Introduction to Statistical Physics

Rigorous and comprehensive, this textbook introduces undergraduate students to simulation methods in statistical physics.

The book covers a number of topics, including the thermodynamics of magnetic and electric systems; the quantum mechanical basis of magnetism; ferrimagnetism, antiferromagnetism, spin waves and magnons; liquid crystals as a non-ideal system of technological relevance; and diffusion in an external potential. It also covers hot topics such as cosmic microwave background, magnetic cooling and Bose–Einstein condensation.

The book provides an elementary introduction to simulation methods through algorithms in pseudocode for random walks, the 2d Ising model and a model liquid crystal. Any formalism is kept simple and derivations are worked out in detail to ensure the material is accessible to students from subjects other than physics.

João Paulo Casquilho is an Associate Professor at the Universidade Nova de Lisboa, Portugal. His research work includes experimental and theoretical studies in liquid crystals rheology under applied electric or magnetic fields and dynamical systems.

Paulo Ivo Cortez Teixeira is an Adjunct Professor at the Instituto Superior de Engenharia de Lisboa and a research associate at the Universidade de Lisboa, Portugal. He is a theoretical soft matter physicist who has worked on colloids, elastomers, foams and liquid crystals.

Introduction to Statistical Physics
and to Computer Simulations

João Paulo Casquilho

Universidade Nova de Lisboa, Portugal

Paulo Ivo Cortez Teixeira

Instituto Superior de Engenharia de Lisboa, Portugal

CAMBRIDGE
UNIVERSITY PRESS

University Printing House, Cambridge CB2 8BS, United Kingdom

One Liberty Plaza, 20th Floor, New York, NY 10006, USA

477 Williamstown Road, Port Melbourne, VIC 3207, Australia

314-321, 3rd Floor, Plot 3, Splendor Forum, Jasola District Centre, New Delhi - 110025, India

79 Anson Road, #06-04/06, Singapore 079906

Cambridge University Press is part of the University of Cambridge.

It furthers the University's mission by disseminating knowledge in the pursuit of
education, learning and research at the highest international levels of excellence.

www.cambridge.org
Information on this title: www.cambridge.org/9781107053786

First published 2015

A catalogue record for this publication is available from the British Library

Library of Congress Cataloging in Publication data
Casquilho, João Paulo, 1951– author.
[Introducão à física estatística. English]
Introduction to statistical physics: and to computer simulations / by João Paulo Casquilho,
Paulo Ivo Cortez Teixeira.
pages   cm
A translation from the Portuguese language edition Introducão à física estatística.
Includes bibliographical references and index.
ISBN 978-1-107-05378-6 (hardback)
1. Statistical physics.   I. Teixeira, Paulo Ivo Cortez, 1965– author.   II. Title.
QC174.8.C3713   2015
530.15′95–dc23
2014032049

ISBN 978-1-107-05378-6 Hardback

J. P. Casquilho dedicates this book to the memory of his parents.
P. I. C. Teixeira dedicates this book to his parents.

Contents

Preface

Statistical physics is a core subject in any degree course in physics, engineering physics, or physics (or chemistry) for education. Its role is on a par with that of introductory quantum mechanics: both provide an essential background in the fundamentals of physics and are prerequisites for more advanced subjects such as atomic and molecular physics, condensed matter physics or solid state physics. On the other hand, statistical physics plays a central part in laying the foundations and enabling the interpretation of classical thermodynamics, the derivation of its laws and of results for model systems such as the classical ideal gas.

Statistical physics is also an excellent vehicle for introducing numerical simulation methods, which are ever more prevalent in physics and engineering. Indeed, computer simulations in statistical physics are playing an increasingly important role in the understanding of the properties and phase transitions of physical systems. Moreover, the computational techniques of statistical physics have been fruitfully applied to problems in many other fields, such as optimisation of assembly lines in an engineering context. Monte Carlo simulations of a number of model systems are, therefore, implemented in this course.

This book grew out of a set of lecture notes for the undergraduate statistical physics course and the graduate computer simulation methods course taught by one of us (J.P.C.) to physical engineering students at the School of Science and Technology of the New University of Lisbon, Portugal (FCT/UNL), in the years 2001–2008. In its final form, the book clearly comprises too much material for a one-semester course. This enables instructors to first cover the foundations of the subject, and then make a selection of more advanced topics, on the basis of their personal preferences and those of the group being taught. This English edition is a thoroughly revised and expanded translation, by the authors, of the Portuguese edition (IST Press, Lisbon, 2011); some of the original chapters have been broken up into shorter chapters, for a sharper focus and greater clarity.

We have deliberately kept our formalism (almost) elementary; no extensive knowledge of, e.g., quantum mechanics or analytical mechanics is presupposed, which should make the book accessible to students of subjects other than physics, such as materials science, materials engineering, chemistry, chemical engineering, or biomedical engineering. A sound basis in general physics, namely classical mechanics, classical thermodynamics, electromagnetism, and some knowledge of modern physics is, however, required. We do work out most derivations in considerable detail, and provide mathematical background material in a number of appendices. Sections marked '*' contain more advanced material, or a more detailed discussion of, or further elaboration on, particular subjects.

This book is organised into five parts, as we now describe.

Random walks. In the first part we give an introduction to statistical methods in physics, and to Monte Carlo simulation methods, through the study of random walks (one chapter).

Statistical thermodynamics. In the second part we present and discuss the basic postulates of statistical physics. We then develop the statistical ensemble formalism and establish the connection with thermodynamics (three chapters).

The ideal gas. In the third part we treat the classical and quantum ideal gases using the statistical ensemble formalism (two chapters). As an extension of the classical ideal gas we study the classical real gas, with a view to applications to non-ideal systems. As applications of the quantum ideal gas we discuss the free electron model in metals, Bose–Einstein condensation, thermal vibrations in crystals and blackbody radiation.

Non-ideal systems, phase transitions and critical phenomena. In the fourth part (four chapters), one chapter is devoted to the mean-field and Landau theories of magnetic phase transitions, and to spin waves. The theory and Monte Carlo simulations of the Ising model are presented in a separate chapter. There follows a chapter on liquid crystals (mostly nematic), where we discuss the analogies between mean-field and Landau-type theories of liquid crystals, and those of ferromagnetism. We introduce the Onsager theory of the nematic phase in solutions and implement Monte Carlo simulations of confined liquid crystals using the Lebwohl–Lasher model. Finally, some of the results previously obtained for phase transitions are recapitulated in the more general context of critical phenomena in a separate chapter.

Brownian motion and diffusion. In the fifth and final part (one chapter) we briefly address irreversible processes through the study of Brownian motion and diffusion, which are related phenomena. Appropriately, we start and finish this book with the random walk, as both paradigm and metaphor.

Acknowledgements

The authors thank Professors Assis Farinha Martins and Grégoire Bonfait, of FCT/UNL, for reading parts of the original manuscript and for their valuable criticism and suggestions. We also acknowledge the invaluable assistance of IST Press and Cambridge University Press, at various stages of manuscript preparation. Thanks are due to some of our anonymous referees, for their insightful and constructive criticisms and suggestions: we believe that the book is better as a result. Finally, we acknowledge the financial support of Fundação para a Ciência e Tecnologia of Portugal, in the form of Projects no. PEst-OE/FIS/UI0618/2011 and PEst-C/CTM/LA0025/2011.

1 Random walks

1.1 Introduction

We shall start our study of statistical physics with *random walks*. As their name suggests, random walks are sequences of steps taken in random directions, i.e., where the direction of a step is chosen according to a given probability distribution. In the most general case, the length of a step is also random. The probability that each step in a sequence of N steps has a given length and is taken in a given direction is independent of the lengths and directions of all other steps in the sequence; in other words, the N steps are statistically independent. We shall study random walks to find out how probable it is that a random walker ends up a certain distance from its starting point after N steps, and how far on average the random walker strays from its starting point.

Random walks are important for a number of reasons. Firstly, they illustrate some basic results of probability theory. Secondly, their statistics are formally identical to those of many key problems in physics and other disciplines, such as diffusion in fluids and solids and the conformations of polymer molecules, some of which will be addressed later in this book. Finally, random walks are particularly suitable vehicles for introducing Monte Carlo simulation methods into statistical physics as they are simple but require many techniques employed in the study of more complex systems.

1.2 Probability: basic definitions

When seeking to describe a system statistically (i.e., in probabilistic terms), it is often useful to consider a *statistical ensemble*. This consists of a very large number N of virtual copies of the system under study. The probability of some particular event occurring is then defined with respect to the ensemble: it is the fraction of systems in the ensemble in which the event actually occurs. Consider, for example, a die throw. One possible statistical description is to assume that the die is thrown N times in succession, under identical conditions. Alternatively, we can imagine a very large number N of identical dice (the representative ensemble of the die) and that each of these is thrown once, all under identical conditions.

We define the probability P_r of a random event r as the limiting value of the relative frequency of r, when the number of trials $N \to \infty$:

$$P_r = \lim_{N \to \infty} f(r), \qquad f(r) = \frac{N_r}{N}, \tag{1.1}$$

where N_r is the number of times that r occurs in N trials. This is the so-called 'frequencist' definition of probability, which is the one most commonly used in the physical sciences. Coming back to the preceding example, the probability of a certain outcome is given by the fraction of trials with that particular outcome. This *ensemble probability* is always defined with respect to a specific ensemble. For example, the probability that the outcome of a die throw is 6, defined with respect to the ensemble of perfect six-sided dice, is different from the probability that the outcome is 6, defined with respect to the ensemble of perfect ten-sided dice.

Consider now an experiment with L mutually exclusive possible outcomes (e.g., a die throw), and label each of these outcomes with the index r: thus $r = 1, 2, \ldots, L$. After $N \gg 1$ trials, outcome 1 was found N_1 times, outcome 2 N_2 times, $\ldots$, and outcome L N_L times. Because the outcomes are mutually exclusive, we must have $N_1 + N_2 + \cdots + N_L = N$. Dividing both sides of this equation by N and using Eq. (1.1) we obtain

$$\sum_{r=1}^{L} P_r = 1, \tag{1.2}$$

i.e., the probability distribution is *normalised to unity*. If the outcomes are *equally probable*, it follows straightforwardly from Eq. (1.2) that

$$P_r = \frac{1}{L}, \quad \text{for all } r. \tag{1.3}$$

The probability of that the outcome is r **or** s is

$$P(r \text{ or } s) = \frac{N_r + N_s}{N} = P_r + P_s, \tag{1.4}$$

which is straightforwardly generalised to more than two outcomes. For example, when throwing a die the probability that the outcome is 4 or 5 is $1/6 + 1/6 = 1/3$

Now suppose that two separate experiments are performed: one with L mutually exclusive possible outcomes $r = 1, 2, \ldots, L$, and the other with M mutually exclusive possible outcomes $s = 1, 2, \ldots, M$. An example might be throwing a die ($L = 6$) and flipping a coin ($M = 2$) simultaneously. The probability that outcomes r and s occur simultaneously is called the *joint probability* of r and s. For each of the N_r possible trials of the first experiment with outcome r, there are N_s possible trials of the second experiment with outcome s, so the total number of trials with outcomes r and s is $N_{rs} = N_r N_s$. The joint probability is then

$$P_{rs} = \frac{N_{rs}}{N} = \frac{N_r N_s}{N}. \tag{1.5}$$

Two events are said to be *statistically independent* when the probability that one event has a particular outcome does not depend on the probability of the other event having a particular outcome. In this case, for all N_r trials with outcome r of the first event, there will be a fraction P_s for which the outcome of the second event is s, so that $N_{rs} = N_r P_s$, leading to

$$P_{rs} = \frac{N_{rs}}{N} = \frac{N_r}{N} P_s = P_r P_s. \tag{1.6}$$

This is easily generalised to more than two independent events as: *The joint probability of statistically independent events is the product of the probabilities of the individual events.*

1.3 Random variables and distribution functions

A *random variable* is a variable X which takes values x in a set B according to a given *probability law* or *distribution function* F_X such that

$$F_X(x) = P[X \le x], \quad \text{for all } x \in B. \tag{1.7}$$

In other words, the distribution function of X evaluated at x equals the probability that $X \le x$. Two random variables are said to be *equal in distribution* if they have the same distribution function.

Random variables may be discrete or continuous.

For a *discrete random variable*, there exists a function p_X, called *probability mass function*, such that

$$P[X \le x] = \sum_{t \le x} p_X(t), \tag{1.8}$$

$$P[X = x] = p_X(x), \tag{1.9}$$

i.e., $p_X(x)$ equals the probability that $X = x$ and the sum is over all possible values t of X not greater than x. In this case, normalisation gives

$$\sum_{x \in B} p_X(x) = 1, \tag{1.10}$$

where the sum is over all possible values x of X.

For a *continuous random variable* there exists a function $f_X(x)$ called *probability density function* such that

$$P[X \le x] = \int_{-\infty}^{x} f_X(t)dt, \tag{1.11}$$

$$P[X \in (x, x + dx)] = f_X(x)dx, \tag{1.12}$$

where in this case we have assumed that X may take any value between $-\infty$ and $+\infty$. Normalisation now gives

$$\int_{-\infty}^{+\infty} f_X(x)dx = 1. \tag{1.13}$$

We define the *most probable value* of a random variable X as that for which the probability mass function of X, or the probability density function of X, is maximised (for discrete and continuous X, respectively).

The *mean*, *mean value*, *expected value* or *mathematical expectation* of a random variable X is defined as

$$E[X] = \sum_{k=1}^{\infty} x_k p_k, \qquad p_k = P[X = x_k], \tag{1.14}$$

if X is discrete, and

$$E[X] = \int_{-\infty}^{+\infty} x f_X(x) dx, \tag{1.15}$$

if X is continuous. The values of X cluster around its mean, which is therefore a measure of the localisation of X, or of its distribution. Note that the mean is the weighted average of all values of X.

Henceforth we shall often use x to denote both the random variable X and its values. The mean of X will then be denoted $\bar{x}$. For consistency we shall employ this notation for discrete as well as continuous variables, although in probability theory $\bar{x}$ usually denotes the arithmetic mean of a sample of X. Thus

$$E[X] = \bar{x}. \tag{1.16}$$

The nth *central moment* of a random variable X is defined as $E[(X - E[X])^n]$. If the distribution function of X is known, then the mean and all moments of X can be calculated. The converse is also true in the case of smooth distributions: this result is often used when the mean and a few moments of a distribution are known, experimentally or from computer simulations ('numerical experiments'), to extract an approximation to the probability distribution function.

The *deviation from the mean* is defined as $X - E[X]$, which in the notation of Eq. (1.16) may also be written

$$X - E[X] = x - \bar{x} \equiv \Delta x. \tag{1.17}$$

If a random variable X deviates very little from its mean, then the true value of the quantity X may be replaced by the mean of X, with a very small error. This is a key issue in statistical physics: in order to quantify how much the true value of X deviates from its mean, we shall calculate the first and second centred moments of a continuous random variable. The same results can easily be obtained for a discrete random variable. Because $X - E[X]$ is also a random variable, we find, from Eqs (1.13)–(1.17):

$n = 1$ (*mean deviation from the mean*):

$$E[X - E[X]] \equiv \overline{\Delta x} = \int_B (x - \bar{x}) f_X(x) dx = 0. \tag{1.18}$$

$n = 2$ (*variance, mean square deviation or scatter*):

$$\mathrm{Var}(X) \equiv \sigma^2(X) = E\left[(X - [X])^2\right] = \overline{(\Delta x)^2} = \int_B (x - \bar{x})^2 f_X(x) dx$$

$$= \int x^2 f_X(x) dx - 2\bar{x} \int x f_X(x) dx + \bar{x}^2 \int f_X(x) dx$$

$$= \overline{x^2} - \bar{x}^2. \tag{1.19}$$

Clearly the variance cannot be negative, as $(\Delta x)^2 \geq 0$. Equation (1.19) then implies that $\overline{x^2} \geq \bar{x}^2$. The variance vanishes only if $x = \bar{x}$ for all x. The larger the variance, the greater the scatter of x about its mean. If $\sigma^2(X)$ is small, then X will always be close to $\bar{x}$, and we can replace the true value of X by its mean $\bar{x}$. The relative error of this procedure is the *relative fluctuation*:

$$\delta(X) = \sigma(X)/E[X], \tag{1.20}$$

where $\sigma(X) = \sqrt{\sigma^2(X)}$ is the *standard deviation* of X.

If a discrete random variable X can take only a finite number N of values all with the same probability p_k, normalisation implies that $p_k = 1/N$ as in Eq. (1.3). By Eq. (1.14) the statistical mean then coincides with the arithmetic mean:

$$\bar{x} = \frac{1}{N} \sum_{k=1}^{N} x_k. \tag{1.21}$$

In this case the discrete version of Eq. (1.18) is just that *the deviations from the mean add up to zero*.

In the limit $N \to \infty$, the statistical mean and the arithmetic mean of a discrete random variable are related through a theorem known as the *law of large numbers*, which can be stated as follows (Apostol, 1969): Let $X_1, X_2, \ldots, X_N$ be a sequence of N independent random variables, all equal in distribution to some random variable X of finite mean and variance. If we define a new random variable $\overline{X}$, the *arithmetic mean* of $X_1, X_2, \ldots, X_N$, as

$$\overline{X} = \frac{1}{N} \sum_{k=1}^{N} X_k, \tag{1.22}$$

then

$$\lim_{N \to \infty} \overline{X} = E[X]. \tag{1.23}$$

We close this section with a brief reference to *functions of random variables*. For simplicity we shall drop the X subscript. The mean of a function $g(X)$ of random variable X is then defined as

$$\overline{g(X)} = \sum_x g(x)p(x) \quad \text{or} \quad \overline{g(X)} = \int g(x)f(x)dx, \tag{1.24}$$

where $p(x)$ and $f(x)$ are given by Eqs. (1.9) and (1.12), respectively.

1.4 The simple random walk

1.4.1 The binomial distribution

We shall start by deriving the probability distribution for a one-dimensional random walk. Here steps are all the same length ℓ and can be taken either to the right or to the left. Out of N steps, n_1 are rightward steps and $n_2 = N - n_1$ are leftward steps. Assuming that consecutive steps are statistically independent, then at each moment in time the probability of taking a step right is p and the probability of taking a step left is $q = 1 - p$. If we view a random walk as N trials of a *Bernoulli experiment* where a step right is a success and a step left is a failure, then the probability of n_1 successes out of N trials is given by the *binomial distribution*:

$$P(X = n_1) \equiv P(n_1) = \binom{N}{n_1} p^{n_1} q^{n_2}, \qquad n_1 = 0, 1, \ldots, N,$$
$$P(n_1) = 0, \qquad\qquad\qquad\qquad \text{any other value of } n_1, \qquad (1.25)$$

where X is the random variable equal to the number of successes in N independent Bernoulli experiments, $n_2 = N - n_1$ is the number of failures, and

$$\binom{N}{n_1} = \frac{N!}{n_1!(N - n_1)!} \qquad (1.26)$$

is the binomial coefficient.

Proof of Eq. (1.25)   For N Bernoulli experiments, the probability of n_1 consecutive successes is p^{n_1} (because all outcomes are independent). Likewise, the probability of n_2 consecutive failures is q^{n_2}. Hence the probability that n_1 consecutive successes are followed by n_2 consecutive failures is $p^{n_1} q^{n_2}$. The probability of n_1 successes and n_2 failures in any other order is also $p^{n_1} q^{n_2}$, since each of the N outcomes is either a success or a failure. Then the probability of n_1 successes and n_2 failures regardless of order equals the probability of n_1 successes and n_2 failures in some order, times the number of possible ways in which n_1 out of N experiments are successes, which is given by the binomial coefficient.

The mean of n_1 can be found using Eqns. (1.14) and (1.25):

$$\overline{n_1} = \sum_{n_1=0}^{N} n_1 P(n_1) = \sum_{n_1=0}^{N} n_1 \binom{N}{n_1} p^{n_1} q^{n_2}. \qquad (1.27)$$

Noting that

$$n_1 p^{n_1} q^{n_2} = p \frac{\partial}{\partial p} \left(p^{n_1} q^{n_2} \right), \qquad (1.28)$$

that the binomial theorem implies

$$\sum_{n_1=0}^{N} P(n_1) = (p + q)^N, \qquad (1.29)$$

and that $P(n_1)$ is normalised to unity, Eq. (1.10), we obtain

$$\overline{n_1} = \sum_{n_1=0}^{N} p \frac{\partial}{\partial p} P(n_1) = p \frac{\partial}{\partial p} \left(\sum_{n_1=0}^{N} P(n_1) \right)$$

$$= p \frac{\partial}{\partial p} \left[(p + q)^N \right] = Np(p + q)^{N-1} = Np. \qquad (1.30)$$

Along the same lines, it may be easily shown that

$$\overline{n_1^2} = Np + N(N - 1)p^2, \qquad (1.31)$$

whence from Eq. (1.19) with Eqs. (1.30) and (1.31) we find for the variance:

$$\sigma^2(n_1) = \overline{n_1^2} - \overline{n_1}^2 = Npq. \tag{1.32}$$

Up till now we have concentrated on a description of the one-dimensional random walk in terms of the random variable n_1, the number of rightward steps out of N total steps. If, however, we wish to study the *displacement*, it is convenient to introduce a new variable, the *effective number of rightward steps*, as (Reif, 1985)

$$n = n_1 - n_2, \tag{1.33}$$

where $n_1 + n_2 = N$ (recall that n_2 is the total number of leftward steps). We can then define the *effective rightward displacement* as

$$L = nl, \tag{1.34}$$

where l is the step length.

The mean of n is easily calculated from Eq. (1.30):

$$\overline{n} = \overline{n_1 - n_2} = \overline{n_1} - \overline{n_2} = N(p - q). \tag{1.35}$$

Because $n = n_1 - n_2 = 2n_1 - N$, n takes only integer values with spacing $\delta n = 2$, whence

$$\Delta n \equiv n - \overline{n} = (2n_1 - N) - (2\overline{n_1} - N) = 2\Delta n_1, \tag{1.36}$$

where $\Delta n_1 = n_1 - \overline{n_1}$. From Eq. (1.32) we then get the variance of n:

$$\sigma^2(n) = \overline{\Delta n^2} = 4Npq. \tag{1.37}$$

The mean and variance of the displacement then follow easily from Eqs. (1.34)–(1.37):

$$\overline{L} = Nl(p - q), \tag{1.38}$$

$$\sigma^2(L) = 4Nl^2pq. \tag{1.39}$$

The variance of n is thus quadratic in l, whence the standard deviation $\sigma(L)$ is a measure of the linear scatter of L values.

As mentioned before, a suitable measure of the width of a distribution is its relative fluctuation, given by Eq. (1.20). For n_1 this is, from Eqs. (1.30) and (1.32):

$$\delta(n_1) \equiv \frac{\sigma(n_1)}{\overline{n_1}} = \sqrt{\frac{q}{p}} \frac{1}{\sqrt{N}}, \tag{1.40}$$

i.e., the relative fluctuation decays with $N^{-1/2}$ when N increases (see Figure 1.1).

In the special case where $p = q = 1/2$, i.e., rightward and leftward steps are equally probable, we have the following results:

$$\overline{n} = 0; \qquad \sigma^2(n) = N; \qquad \sigma(n) = \sqrt{N}; \tag{1.41}$$

$$\overline{L} = 0; \qquad \sigma^2(L) = Nl^2; \qquad \sigma(n) = \sqrt{N}l; \tag{1.42}$$

$$\delta(n_1) = \frac{1}{\sqrt{N}}. \tag{1.43}$$

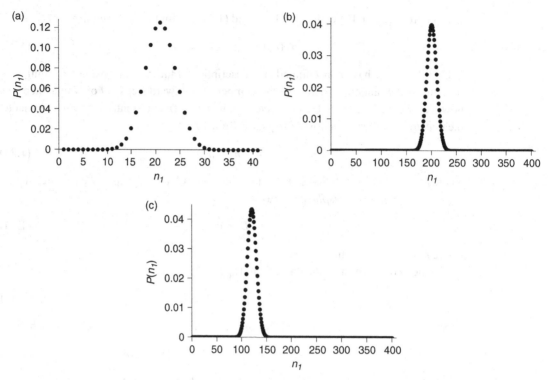

Figure 1.1 Examples of binomial distributions, Eq. (1.25.) (a) $N = 40$; $p = 0.5$; $\overline{n}_1 = 20$; $\sigma^2(n_1) = 10$; $\delta(n_1) = 0.16$. (b) $N = 400$; $p = 0.5$; $\overline{n}_1 = 200$; $\sigma^2(n_1) = 100$; $\delta(n_1) = 0.05$. (c) $N = 400$; $p = 0.3$; $\overline{n}_1 = 120$; $\sigma^2(n_1) = 84$; $\delta(n_1) = 0.076$.

We thus conclude that, for the symmetric one-dimensional random walk, N is a measure of the variance, $N^{1/2}$ is a measure of the standard deviation, and $N^{-1/2}$ is a measure of the relative fluctuation, with l a scale factor. Moreover, because rightward and leftward steps are equally probable, the mean effective displacement vanishes, as might be intuitively expected.

1.4.2 The Gaussian distribution

For large N, the binomial distribution $P(n_1)$ given by Eq. (1.25) acquires a very pronounced maximum at $n_1 = \overline{n}_1$, dropping rapidly as n_1 moves away from $\overline{n}_1$ (see Figure 1.1). This enables us to find an approximate expression for $P(n_1)$ when $N \to \infty$:

$$P(n_1) = \frac{1}{\sqrt{2\pi \sigma^2(n_1)}} \exp\left[-\frac{1}{2}\frac{(n_1 - \overline{n}_1)^2}{\sigma^2(n_1)}\right], \tag{1.44}$$

where $\overline{n}_1$ and $\sigma^2(n_1)$ are given by Eqs. (1.30) and (1.32), respectively. Equation (1.44) is known as the *Gaussian distribution* or *standard normal distribution*. Gaussian distributions occur very often (but by no means in all cases!) in statistics when one has to deal with large numbers of trials.

Proof of Eq. (1.44)   Start by making the change of variables $n_1 = \overline{n_1} + k$, with $\overline{n_1}$ given by Eq. (1.30), in Eq. (1.25):

$$P(k) = \frac{N!}{(Np+k)!(Nq-k)!} p^{Np+k} q^{Nq-k}.$$

Now use Stirling's formula, Eq. (1.126), for $N!$ when $N \gg 1$:

$$N! \simeq \sqrt{2\pi N} N^N e^{-N},$$

with the result

$$P(k) = \sqrt{2\pi N} N^N e^{-N} \times \frac{p^{Np+k}}{\sqrt{2\pi(Np+k)}(Np+k)^{Np+k} e^{-(Np+k)}}$$

$$\times \frac{q^{Nq-k}}{\sqrt{2\pi(Nq-k)}(Nq-k)^{Nq-k} e^{-(Nq-k)}}$$

$$= \frac{1}{\sqrt{2\pi N(p+k/N)(q-k/N)}(1+k/Np)^{Np+k}(1-k/Nq)^{Nq-k}},$$

whence

$$\ln P(k) = -\frac{1}{2}\ln\left[2\pi N(p+k/N)(q-k/N)\right]$$

$$- (Np+k)\ln(1+k/Np) - (Nq-k)\ln(1-k/Nq).$$

For $k/Np \ll 1$ and $k/Nq \ll 1$ we can truncate the series expansions of the logarithms at low order:

$$\ln\left(1+\frac{k}{Np}\right) \simeq \frac{k}{Np} - \frac{k^2}{2N^2p^2} + \frac{k^3}{3N^3p^3},$$

$$\ln\left(1-\frac{k}{Nq}\right) \simeq -\frac{k}{Nq} - \frac{k^2}{2N^2q^2} - \frac{k^3}{3N^3q^3}.$$

Substitution into the expression for $\ln P(k)$ then yields:

$$\ln P(k) = -\frac{1}{2}\ln\left\{2\pi N\left[pq - \frac{k}{N}(p-q) - \frac{k^2}{N^2}\right]\right\}$$

$$- \frac{k^2}{2Np} - \frac{k^2}{2Nq} + \frac{k^3}{6N^2p^2} - \frac{k^3}{6N^2q^2},$$

or, equivalently,

$$P(k) = \frac{\exp\left(-\frac{k^2}{2Npq} - \frac{k^3}{6N^2}\frac{p-q}{p^2q^2}\right)}{\sqrt{2\pi N\left[pq - \frac{k}{N}(p-q)\frac{k^2}{N^2}\right]}}.$$

Further neglecting terms of order k/N or higher, we find:

$$P(k) = \frac{\exp\left(-\frac{k^2}{2Npq}\right)}{\sqrt{2\pi Npq}}$$

Changing variables back to n_1 then gives Eq. (1.44).

The probability that, after a very large number of steps N, the effective number of right-ward steps, as defined by Eq. (1.33), is n, follows from Eq. (1.44) combined with Eqs. (1.30) and (1.32), noting that $n_1 = (N+n)/2$ (Reif, 1985):

$$P(n) = \frac{1}{\sqrt{2\pi Npq}} \exp\left\{-\frac{[n-N(p-q)]^2}{8Npq}\right\}, \tag{1.45}$$

where we have used the fact that $n_1 - Np = (N+n-2Np)/2 = [n-N(p-q)]/2$. This result may be seen as a special case of the *central limit theorem*, to which we shall come back later: if a random variable (the effective number of rightward steps) is the sum of a very large number of other, statistically independent, random variables (all individual steps), then its distribution is Gaussian.

The above distribution can be re-expressed in terms of the effective rightward displacement, defined by Eq. (1.34). If the step length l is much smaller than the relevant lengthscale of the system under study (e.g., its linear dimension), the discrete variable L with increment $\delta L = 2l$ may be replaced by a continuous, 'macroscopic' variable x with increment $dx \gg l$. Then the probability that after $N \gg 1$ steps the effective rightward displacement lies between x and $x + dx$ is, by definition (1.11), the sum of $P(n)$ over all n in dx, of which there are $dx/2l$. Because $P(n)$ is approximately constant in such a narrow interval, to a good approximation the probability we seek is just $P(n)$ times $dx/2l$, whence we can write

$$f(x)dx = P(n)\frac{dx}{2l}, \tag{1.46}$$

where $f(x)$ is the probability density, according to definition (1.12). Equations (1.45) and (1.46) finally yield

$$f(x)dx = \frac{1}{\sqrt{2\pi\sigma^2}} \exp\left[-\frac{1}{2}\frac{(x-\mu)^2}{\sigma^2}\right] dx, \tag{1.47}$$

where

$$\mu = Nl(p-q), \tag{1.48}$$

$$\sigma^2 = 4Nl^2pq. \tag{1.49}$$

Equation (1.47) is the most common form of the Gaussian distribution for a continuous variable. It gives *the probability that after N steps of size l each, the random walker will find itself at a distance between x and x + dx from its starting point.* This distribution is symmetric about μ, which as we shall see shortly is its mean value. Figure 1.2 shows some exemplary Gaussian distributions.

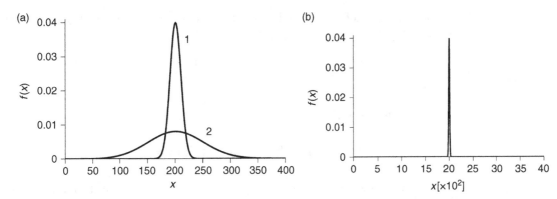

Figure 1.2 Examples of Gaussian distributions, Eq. (1.47). Here $\bar{x} = \mu$ and $\delta = \sigma/\mu$. (a) (1) $\bar{x} = 200$; $\sigma = 10$; $\delta = 0.05$; (2) $\bar{x} = 200$; $\sigma = 50$; $\delta = 0.25$. (b) $\bar{x} = 2000$; $\sigma = 10$; $\delta = 5 \times 10^{-3}$.

It is also easy to check that $f(x)$ given by (1.47) is normalised to unity:

$$
\begin{aligned}
\int_{-\infty}^{+\infty} f(x)dx &= \frac{1}{\sqrt{2\pi\sigma^2}} \int_{-\infty}^{+\infty} e^{-(x-\mu)^2/2\sigma^2} dx \\
&= \frac{1}{\sqrt{2\pi\sigma^2}} \int_{-\infty}^{+\infty} e^{-y^2/2\sigma^2} dy \\
&= \frac{1}{\sqrt{2\pi\sigma^2}} \sqrt{2\pi\sigma^2} \\
&= 1,
\end{aligned}
$$

where we have made the change of variables $y = x - \mu$ and used the results of Appendix 1.A.2. The limits of integration can be taken to be $\pm\infty$ because $f(x)$ is negligibly small for $x \gg \mu$ or $x \ll \mu$ (see Figure 1.2): after $N \gg 1$ steps, it is very unlikely that the displacement will deviate substantially from its mean, which we shall now calculate:

$$
\begin{aligned}
\bar{x} = \int_{-\infty}^{+\infty} xf(x)dx &= \frac{1}{\sqrt{2\pi\sigma^2}} \int_{-\infty}^{+\infty} xe^{-(x-\mu)^2/2\sigma^2} dx \\
&= \frac{1}{\sqrt{2\pi\sigma^2}} \left[\int_{-\infty}^{+\infty} ye^{-y^2/2\sigma^2} dy + \mu \int_{-\infty}^{+\infty} e^{-y^2/2\sigma^2} dy \right] \\
&= \frac{1}{\sqrt{2\pi\sigma^2}} \mu\sqrt{2\pi\sigma^2} \\
&= \mu, \quad\quad\quad\quad\quad\quad\quad\quad\quad\quad\quad\quad\quad\quad\quad\quad\quad (1.50)
\end{aligned}
$$

where we made the same change of variables as before and the first integral in the square brackets vanishes because the integrand is an odd function of y.

Using the results of Appendix 1.A.2, the variance may likewise be found to be

$$
\sigma^2(x) = \sigma^2, \quad\quad\quad\quad\quad\quad\quad\quad\quad\quad\quad (1.51)
$$

the proof of which is left as an exercise for the reader. It then follows from Eqs. (1.48) and (1.49) that

$$\bar{x} = Nl(p - q), \tag{1.52}$$

$$\sigma^2(x) = 4Nl^2 pq, \tag{1.53}$$

which agree with Eqs. (1.38) and (1.39) derived for arbitrary N.

1.4.3 The Poisson distribution

For $p \ll 1$, the binomial distribution, Eq. (1.25), can be approximated by the *Poisson distribution*,

$$P(n_1) = \frac{\lambda^{n_1}}{n_1!} e^{-\lambda}, \tag{1.54}$$

with $\lambda = \overline{n_1}$, as we shall now show.

Rewrite Eq. (1.25) with $n_2 = N - n_1$ and $q = 1 - p$:

$$P(n_1) = \frac{N!}{n_1!(N - n_1)!} p^{n_1}(1 - p)^{N - n_1}$$

$$= \frac{N(N - 1)(N - 2)\dots(N - n_1 + 1)}{n_1!} p^{n_1}(1 - p)^N (1 - p)^{-n_1}$$

$$= \frac{N^{n_1}(1 - 1/N)(1 - 2/N)\dots(1 - (n_1 - 1)/N) p^{n_1}(1 - p)^N}{n_1!(1 - p)^{n_1}}.$$

Because p is small, $P(n_1)$ will only be appreciably non-zero for $n_1 \ll N$. In this case $(1 - 1/N)(1 - 2/N)\dots(1 - (n_1 - 1)/N) \simeq 1$. Further noting that $\ln(1 - p)^N N \ln(1 - p) \simeq -Np$, we have $(1 - p)^N \simeq e^{-Np}$ and $(1 - p)^{n_1} \approx 1$, whence

$$P(n_1) \simeq \frac{(Np)^{n_1} e^{-Np}}{n_1!}.$$

Since by Eq. (1.30) $\overline{n_1} = Np$, Poisson's law, Eq. (1.54), immediately follows. This law tells us something almost trivial: that if the probability of taking a step right is small ($p \ll 1$), then the probability of taking n_1 steps right falls off very rapidly with n_1, i.e., large numbers of rightward steps have very low probability.

1.5 The generalised random walk

1.5.1 Joint distributions of several random variables

When dealing with more than one random variable, the definitions of Section 1.3 need to be generalised. For simplicity, let us consider the case of two random variables, X_1 and X_2; the results we shall obtain are easily generalised to any number of random variables

The *joint distribution function* is defined as

$$F_{X_1,X_2}(x_1,x_2) = P[X_1 \leq x_1, X_2 \leq x_2], \qquad \text{for all } x_1 \in B_1, x_2 \in B_2. \tag{1.55}$$

Henceforth we shall use the language of continuous random variables, but the same procedure holds for discrete random variables as well. The definition of probability density, Eq. (1.12), generalises to two variables as

$$P[X_1 \leq x_1, X_2 \leq x_2] = \int_{t_1 \in B_1, t_1 \leq x_1} \int_{t_2 \in B_2, t_2 \leq x_2} f_{X_1,X_2}(t_1,t_2)dt_1 dt_2, \tag{1.56}$$

$$P[X_1 \in (x_1, x_1 + dx_1), X_2 \in (x_2, x_2 + dx_2)] = f_{X_1,X_2}(x_1,x_2)dx_1 dx_2, \tag{1.57}$$

where $f_{X_1,X_2}(x_1,x_2)$ is the *joint probability density*.

We further define the *marginal density* of X_1 as

$$f_{X_1}(x_1) = \int_{B_2} f_{X_1,X_2}(x_1,x_2)dx_2, \tag{1.58}$$

which is the probability density that X_1 lies between x_1 and $x_1 + dx_1$ for any value of X_2 (the marginal density of X_2 is defined likewise). Normalisation then requires that each individual marginal density must also be normalised to unity:

$$1 = \int_{B_1} \int_{B_2} f_{X_1,X_2}(x_1,x_2)dx_1 dx_2 = \int_{B_1} f_{X_1}(x_1)dx_1 = \int_{B_2} f_{X_2}(x_2)dx_2, \tag{1.59}$$

where we have used Eq. (1.58) and its counterpart for $f_{X_2}(x_2)$.

Two random variables are said to be *statistically independent* or *uncorrelated* if

$$f_{X_1,X_2}(x_1,x_2) = f_{X_1}(x_1)f_{X_2}(x_2), \tag{1.60}$$

i.e., if the probability that X_1 takes the value x_1 does not depend on the value of X_2, and vice versa. Clearly in this case we also have

$$f_{X_1,X_2}(x_1,x_2)dx_1 dx_2 = \left[f_{X_1}(x_1)dx_1 \right] \left[f_{X_2}(x_2)dx_2 \right], \tag{1.61}$$

whence

$$\begin{aligned}
E[X_1, X_2] &= \int_{B_1} \int_{B_2} x_1 x_2 f_{X_1,X_2}(x_1,x_2)dx_1 dx_2 \\
&= \left[\int_{B_1} x_1 f_{X_1}(x_1)dx_1 \right] \left[\int_{B_2} x_2 f_{X_2}(x_2)dx_2 \right] \\
&= E[X_1]E[X_2],
\end{aligned} \tag{1.62}$$

i.e., the mean of the product of two statistically independent random variables is the product of the means of the two variables.

The *covariance* of two random variables X_1 and X_2 is defined as

$$\text{Cov}(X_1,X_2) = E[(X_1 - E[X_1])(X_2 - E[X_2])], \tag{1.63}$$

which is the mean of the products of the deviations of the two variables from their respective means. If the two variables are statistically independent, it follows from Eqs. (1.18) and (1.62) that

$$\text{Cov}(X_1, X_2) = E[X_1 - E[X_1]]\, E[X_2 - E[X_2]] = 0. \tag{1.64}$$

We are now in a position to find the variance of the sum of two random variables:

$$\begin{aligned}
\text{Var}(X_1 + X_2) &= E\left[\{(X_1 + X_2) - E[X_1 + X_2]\}^2\right] \\
&= E\left[\{(X_1 - E[X_1]) + (X_2 - E[X_2])\}^2\right] \\
&= E\left[(X_1 - E[X_1])^2\right] + E\left[(X_2 - E[X_2])^2\right] \\
&\quad + 2E[(X_1 - E[X_1])(X_2 - E[X_2])] \\
&= \text{Var}(X_1) + \text{Var}(X_2) + 2\text{Cov}(X_1, X_2). \tag{1.65}
\end{aligned}$$

From Eq. (1.64) we conclude that *if the two variables are statistically independent, then the variance of their sum equals the sum of their variances.* The converse is true only under certain conditions: *if X_1 and X_2 are standard-normal distributed and $\text{Cov}(X_1, X_2) = 0$, then X_1 and X_2 are statistically independent.* The above results are readily generalised to N random variables.

It is also straightforward to generalise Eq. (1.24) to find *the mean of a function of two random variables*:

$$\overline{g(X_1, X_2)} = \int_{B_1} \int_{B_2} g(x_1, x_2) f(x_1, x_2)\, dx_1 dx_2, \tag{1.66}$$

with $f(x_1, x_2)$ given by Eq. (1.57), where we have dropped the indices for simplicity. In general it can be shown that, for any two functions $g(X_1, X_2)$ and $h(X_1, X_2)$ of the random variables X_1 and X_2,

$$\overline{(g + h)} = \overline{g} + \overline{h}. \tag{1.67}$$

1.5.2 General results for random walks in one dimension

In Section 1.4 we studied the random walk using combinatorial methods. Although we were able to derive a number of important results, this is not the most convenient approach if we wish to consider variable step lengths, or walks in more than one dimension.

Start by considering a one-dimensional, N-step random walk where steps can have arbitrary sizes. Let S_i be the (positive or negative) displacement along OX at step i, and let the probability that S_i lies between s_i and $s_i + ds_i$ be given by

$$P[S_i \in (s_i, s_i + ds_i)] = f(s_i) ds_i.$$

We shall assume, as before, that all individual displacements (steps) are statistically independent. We further assume that the probability densities are the same for all steps, i.e., that they are equal in distribution to some random variable s:

$$f(s_i) = f(s), \qquad \text{for all } i. \tag{1.68}$$

We want to know the total displacement after N steps:

$$x = \sum_{i=1}^{N} s_i. \tag{1.69}$$

There will be a probability density $f(x)$ such that the probability that x lies between x and $x + dx$ is $f(x)dx$ (see note at the end of this section). However, we can find the moments of x without knowledge of $f(x)$, using the results of Section 1.5.1.

- *Mean of x*: From Eq. (1.69) we get

$$\bar{x} = \overline{\left(\sum_{i=1}^{N} s_i \right)} = \sum_{i=1}^{N} \bar{s_i}, \tag{1.70}$$

and Eq. (1.68) then implies

$$\bar{x} = N\bar{s}, \tag{1.71}$$

where $\bar{s}$ is just the mean displacement associated with an individual step:

$$\bar{s} = \int_{B_i} s f(s) ds. \tag{1.72}$$

- *Variance of x*: Because the s_i are statistically independent, we have, from Eqs. (1.64) and (1.65):

$$\sigma^2(x) = \sigma^2 \left(\sum_{i=1}^{N} s_i \right) = \sum_{i=1}^{N} \sigma^2(s_i). \tag{1.73}$$

From Eq. (1.68) we again have

$$\sigma^2(x) = N\sigma^2(s_i), \tag{1.74}$$

where $\sigma^2(s_i)$ is just the variance of the displacement associated with an individual step:

$$\sigma^2(s_i) \equiv \sigma^2(s) = \int_{B_i} (\Delta s)^2 f(s) ds, \tag{1.75}$$

where $\Delta s = s - \bar{s}$. Equations (1.71) and (1.74) are very general, and therefore very important results. The relative fluctuation of x, defined through Eq. (1.20), is

$$\delta(x) = \frac{\sigma(x)}{\bar{x}} = \frac{\sigma(s)}{\bar{s}} \frac{1}{\sqrt{N}}, \tag{1.76}$$

whence, if $\bar{s} \neq 0$ and $\sigma(s)$ is finite, we have

$$\delta(x) \propto \frac{1}{\sqrt{N}}, \tag{1.77}$$

which implies that the relative deviation of x from $\bar{x}$ drops with $N^{-1/2}$ as N increases. We have thus rederived, by a more general method, a result that had already been arrived at in Section 1.4.1 using combinatorial arguments. This result is of paramount importance in statistical physics: *For macroscopic systems, where the number of particles N is of the order of Avogadro's number $N_A \simeq 6 \times 10^{23}$ mol^{-1}, the error incurred when replacing the* *true (unknown) value of a macroscopic quantity associated with the random variable X by its mean value $\bar{x}$ is negligible. In the limit $N \to \infty$ the relative fluctuations of extensive quantities*[1] *vanish.* The mean $\bar{x}$ is found by the methods of statistical physics that will be developed in the next few chapters.

In the special case where all steps have the same size l, p is the probability of a positive displacement and $q = 1 - p$ is the probability of a negative displacement. We then have

$$\bar{s} = lp + (-l)q = l(p - q), \tag{1.78}$$

$$\sigma^2(s) = \overline{s^2} - \bar{s}^2 = l^2(p + q) - l^2(p - q)^2 = 4l^2 pq, \tag{1.79}$$

which, together with Eqs. (1.71) and (1.74), again lead to Eqs. (1.52) and (1.53), repeated below:

$$\bar{x} = Nl(p - q),$$

$$\sigma^2(x) = 4Nl^2 pq.$$

Note that *these results have been obtained without any knowledge of the probability distribution for x.*

1.5.3 Random walks in three dimensions

The three-dimensional random walk can be studied using the results derived in the preceding subsection for the one-dimensional random walk with a distribution of step lengths. Basically the problem involves summing vectors all of the same length, but random directions. In other words, we wish to know the probability that the total three-dimensional displacement $\mathbf{r} = \mathbf{x} + \mathbf{y} + \mathbf{z}$ has some magnitude and some direction, where its components along each of the coordinate axes are themselves one-dimensional random variables. That is, x, y and z are each given by an equation like (1.69) and therefore satisfy Eqs. (1.71) and (1.74). When evaluating $\overline{x_\alpha}$ and $\sigma^2(s_\alpha)$ ($\alpha = x, y, z$) care must be taken that in three dimensions there are six possible directions, so the two probabilities p and q satisfying $p + q = 1$ of the one-dimensional case must now be replaced by six probablities p_α and q_α, with normalisation $\sum_{\alpha=1}^{3}(p_\alpha + q_\alpha) = 1$. In particular, Eq. (1.79), which was derived on the assumption that $p + q = 1$, is no longer valid. However, we can still treat each dimension separately and obtain (see Problems 1.6, 1.8 and 1.9):

$$\sigma^2(s_\alpha) = l_\alpha^2(p_\alpha + q_\alpha) - l_\alpha^2(p_\alpha - q_\alpha)^2, \qquad \alpha = x, y, z. \tag{1.80}$$

For the displacement $\mathbf{r}$, we have $\bar{\mathbf{r}} = \bar{\mathbf{x}} + \bar{\mathbf{y}} + \bar{\mathbf{z}}$, whence

$$\bar{\mathbf{r}}^2 = \bar{\mathbf{r}} \cdot \bar{\mathbf{r}} = \bar{x}^2 + \bar{y}^2 + \bar{z}^2, \tag{1.81}$$

$$\overline{r^2} = \overline{\mathbf{r} \cdot \mathbf{r}} = \overline{x^2} + \overline{y^2} + \overline{z^2}, \tag{1.82}$$

$$\Rightarrow \sigma^2(r) = \sigma^2(x) + \sigma^2(y) + \sigma^2(z). \tag{1.83}$$

---

[1] To be defined in Section 2.2.1.

We show at the end of this subsection that x, y and z are standard-normal distributed. From this and Eq. (1.83) we conclude that x, y and z are statistically independent variables (recall Eq. (1.65) and the discussion thereon).

If $\bar{\mathbf{r}} = \mathbf{0}$ the variance equals the *mean square radius*, which thus measures the scatter in r-values:

$$\sigma^2(r) = \overline{r^2}. \tag{1.84}$$

From Eqs. (1.74), (1.83) and (1.84) we find the important result

$$\overline{r^2} = N\left[\sigma^2(s_x) + \sigma^2(s_y) + \sigma^2(s_z)\right] \propto N. \tag{1.85}$$

If steps in the positive and negative directions of all axes are equally probable, i.e., if there is no polarity along any of the axes, it follows from Eq. (1.78) with $p_\alpha = q_\alpha$ that $\bar{x} = \bar{y} = \bar{z} = 0$, whence Eq. (1.79) must be replaced by

$$\sigma^2(s_x) = \overline{x^2} = l_x^2(p_x + q_x), \tag{1.86}$$

$$\sigma^2(s_y) = \overline{y^2} = l_y^2(p_y + q_y), \tag{1.87}$$

$$\sigma^2(s_z) = \overline{z^2} = l_z^2(p_z + q_z). \tag{1.88}$$

The above equations immediately suggest how a random walk can be made anisotropic: either the step lengths along x, y and z are different ($l_x \neq l_y \neq l_z$), or the probabilities of taking steps along x, y and z are different ($p_x + q_x \neq p_y + q_y \neq p_z + q_z$). Such a random walk is known generally as a *biased random walk*. Consider the case where the probabilities of taking steps along different axes are the same ($p_x + q_x = p_y + q_y = p_z + q_z = 1/3$). Then for a general anisotropic random walk we have, from Eqs. (1.86)–(1.88):

$$\overline{x^2} = l_x^2/3, \quad \overline{y^2} = l_y^2/3, \quad \overline{z^2} = l_z^2/3, \tag{1.89}$$

whereas for an isotropic random walk $l_x = l_y = l_z \equiv l$ and

$$\overline{x^2} = \overline{y^2} = \overline{z^2} = l^2/3 = \overline{r^2}/3. \tag{1.90}$$

Below we discuss a few exemplary physical systems that can be modelled as random walks.

1. *Diffusing gas molecule.* A gas molecule travels in three-dimensional space with a mean free path l between consecutive collisions with other gas molecules. What is its mean displacement after N such collisions? This is studied in Chapter 11 (see also Problem 1.6).

2. *Conformations of a polymer chain in solution.* A polymer is a macromolecule composed of N repeating units called monomers, which may or may not be all identical. In dilute solutions the interactions between polymer molecules can be neglected. What is then the mean square size of such a molecule? An isotropic polymer will be spherical on average, whereas a liquid crystalline polymer will on average be either elongated or flattened along a specific direction (different mean square radii along three mutually perpendicular directions, corresponding to $p_x + q_x = p_y + q_y = p_z + q_z = 1/3$, and $l_x \neq l_y \neq l_z$. This is illustrated in Problems 1.14, 3.3 and 4.7.

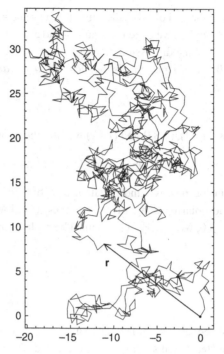

Figure 1.3 Two-dimensional random walk composed of $N = 1000$ steps. Each step has unit size and random orientation. $\mathbf{r}$ is the random walker's position after N steps. Note that a *single* random walk has a very irregular shape: it is only the *mean square displacement* that is the same in all directions.

3. *Magnetism*. If N atoms of spin magnetic moment 1/2 are placed in a magnetic field, we know from quantum mechanics that their magnetic moments can only align either parallel or antiparallel to the field. What is then the total mean magnetic moment of the N-atom sample per unit volume? This is studied in Chapters 3, 4 and 7. See also Problem 1.4.

4. *Brownian motion*. This is the motion performed by a macroscopic particle, e.g., a speck of dust, suspended in a liquid, as a result of its collisions with the liquid molecules. What is the particle's mean square displacement? Figure 1.3 shows a simulation of Brownian motion modelled by a two-dimensional random walk. This is studied in Chapter 11.

The central limit theorem*

The probability law for the total displacement of a random walker after N steps of random size can be derived from a very general result of probability theory known as the *central limit theorem*, which may be stated as follows:

> *Consider a sequence of independent random variables X_n, $n = 1, 2, \ldots, N$, each with finite mean value and variance. Define a new random variable Y as $Y = \sum_{n=1}^{N} X_n$. In the limit $N \to \infty$, the probability distribution of Y is the standard normal distribution.*

This theorem implies that, whatever the probability distribution $f(s)ds$ of each individual step, if the steps are statistically independent and if $f(s)$ drops sufficiently fast as $|s| \to \infty$, then for a sufficiently large number of steps N the total displacement will be standard-normally distributed, cf. Eq. (1.47):

$$f(x)dx = \frac{1}{\sqrt{2\pi\sigma^2}} \exp\left[-\frac{1}{2}\frac{(x-\mu)^2}{\sigma^2}\right] dx, \tag{1.91}$$

where

$$\mu = N\bar{s}, \tag{1.92}$$

$$\sigma^2 = N\sigma^2(s). \tag{1.93}$$

These, together with Eqs. (1.50) and (1.51), again lead to Eqs. (1.71) and (1.74), which we derived earlier by direct calculation of the moments of the probability distribution. The central limit theorem, whose proof can be found, for example, in Reif (1985), is a very general result that explains why so many phenomena in nature are approximately Gaussian, or standard-normally distributed: they are the sum of a large number of statistically independent effects, e.g., random collisions between molecules, uncorrelated changes in the orientation of a macromolecular backbone...

For the isotropic, three-dimensional random walk discussed earlier, x, y and z are statistically independent, so the probability that the total displacement will lie between $\mathbf{r}$ and $\mathbf{r} + d\mathbf{r}$ is $f(\mathbf{r})d\mathbf{r} = f(x)dxf(y)dyf(z)dz$. From Eq. (1.91) we thus have

$$f(x)dxf(y)dyf(z)dz \propto \exp\left(-\frac{x^2}{\overline{x^2}}\right) dx \exp\left(-\frac{y^2}{\overline{y^2}}\right) dy \exp\left(-\frac{z^2}{\overline{z^2}}\right) dz. \tag{1.94}$$

If in addition $\bar{\mathbf{r}} = \mathbf{0}$, it follows from Eq. (1.90) that

$$f(\mathbf{r})d^3r \propto \exp\left(-\frac{3r^2}{2\overline{r^2}}\right) d^3r, \tag{1.95}$$

where $d^3r = dxdydz = d\mathbf{r}$, and the proportionality constant is found by imposing normalisation to unity.

1.6 Monte Carlo sampling of random walks*

Suppose that some physical system at equilibrium can be modelled as a d-dimensional random walk of N steps of size l each. In three dimensions this could be, for example, a polymer chain under certain conditions. Such a chain can adopt a very large number of configurations. It is conceivable that the chain's instantaneous configuration will change in time, so that after a sufficiently long time it will have adopted all possible configurations. The sequence of chain configurations will be associated with a sequence of random walks. Instead of evolving the chain – or, in general, any system undergoing transitions between different configurations – in time, and computing 'time averages' of the physical quantities of interest, we consider instead, at some instant, a set of virtual copies of the system, each in a given configuration chosen at random. The set is chosen large enough that it contains all possible configurations of the system. In the case of a polymer chain, this set would consist of chains all with the same number N of steps, i.e., monomers, but different shapes. This is a *statistical ensemble* as defined at the start of this chapter. Now assume the ensemble contains Ω virtual copies of the system, of which Ω_i are in configuration i. The probability that the system will be found in state i when performing a measurement is then, by Eq. (1.1),

$$P_i = \lim_{\Omega \to \infty} \frac{\Omega_i}{\Omega}. \tag{1.96}$$

The probability thus defined is known as *ensemble probability*. The mean of a quantity such as the end-to-end distance of a random walk will now be written in terms of an ensemble average:

$$\bar{r} = \sum_{i=1}^{\Omega} r_i P_i, \tag{1.97}$$

where r_i is the end-to-end distance in the ith configuration. Equation (1.97) is rather impractical, as it requires knowledge of all possible configurations, which for many systems is not an easy task. The key idea of the *Monte Carlo method* is to approximate the sum over all configurations in Eq. (1.97) by a sum over *some* configurations, taken as a statistical sample.

We shall now apply the Monte Carlo method to the random walk and show that we can reproduce approximately, by numerical means, some of the theoretical results derived earlier. We thus need to generate M random walks of N steps each, where each walk is in some configuration i (there may be more than one walk in the same configuration), for which the end-to-end distance is r_i (there may be more than one configuration with the same r_i). In order to get a good distribution of configurations (i.e., a uniform distribution of statistically independent configurations), we shall generate the M configurations randomly using a sequence of computer-generated *random numbers*. Then for a sufficiently large sample size M we have, by the law of large numbers, Eq. (1.23):

$$\bar{r} = \frac{1}{N} \sum_{i=1}^{M} r_i, \tag{1.98}$$

i.e., the mean of r is just the arithmetic mean of r_i for all configurations generated. This is the *simple sampling* Monte Carlo method. Its accuracy therefore relies heavily on the quality of the random number generator used: after sampling M walks of N steps each, no correlations between random numbers must be apparent. Note that the Monte Carlo method as applied to the random walk may be used as a test of the random number generator: if the theoretical predictions are not verified after all other possible sources of error have been ruled out, then the random number generator must be at fault.

Let us consider a special case of the problem treated in Section 1.5: a random walk on an isotropic two-dimensional (square) lattice, i.e., where there are no privileged directions. Now there are four vectors connecting any given lattice site with its four nearest neighbours. If we take the lattice spacing to be unity, these vectors are (Figure 1.4):

$$\mathbf{v}_1 = (1,0); \quad \mathbf{v}_2 = (0,1); \quad \mathbf{v}_3 = (-1,0); \quad \mathbf{v}_4 = (0,-1). \tag{1.99}$$

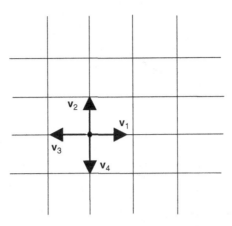

Figure 1.4   Illustration of vectors connecting a lattice site to its four nearest neighbours in a square lattice, Eq. (1.99).

We are now ready to write the following algorithm for generating and sampling random walks.

Algorithm SRW.   This algorithm generates and performs a simple sampling of M random walks of N steps each. It computes the mean end-to-end distance and the mean square radius of the walk [Binder and Heermann, 1992].

begin sampling loop from $i = 1$ to $i = M$
     starting site: $x = 0, y = 0$
     begin stepping loop $j = 1$ to N
          generate random integer n such that $0 \le n \le 3$
               $n = 0 :$ $x = x + 1$
               $n = 1 :$ $x = x - 1$
               $n = 2 :$ $y = y + 1$
               $n = 0 :$ $y = y - 1$
     end stepping loop
     compute $x_i(N)$ and $y_i(N)$
end sampling loop
compute $\bar{r}$ and $\overline{r^2}$

Figure 1.5 shows the results of running a code based on this algorithm on a personal computer. Here $M = 10\,000$ and $N = 2000$. Note that in order to check laws such as Eqs. (1.86)–(1.88) it is not enough to simulate walks of a fixed number of steps: instead, one must sample all walks with numbers of steps ranging from 1 to N.

The statistical character of, e.g., Eqs. (1.86)–(1.88) and (1.89) is now apparent. Each individual walk has some end-to-end distance (see Figure 1.6): the statistical laws only describe the behaviour of averages over a sufficiently large number of walks.

It must be emphasised that Monte Carlo methods play an important role in statistical physics not because they allow us to verify already known analytical results, but because they provide a means to obtain results that cannot be derived analytically. We should

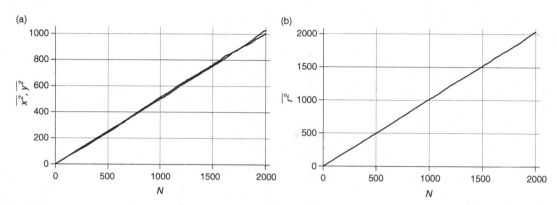

Figure 1.5 Results of simple sampling of $M = 10\,000$ random walks of length $N = 2000$. In the left panel it can be seen that $\overline{x^2}$ and $\overline{y^2}$ both approximate their expected behaviours, $\overline{x^2} = \overline{y^2} = N/2$. The right panel shows that $\overline{r^2} = \overline{x^2} + \overline{y^2} \simeq N$, also as expected.

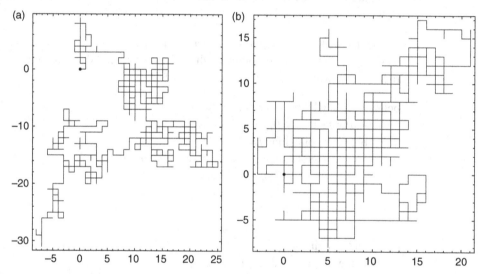

Figure 1.6 Two examples of random walks with $N = 1000$ steps, generated using algorithm SRW.

not, however, underestimate the relevance of reproducing known results as a check on the soundness of methods that may then be employed to study other systems. Coming back to the example of a polymer chain, a more realistic model would be a *self-avoiding random walk* (SAW), as a polymer chain cannot intersect itself due to excluded-volume effects.

An algorithm for generating a SAW must check, at every step, whether the random walker has already visited the site it wants to visit next. If not, then the random walker may proceed. Otherwise, the walk will terminate. It is apparent that most attempts at generating

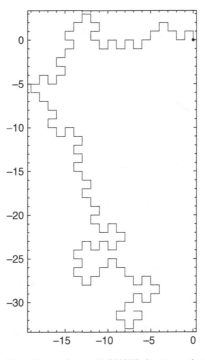

Figure 1.7   Example of indefinitely growing self-avoiding random walk (IGSAW). Starting at the top right-hand corner, this walk terminated after 146 steps.

SAWs with large numbers of steps will be unsuccessful.[2] A possible solution is to generate 'clever walks' that find their way around sites that have been visited before. One such walk is the *indefinitely growing self-avoiding random walk* (IGSAW), which can be generated by an algorithm similar to the SAW's, except that each new step is attempted not once, but l times, until a site is found that has not been visited before. The number of attempts l should be optimised taking into account the lattice dimensionality d (note that for $d = 1$ a SAW cannot turn back, and that finding a way around previously visited sites is easier the larger d is). Even so, a two-dimensional IGSAW has a non-negligible probability of getting trapped, as shown in Figure 1.7. Figures 1.8a and 1.8b plot, respectively, the mean square radius $\overline{r^2}$ and the sample size M vs N, for an IGSAW. The sample size M drops as N increases; at large N the dependence is exponential, as can be seen from Figure 1.8b or derived analytically (Binder and Heermann, 1992). Consequently, simple sampling is only reliable in a range of N whose width depends on the value of M for very small N. More sophisticated algorithms exist that allow the generation of SAWs of a fixed number of steps, and therefore to sample at constant N (see Problem 1.14).

In an IGSAW, unlike in a simple random walk, even steps very far apart are no longer statistically independent, so it is difficult to derive a statistical law for, e.g., the dependence

---

[2] Especially in low dimensions.

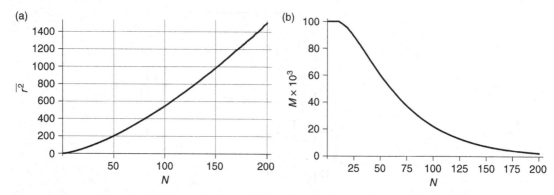

Figure 1.8 For an IGSAW, N dependence of (a) the mean square radius $\overline{r^2}$ and (b) the sample size M. For instance, for $N = 200$, $M = 2339$.

of the mean square radius on the number of steps. Recall that for a simple random walk we had Eq. (1.85), which can be written in the form

$$\overline{r^2} \propto N^\nu, \tag{1.100}$$

with $\nu = 1$. An IGSAW also obeys a scaling law of this type at large N; ν is found from the slope of a log-log plot of $\overline{r^2}$ vs N:

$$\log \overline{r^2} = \nu \log N + \text{const..} \tag{1.101}$$

The data in Figure 1.8a yield $\nu = 1.45$, implying that an IGSAW is 'swollen' relative to a simple random walk.[3] The exponent ν is termed a *critical exponent*.[4] Critical exponents are the same within each *universality class*: in the case of random walks, which universality class a walk belongs to depends on the type of walk and the lattice dimensionality, but not on the lattice structure. It is known that in the limit $N \to \infty$, exponent ν depends on dimensionality d as (Landau and Binder, 2000):

$$\nu \simeq \frac{6}{(d+2)}, \quad d \leq 4, \tag{1.102}$$

whence we conclude that as d increases, an IGSAW becomes less and less swollen relative to a simple random walk, until the two become identical at $d = 4$ (for which $\nu = 1$).

Monte Carlo methods are very useful for studying this type of problems (see, e.g., Binder and Heermann (1992), Landau and Binder (2000)), of which we shall see more examples in Chapters 8 and 9. Note that sample sizes of order 10^6 are often necessary to obtain good statistics.

---

[3] More rigorously, one may perform a fit of the form $\overline{r^2} = aN^\nu \left(1 + bN^{-\Delta}\right)$, where the second term in brackets is a correction term.

[4] Note that this particular critical exponent is associated with a geometrical quantity (the linear size of the walk) and not with the divergence of any thermodynamic quantity, as is the case with the critical exponents for continuous phase transitions that we shall encounter in Chapter 10.

1.A Some useful integrals

1.A.1 Integrals of the form $\int_0^{+\infty} e^{-bx}x^n dx$

Here we follow (Reif, 1985, App. A3). Start by taking $b = 1$. For $n = 0$ it is trivially found that

$$\int_0^{+\infty} e^{-x}dx = -\left[e^{-x}\right]_0^{+\infty} = 1. \tag{1.103}$$

For $n > 0$ we integrate by parts:

$$\int_0^{+\infty} e^{-x}x^n dx = -\int_0^{+\infty} x^n d\left(e^{-x}\right) = -\left[x^n e^{-x}\right]_0^{+\infty} + n\int_0^{+\infty} x^{n-1}e^{-x}dx. \tag{1.104}$$

The first term on the right-hand side of the above equation vanishes, leading to the recursion relation

$$\int_0^{+\infty} e^{-x}x^n dx = n\int_0^{+\infty} e^{-x}x^{n-1}dx. \tag{1.105}$$

If n is a positiver integer, we may iterate the above relation to get

$$\int_0^{+\infty} e^{-x}x^n dx = n(n-1)(n-2)\ldots(2)(1) = n!. \tag{1.106}$$

For non-integer n the integral will still converge if $n > -1$. In this case we define the *gamma function* as

$$\Gamma(n) = \int_0^{+\infty} e^{-x}x^{n-1}dx. \tag{1.107}$$

It follows from Eq. (1.106) that, if n is an integer,

$$\Gamma(n) = (n-1)!, \tag{1.108}$$

and from Eq. (1.105) that

$$\Gamma(n) = (n-1)\Gamma(n-1). \tag{1.109}$$

Finally, making the change of variables $y = bx$ with $b > 0$ in Eq. (1.107), we easily obtain the more general result

$$\int_0^{+\infty} e^{-bx}x^n dx = b^{-(n+1)}\Gamma(n+1). \tag{1.110}$$

1.A.2 Integrals of the form $\int_0^{+\infty} e^{-ax^2}x^n dx$

Here we follow (Reif, 1985, App. A4). Let

$$I(n) = \int_0^{+\infty} e^{-ax^2}x^n dx, \quad \text{with } n \geq 0. \tag{1.111}$$

Making the change of variables $y \equiv \sqrt{a}x$ we obtain, for $n = 0$,

$$I(0) = \frac{1}{\sqrt{a}} \int_0^{+\infty} e^{-y^2} dy = \frac{1}{2}\sqrt{\frac{\pi}{a}}, \tag{1.112}$$

where we used the following result, to be proved at the end of this appendix:

$$\mathcal{I} = \int_{-\infty}^{+\infty} e^{-y^2} dy = \sqrt{\pi}. \tag{1.113}$$

For $n = 1$,

$$I(1) = a^{-1} \int_0^{+\infty} y e^{-y^2} dy = a^{-1} \left[-\frac{1}{2}e^{-y^2} \right]_0^{+\infty} = \frac{1}{2a}. \tag{1.114}$$

By differentiating with respect to a, we derive the following recursion relation:

$$I(n) = -\frac{\partial}{\partial a} \left(\int_0^{+\infty} e^{-ax^2} x^{n-2} dx \right) = -\frac{\partial I(n-2)}{\partial a}, \tag{1.115}$$

which allows all integrals $I(n)$ for $n > 1$ to be written in terms of $I(0)$ or $I(1)$. For example,

$$I(2) = -\frac{\partial I(0)}{\partial a} = -\frac{\sqrt{\pi}}{2} \frac{\partial}{\partial a} \left(a^{-1/2} \right) = \frac{\sqrt{\pi}}{4} a^{-3/2}. \tag{1.116}$$

Alternatively, $I(n)$ can be written in terms of the gamma function. Making the change of variables $x = (u/a)^{1/2}$ in Eq. (1.111), we obtain

$$dx = \frac{1}{2} a^{-1/2} u^{-1/2} du, \tag{1.117}$$

and $I(n)$ thus becomes

$$I(n) = \frac{1}{2} a^{-(n+1)/2} \int_0^{+\infty} e^{-u} u^{(n-1)/2} du. \tag{1.118}$$

Using the definition of gamma function, Eq. (1.107), this becomes

$$I(n) \equiv \int_0^{+\infty} e^{-ax^2} x^n dx = \frac{1}{2} \Gamma\left(\frac{n+1}{2} \right) a^{-(n+1)/2}. \tag{1.119}$$

From Eqs. (1.103) or (1.108) we find

$$\Gamma(1) = 1. \tag{1.120}$$

On the other hand, setting $n = 1/2$ in Eq. (1.107) and making the change of variables $x = y^2$ we obtain

$$\Gamma\left(\frac{1}{2} \right) = \int_0^{+\infty} e^{-x} x^{-1/2} dx = \int_0^{+\infty} e^{-y^2} dy, \tag{1.121}$$

which combined with Eq. (1.113) yields

$$\Gamma\left(\frac{1}{2} \right) = \sqrt{\pi}. \tag{1.122}$$

Equations (1.109) and (1.119)–(1.122) allow us to construct the following table of integrals (Table 1.1). They should be checked as an exercise.

Table A.1 Some useful integrals

$I(0) = \frac{1}{2}\sqrt{\pi}a^{-1/2}$	$I(3) = \frac{1}{2}a^{-2}$
$I(1) = \frac{1}{2}a^{-1}$	$I(4) = \frac{3}{8}\sqrt{\pi}a^{-5/2}$
$I(2) = \frac{1}{4}\sqrt{\pi}a^{-3/2}$	$I(5) = a^{-3}$

Proof of Eq. (1.113)   Start by writing

$$\mathcal{I}^2 = \int_{-\infty}^{+\infty} e^{-x^2}\,dx \int_{-\infty}^{+\infty} e^{-y^2}\,dy = \int_{-\infty}^{+\infty}\int_{-\infty}^{+\infty} e^{-(x^2+y^2)}\,dx\,dy. \tag{1.123}$$

This is an integral over the whole xy plane. Changing to polar coordinates, the infinitesimal element of area is now $r\,dr\,d\theta$ and we have

$$\mathcal{I}^2 = \int_0^{+\infty}\int_0^{2\pi} e^{-r^2} r\,dr\,d\theta = 2\pi \int_0^{+\infty} e^{-r^2} r\,dr = 2\pi \int_0^{+\infty}\left(-\frac{1}{2}\right)d\left(e^{-r^2}\right)$$

$$= -\pi \left[e^{-r^2}\right]_0^{+\infty} = \pi, \tag{1.124}$$

whence $\mathcal{I} = \sqrt{\pi}$.

1.A.3 Integrals of the form $\int_{-\infty}^{+\infty} e^{-(ax^2+bx+c)}\,dx$

Integrals of this form can be easily calculated by completing the square in the exponent:

$$ax^2 + bx + c = a\left(x^2 + \frac{b}{a}x + \frac{c}{a}\right)$$

$$= a\left(x^2 + \frac{b}{a}x + \frac{b^2}{4a^2} + \frac{c}{a} - \frac{b^2}{4c^2}\right)$$

$$= a\left(x + \frac{b}{2a}\right)^2 + c - \frac{b^2}{4a},$$

and then using Eq. (1.113):

$$\int_{-\infty}^{+\infty} e^{-(ax^2+bx+c)}\,dx = e^{-c+b^2/4a} \int_{-\infty}^{+\infty} e^{-a(x+b/2a)^2}\,dx$$

$$= \frac{e^{-c+b^2/4a}}{\sqrt{a}} \int_{-\infty}^{+\infty} e^{-y^2}\,dy$$

$$= e^{-c+b^2/4a}\sqrt{\frac{\pi}{a}}, \tag{1.125}$$

where we have made the change of variables $y = \sqrt{a}\,(x + b/2a)$.

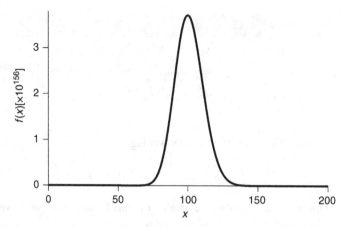

Figure 1.9   Plot of $F_n(x) = e^{-x}x^n$ for $n = 100$. Statistical physics typically deals with $n \sim 10^{24}$ (Avogadro's number), for which $F_n(x)$ would be quite considerably sharper.

1.B Stirling's formula

Computing $n!$ is hard for large n. We shall now derive an approximation for $n!$ in this limit, starting from the integral representation derived in Appendix 1.A.1, Eq. (1.106) along the lines of (Reif, 1985, App. A6).

Consider the integrand, $F_n(x) = e^{-x}x^n$. For large n, x^n is a steeply increasing function of x, whereas e^{-x} is a steeply decreasing function of x. Then $F_n(x)$ has a sharp maximum at some $x = x_0$ and decreases abruptly as x deviates from x_0: see Figure 1.9.

Finding x_0 is most conveniently done in terms of the logarithm of $F_n(x)$, which obviously is also maximised when $x = x_0$:

$$\frac{d\ln F_n(x)}{dx} = 0 \Rightarrow \frac{d}{dx}(n\ln x - x) = \frac{n}{x} - 1 = 0,$$

whence $x_0 = n$ (see Figure 1.9). Because this maximum is rather sharp, the only contributions to the integral in Eq. (1.106) will come from x in the vicinity of x_0. We shall therefore obtain an approximation for this integral by expanding $F_n(x)$ in powers of $\xi = x - n \ll 1$. Actually, we shall expand $\ln F_n(x)$, which varies more smoothly than $F_n(x)$ and does not have very large high-order derivatives for $x \approx x_0$, which will allow us to truncate the expansion at low order. We thus have

$$\ln F_n(x) = n\ln x - x = n\ln(n + \xi) - (n + \xi).$$

Expanding the logarithm, we get

$$\ln(n + \xi) = \ln n + \ln\left(1 + \frac{\xi}{n}\right) = \ln n + \frac{\xi}{n} - \frac{1}{2}\left(\frac{\xi}{n}\right)^2 + \cdots$$

whence

$$\ln F_n(x) \approx n\ln n - n + \frac{1}{2}\frac{\xi^2}{n} \Rightarrow F_n(x) \approx n^n e^{-n} e^{-\xi^2/2n}.$$

The last exponential factor clearly shows that $F_n(x)$ has a maximum at $\xi = 0$ and becomes very small when $|\xi| \gg \sqrt{n}$ (see Figure 1.9). Substituting this into Eq. (1.106) yields

$$n! \approx \int_{-n}^{+\infty} n^n e^{-n} e^{-\xi^2/2n} d\xi = n^n e^{-n} \int_{-\infty}^{+\infty} e^{-\xi^2/2n} d\xi,$$

where we have replaced the lower limit of integration by $-\infty$. This is permissible because the argument is negligibly small for $\xi < -n$. By Eq. (1.112) this equals

$$n! \approx \sqrt{2\pi n} n^n e^{-n}, \quad \text{for } n \gg 1, \tag{1.126}$$

which can be written in the alternative form:

$$\ln n! \approx n \ln n - n + \frac{1}{2} \ln(2\pi n). \tag{1.127}$$

Equation (1.126) or (1.127) is known as *Stirling's formula*. For very large n, $n \gg \ln n$ and Eq. (1.127) simplifies to

$$\ln n! \approx n \ln n - n. \tag{1.128}$$

In most situations of interest in statistical physics the numbers involved are large enough that Eq. (1.128) is a good approximation (e.g., if $n = 6 \times 10^{23}$ – of the order of Avogadro's number – then $\ln n = 55$).

Problems

1.1 Prove Eq. (1.31).
1.2 Consider an ideal gas of N_0 molecules in equilibrium, in a volume V_0.
   (a) What is the probability p that a gas molecule finds itself in a subvolume V of V_0?
   (b) Let N be the number of molecules in volume V. Find $\overline{N}$ in terms of N_0, V_0 and V.
   (c) Compute the relative fluctuation of N for $V = V_0/2$ and (i) $N_0 = 100$; (ii) $N_0 = 10^{24}$.
1.3 Prove Eq. (1.51).
1.4 Consider an ideal system of n spins 1/2 in equilibrium, in an applied magnetic field **H**. Let μ_i be the component along the field direction of the magnetic moment associated with the ith spin, and let M be the total magnetic moment of the system along the field direction, i.e., $M = \sum_{i=1}^{N} \mu_i$. Each μ_i can only take the values $+\mu$ (magnetic moment 'parallel' to the field) or $-\mu$ (magnetic moment 'antiparallel' to the field), with probabilities p and q, respectively.
   (a) If **H** $\neq$ **0**, is $p > q$ or $p < q$? Why?
   (b) Find $\overline{M}$ and $\sigma^2(M)$.
   (c) Assume $p = 0.51$. Find $\overline{M}$ and the relative fluctuation of M for (i) $N = 100$; (ii) $N = 10^{24}$. What can you conclude?

1.5 A particle is free to move at random along OX. The particle displacement s (which can be positive or negative) is a Gaussian random variable:

$$f(s) = \frac{1}{\sqrt{2\pi\sigma^2}} \exp\left[-\frac{(s-l)^2}{2\sigma^2}\right],$$

where l is the mean displacement. Find the mean and variance of the particle position x after N displacements.

1.6 Consider a molecule diffusing in a gas. Let $\mathbf{s}_i$ be the displacement vector between the $(i-1)$th and ith collisions with other gas molecules. Consecutive displacements are statistically independent and there are no privileged directions. Let $\mathbf{R}$ be the total displacement after N collisions: $\mathbf{R} = \sum_{i=1}^{N} \mathbf{s}_i$.
(a) Find the mean and variance of $\mathbf{R}$.
(b) Repeat the calculation for $||\mathbf{s}_i|| = l$, $i = 1, 2, \ldots, N$.

1.7 Find the normalisation constant in Eq. (1.95).

1.8 Discuss the plots in Figure 1.5.

1.9 Discuss the following plots, which show results for a biased random walk with $p_{x_+} = 3/8$, $p_{x_-} = 1/8$, $p_{y_+} = p_{y_-} = 2/8$ and $M = 10^4$, $N = 10^3$. This may represent the motion of a particle in an applied field (e.g., a charged particle in an electric field).

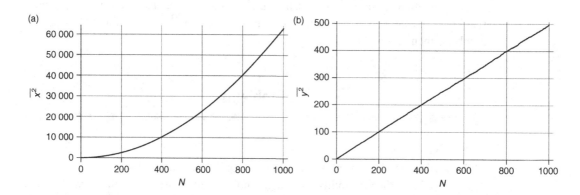

1.10 Code algorithm SRW. Run the code and check that you obtain the expected scaling laws.

1.11 Code an algorithm for the biased random walk. Run the code for different p_x and p_y. Discuss the results.

1.12 Write and run codes for the following: (i) simple random walk with steps of constant size but arbitrary direction; (ii) biased random walk; (iii) random walk with step size given by a Gaussian probability distribution. Compare results for the mean square radius in all these cases.

1.13 Code an algorithm for the indefinitely growing self-avoiding random walk in two and three dimensions. Find its critical exponents from Eq. (1.101), and compare the results with the predictions of Eq. (1.102).

1.14 Code a *reptation* algorithm as follows: (1) generate an N-step, indefinitely growing self-avoiding random walk (IGSAW) ; (2) remove the first step of this walk and add an additional step at the end, using the IGSAW algorithm; (3) repeat (2) M times or until the random walk becomes trapped (it is less likely to be trapped in three dimensions than in two – see Landau and Binder (2000, chap. 4.7) for more efficient IGSAW algorithms). This allows sampling to be done at a constant number of steps N. The reptation model was introduced by de Gennes, and by Doi and Edwards, to study the motion of an entangled polymer molecule in a melt or a concentrated solution [Grosberg and Khokhlov, 1997].

References

Apostol, T. M. 1969. *Calculus*, Vol. 2, 2nd edition. Wiley.
Binder, K., and Heermann, D. W. 1992. *Monte Carlo Simulation in Statistical Physics*, 2nd edition. Springer-Verlag.
Grosberg, A. Y., and Khokhlov, A. R. 1997. *Giant Molecules*. Academic Press.
Landau, D. P., and Binder, K. 2000. *Guide to Monte Carlo Simulations in Statistical Physics*. Cambridge University Press.
Reif, F. 1985. *Fundamentals of Statistical and Thermal Physics*. McGraw-Hill.

2 Review of thermodynamics

2.1 Introduction

When confronted with a physical system, such as a gas in a closed container fitted with a moving piston, or a crystal in a magnetic field, our first task is to identify suitable observable quantities to describe the system. If we opt for a *macroscopic* description, these will be properties of the system as a whole that are, for the most part, readily perceptible to our senses, such as its temperature, pressure, volume, internal energy or entropy. The laws of *thermodynamics* relate these quantities when the system exchanges energy with its surroundings. If, on the other hand, we choose to carry out a *microscopic* description, we shall be using quantities of which we have no direct experience, such as the position, linear momentum, kinetic energy, magnetic moment, etc., of each and every atom, molecule or ion in the system. *Statistical mechanics* allows us to predict the macroscopic properties of the system starting from these microscopic quantities. Before we delve into the principles of statistical mechanics in the next chapter, we shall review here the basic concepts of classical (equilibrium) thermodynamics and how they can be applied to study magnetic systems and dielectrics.

2.2 Basic concepts of equilibrium thermodynamics

2.2.1 The laws of thermodynamics

Thermodynamics is a phenomenological theory concerned with the equilibrium properties of, and energy (heat or work) exchanges between *thermodynamics systems* – macroscopic systems about whose structure thermodynamics makes no assumptions.

The basic hypothesis is that a thermodynamic system at equilibrium has well-defined properties, described by a few macroscopic parameters such as its temperature T, pressure P, etc. A *thermodynamic state* is specified by the values of all thermodynamic parameters describing the system, called thermodynamic variables. A system is said to be in *thermodynamic equilibrium* when its state does not change in time. At equilibrium, thermodynamic variables such as the density or temperature are the same everywhere in the system, i.e., there are no density or temperature gradients as exist in a non-equilibrium state. By 'equilibrium' we mean *stable equilibrium*, in the sense that small fluctuations of the thermodynamic variables around their equilibrium values do not change the state of the system. Larger, non-negligible fluctuations occur in the vicinity of a phase transition (*critical fluctuations*); these belong to the realm of critical phenomena – see Chapter 10.

There are two basic types of thermodynamic variable, *extensive* and *intensive*. An *extensive* variable is proportional to the number of subsystems that make up the system under study, e.g., the number of molecules in a gas. An *intensive* variable, on the other hand, takes the same value independently of how many subsystems there are in the system. A functional relationship involving the thermodynamic variables (usually extensive as well as intensive) of a given system is called an *equation of state* of that system. If, as is often the case in practice, the thermodynamic variables are P, V and T, then the equation of state is of the form $f(P, V, T) = 0$, which tells us that, in this system at equilibrium, only two of the three thermodynamic variables are independent. The equation of state needs to be found experimentally (or derived from statistical mechanical considerations), and is part of the system specification. For instance, the equation of state of an ideal gas is $PV = Nk_BT$, where N is the number of molecules and $k_B = 1.381 \times 10^{-23}\,\mathrm{J\,K^{-1}}$ is Boltzmann's constant.

This equation of state defines a temperature scale: the *ideal-gas temperature T*, whose numerical value will depend on the choice of units. Conventionally this is the kelvin, denoted by the symbol K, and defined in such a way that the triple point of water is at $T = 273.16\,\mathrm{K}$.[1] We shall show later that the absolute temperature, as derived from microscopic considerations, coincides with the ideal-gas temperature. The ideal-gas temperature is universal because at low enough pressures all gases behave in the same way; the ideal gas is an embodiment of this universal limiting behaviour. The existence of a 'thermometer', i.e., a system that can be used to measure temperature, is a consequence of the result known as the *zeroth law of thermodynamics*: any two systems in thermal equilibrium with a third system are in thermal equilibrium with each other.

By definition, a *thermodynamic process* is a change of thermodynamic state. If the initial state is an equilibrium one, then a process may only occur if there is a change in the external conditions imposed on the system. If this change is sufficiently slow that the system can be assumed to be in equilibrium at every instant (i.e., to evolve through a succession of equilibrium states), then the process is termed *quasi-static*. When the external conditions are returned to their initial values, the system may or may not return to its initial state. If it does, the process is termed *reversible*; and *irreversible* if it does not. A reversible process is also quasi-static, but the converse need not be true.

We shall start by studying how a system undergoing a reversible process exchanges energy with its surroundings. We know from mechanics that the work performed by a force F that displaces a body by dx is $d\!\!^-W = Fdx$,[2] where $d\!\!^-$ means that $d\!\!^-W$ is an infinitesimal quantity, but not an exact differential of W, in the sense that it cannot be expressed in terms of the derivative of W (it is an *inexact* differential). Likewise, the work performed *on* a system, such as, a gas at pressure P inside a cylinder fitted with a moving piston, is

$$d\!\!^-W = -PdV, \qquad (2.1)$$

---

[1] The alert reader will not have failed to notice that we need *two* fixed points to define a temperature scale. The missing fixed point is $T = 0\,\mathrm{K}$, at which the pressure of the ideal gas vanishes.

[2] For simplicity, we assume that the displacement is along the direction of the force.

where the minus sign expresses the convention that work performed *on* the system, leading to a decrease in volume ($dV < 0$), is taken to be positive ($đW > 0$). P is the intensive variable conjugate to the extensive variable V.

Consider now a system where the number of particles may change. One possible realisation is a glass of water that loses molecules through evaporation. If the number of particles N in the system changes by dN, the work performed on the system is

$$đW = \mu dN, \tag{2.2}$$

where μ, the *chemical potential*, is the intensive variable conjugate to the number of particles. In general, there are as many chemical potentials as there are types of particle in the system. The name chemical potential comes from the role this quantity plays in chemical reactions but, as we shall see, it is of more general applicability. As an example, the Fermi energy of an ideal Fermi gas is defined in terms of its chemical potential. If there is more than one type of particle, Eq. (2.2) generalises to

$$đW = \sum_i \mu_i dN_i, \tag{2.3}$$

where now μ_i is the chemical potential of species i.

The intensive variables temperature, pressure and chemical potential allow different kinds of equilibrium to be defined between systems, or between parts of the same system. Two systems, call them 1 and 2, are said to be in *thermal equilibrium* when no heat is transferred between them; this requires that $T_1 = T_2$. Likewise, two systems are said to be in *mechanical equilibrium* if neither increases its volume at the expense of the other; this requires that $P_1 = P_2$. Finally, two systems are said to be in *equilibrium with respect to the exchange of particles* when the net exchange of particles between them is nil; this requires that $\mu_1 = \mu_2$. In the example discussed above of a glass of water, this would mean that as many molecules must evaporate as condense, in order that the liquid water be in equilibrium with its vapour.

In the preceding examples, V and N are the external parameters that specify the state of our system; by analogy with mechanics, they are referred to as *generalised coordinates*. Likewise, the intensive variables P and μ are the corresponding *generalised forces* acting on the generalised coordinates to change their values, and therefore performing work. Other examples of pairs of generalised coordinate–generalised force are, in a magnetic system, the total magnetic moment of a sample and the applied magnetic field, respectively; and in an electric system, the total electric moment and the applied electric field, respectively.[3]

The energy of a system may also vary without any change of its generalised coordinates, i.e., even if no work is performed on the system. By definition, the energy transferred to a system that is not work is called *heat*, and is denoted by Q. The heat is related to changes in motion at the molecular scale. A macroscopic equilibrium state of a thermodynamic

---

[3] We use the total magnetic moment and the total electric moment as the extensive quantities conjugate to the magnetic and electric fields, respectively. The magnetisation and the polarisation are, respectively, the total magnetic moment per unit volume and the total electric moment per unit volume, and therefore intensive quantities

system may be specified by its internal energy U; if the system is isolated, it follows from the principle of conservation of energy that $U = $ constant. If on the other hand we allow the system to exchange energy with its surroundings, the infinitesimal change in its internal energy, dU, equals the work performed *on* the system by the 'external forces', $đW$, plus the heat $đQ$ *supplied* to the system:

$$dU = đQ + đW. \tag{2.4}$$

In writing Eq. (2.4) we adhere to the convention that work is positive (negative) if it is performed on (by) the system. As an example, consider a gas in a container with one movable wall. If the gas is compressed, then by Eq. (2.1) the work performed on the gas is $đW = -PdV > 0$. On the other hand, if the gas is at a higher pressure than its surroundings, it will expand, performing the work $đW = -PdV < 0$. In either case, we write $dU = đQ - PdV$.

This is the *first law of thermodynamics* for infinitesimal processes: the internal energy of a given system varies when the system exchanges heat and/or work with its surroundings. This is a consequence of the *principle of conservation of energy*: because energy can be neither created nor destroyed: a change in the energy of a system is only possible if the system exchanges energy with its surroundings. dU is an exact differential, i.e., U is a *state function* or *function of state* of the system. The first law is valid for any thermodynamic process, be it reversible or irreversible.

The *second law of thermodynamics* may be formulated as follows: A macroscopic equilibrium state of any given system may be specified by a state function S, called *entropy*, which has the following properties: (a) in any process whereby an isolated system changes its state, its entropy never decreases, i.e.,

$$\Delta S \geq 0, \tag{2.5}$$

where the equality holds if the process is reversible; (b) if the system is not isolated and undergoes an infinitesimal transformation that involves exchanging a heat $đQ$ with its surroundings, then

$$đQ \leq TdS, \tag{2.6}$$

where, again, the equality holds if the process is reversible. T is the *absolute temperature* of the system, and dS is an exact differential because S is a state function. A possible alternative formulation of the second law is: In an isolated system (hence at constant internal energy), the entropy is maximised in the thermodynamic equilibrium state. We shall see later what the physical meaning of the entropy is; for now, it saves us worrying about the behaviour of individual molecules.

Combining the first and second laws, it follows that the internal energy U of a system is a function of its generalised coordinates, which we shall denote by x_k, and of its entropy S: $U = U(x_1, x_2, \ldots, S)$. The exact differential dU is then

$$dU(x_1, x_2, \ldots, S) = \sum_k \left(\frac{\partial U}{\partial x_k} \right)_{x_{i \neq k}, S} dx_k + \left(\frac{\partial U}{\partial S} \right)_{x_1, x_2, \ldots} dS, \tag{2.7}$$

which may be rewritten as

$$dU = TdS - \sum_k X_k dx_k,$$ (2.8)

where we have defined the generalised forces X_k conjugate to the generalised coordinates x_k as

$$X_k = -\left(\frac{\partial U}{\partial x_k}\right)_{x_{i\neq k},S},$$ (2.9)

and the absolute temperature as

$$T = \left(\frac{\partial U}{\partial S}\right)_{x_1,x_2,\ldots}$$ (2.10)

Furthermore, if only mechanical work is performed on or by the system, then the only relevant generalised coordinate is the volume V, and Eq. (2.8) becomes

$$dU = TdS - PdV,$$ (2.11)

where the pressure is the force conjugate to the volume:

$$P = -\left(\frac{\partial U}{\partial V}\right)_S.$$ (2.12)

Equation (2.8) is the *fundamental thermodynamic relation for an infinitesimal process*. It is valid for any quasi-static process, be it reversible or irreversible, because only in such a process, which is a succession of equilibrium states, is it possible to define, at every moment in time, the state functions U and S.

If the process is reversible, we have, again from the first and second laws, Eqs. (2.4) and (2.6), respectively,

$$TdS = \bar{d}Q,$$ (2.13)

$$-\sum_k X_k dx_k = \bar{d}W.$$ (2.14)

If, on the other hand, the process is irreversible, Eq. (2.8) of course still holds, but we cannot identify $\sum_k X_k dx_k$ with the work performed on the system, nor TdS with the heat supplied to the system by its surroundings. Indeed, if for example a gas is compressed irreversibly, the (purely mechanical) work performed on the system is $\bar{d}W_{\mathrm{irrev}} > -PdV$. In this case, from Eqs. (2.4) and (2.11) we have

$$dU = \bar{d}Q + \bar{d}W = \bar{d}Q - PdV,$$ (2.15)

whence we conclude that $\bar{d}Q < TdS$, i.e., the entropy increase is not just due to the heat the system exchanges with its surroundings: *irreversible work generates entropy*. If $\bar{d}Q = 0$, there is no net heat flux into or out of the system. Such a process is called *adiabatic*.

Equations (2.9), (2.10) and (2.12) relate the generalised coordinates, the generalised forces and the temperature. They are the *equations of state* alluded to earlier. In Eq. (2.8) the entropy may be regarded as a generalised coordinate, and the temperature as a generalised force. Thermodynamics thus introduces a new generalised coordinate – the entropy – that is the collective expression of motion on the microscopic scale, and a 'force' – the absolute temperature – that drives variations of this generalised coordinate. By Eq. (2.10), the absolute temperature coincides with the ideal gas temperature, so we shall no longer make any distinction between the two.

Finally, the *third law of thermodynamics* states that the entropy of a system has the following limiting property:

$$\lim_{T \to 0} S = S_0, \tag{2.16}$$

where S_0 is a constant that does not depend on any system parameters, and is therefore universal. Usually it is set to zero.

We finish this section by noting that the above laws are completely macroscopic in character: nowhere did we refer to the properties of the constituent atoms, molecules or ions, or to their interactions.

2.2.2 Thermodynamic potentials. Maxwell's relations

From the fundamental thermodynamic relation, Eq. (2.8), a number of useful results may be derived. Let us suppose that we want to study a homogeneous fluid, whose only external parameter is the volume V, described by the thermodynamic variables T, S, P and V. Because these four variables are related by two equations of state, (2.10) and (2.12), any two of the four may be chosen as the independent variables specifying the state of the system. The remaining two variables will be given by the equations of state, to be derived from the fundamental thermodynamic relation written in terms of the appropriate thermodynamic potential (to be defined later).

For definiteness, consider a system specified by S and V. Now $U = U(S, V)$ and Eq. (2.8) writes

$$dU = TdS - PdV, \tag{2.17}$$

and the two equations of state are

$$T = \left(\frac{\partial U}{\partial S} \right)_V, \tag{2.18}$$

$$P = - \left(\frac{\partial U}{\partial V} \right)_S. \tag{2.19}$$

Because dU given by Eq. (2.17) is an exact differential, we must have

$$\frac{\partial^2 U}{\partial V \partial S} = \frac{\partial^2 U}{\partial S \partial V} \Rightarrow \left(\frac{\partial T}{\partial V} \right)_S = - \left(\frac{\partial P}{\partial S} \right)_V, \tag{2.20}$$

which expresses the fact that T, S, V and P cannot all vary independently.

Let us now take S and P as the independent variables. The fundamental thermodynamic relation may be rewritten as

$$dU = TdS - PdV = TdS - d(PV) + VdP \Rightarrow d(U + PV) = TdS + VdP, \tag{2.21}$$

which suggests that we define a new state function, called the *enthalpy* and denoted by E^* or H,[4] as

$$E^* = U + PV, \qquad (2.22)$$

in terms of which the fundamental thermodynamic relation becomes

$$dE^* = TdS + VdP. \qquad (2.23)$$

We thus conclude that $E^* = E^*(S, P)$, i.e., that the entropy and the pressure are the natural variables of the enthalpy. The resulting equations of state are thus:

$$T = \left(\frac{\partial E^*}{\partial S}\right)_P, \qquad (2.24)$$

$$V = \left(\frac{\partial E^*}{\partial P}\right)_S. \qquad (2.25)$$

Because dE^* given by Eq. (2.23) is an exact differential, we must have

$$\frac{\partial^2 E^*}{\partial P \partial S} = \frac{\partial^2 E^*}{\partial S \partial P} \Rightarrow \left(\frac{\partial T}{\partial P}\right)_S = \left(\frac{\partial V}{\partial S}\right)_P, \qquad (2.26)$$

Next, take V and T as the independent variables. Proceeding as before, the fundamental thermodynamic relation may be rewritten as

$$dU = TdS - PdV = d(TS) - SdT - PdV \Rightarrow d(U - TS) = -SdT - PdV, \qquad (2.27)$$

which suggests that we define a new state function, called the *Helmholtz free energy* or *Helmholtz potential*, and denoted by F or A, as

$$F = U - TS, \qquad (2.28)$$

in terms of which the fundamental thermodynamic relation becomes

$$dF = -SdT - PdV. \qquad (2.29)$$

We likewise conclude that $F = F(T, V)$, i.e., that the temperature and the volume are the natural variables of the Helmholtz free energy (sometimes referred to simply as the Helmholtz energy), The equations of state follow as before:

$$S = -\left(\frac{\partial F}{\partial T}\right)_V, \qquad (2.30)$$

$$P = -\left(\frac{\partial F}{\partial V}\right)_T. \qquad (2.31)$$

And again, because dF given by Eq. (2.29) is an exact differential, we must have

$$\frac{\partial^2 F}{\partial V \partial T} = \frac{\partial^2 F}{\partial T \partial V} \Rightarrow \left(\frac{\partial S}{\partial V}\right)_T = \left(\frac{\partial P}{\partial T}\right)_V, \qquad (2.32)$$

---

[4] Here we shall always use E^*, as H already denotes the magnetic field.

Finally, take as independent variables T and P. Proceeding as before, the fundamental thermodynamic relation may be rewritten as

$$dU = TdS - PdV = d(TS) - SdT - d(PV) + VdP, \qquad (2.33)$$

which suggests that we define a new state function, called the *Gibbs free energy* or *Gibbs potential*, and denoted by G, as

$$G = U - TS + PV, \qquad (2.34)$$

in terms of which the fundamental thermodynamic relation becomes

$$dG = -SdT + VdP. \qquad (2.35)$$

As before, we conclude that $G = G(T, P)$, i.e., that the temperature and the pressure are the natural variables of the Gibbs free energy (sometimes referred to simply as Gibbs energy), The equations of state are now:

$$S = -\left(\frac{\partial G}{\partial T}\right)_P, \qquad (2.36)$$

$$V = \left(\frac{\partial G}{\partial P}\right)_T. \qquad (2.37)$$

And yet again, because dG given by Eq. (2.35) is an exact differential,,we must have

$$\frac{\partial^2 G}{\partial P \partial T} = \frac{\partial^2 G}{\partial T \partial P} \Rightarrow \left(\frac{\partial S}{\partial P}\right)_T = -\left(\frac{\partial V}{\partial T}\right)_P, \qquad (2.38)$$

Equations (2.20), (2.26), (2.32) and (2.38) relating T, S, P, and V are known as *Maxwell's relations*. They follow immediately from the fact that T, S, P, and V are connected by the fundamental thermodynamic relation, and are therefore not independent. The enthalpy, Helmholtz free energy and Gibbs free energy are the most commonly used *thermodynamic potentials*; they are also functions of state. For each pair of independent variables, the fundamental thermodynamic relation is most conveniently written in terms of the thermodynamic potential that has that pair of variables as its natural variables. We shall discuss the physical meaning of the thermodynamic potentials later, when we derive the general equilibrium conditions of a thermodynamic system. For now, note that each thermodynamic potential is extremised when the infinitesimal variations of its natural variables vanish:

$$dS = 0, \quad dV = 0 \Rightarrow dU = TdS - PdV = 0, \qquad (2.39)$$

$$dS = 0, \quad dP = 0 \Rightarrow dE^* = TdS + VdP = 0, \qquad (2.40)$$

$$dT = 0, \quad dV = 0 \Rightarrow dF = -SdT - PdV = 0, \qquad (2.41)$$

$$dT = 0, \quad dP = 0 \Rightarrow dG = -SdT + VdP = 0. \qquad (2.42)$$

Knowledge of any one of these four state functions, $U = U(S, V), E^* = E^*(S, P), F = F(T, V)$ or $G = G(T, P)$, for all values of its natural variables, allows us to find the other three state functions. For example, if we know F we can get U as follows:

$$U = F + TS = F - T\left(\frac{\partial F}{\partial T}\right)_V = -T^2 \left[\frac{\partial}{\partial T}\left(\frac{F}{T}\right)\right]_V. \qquad (2.43)$$

The above results can be generalised to a homogeneous fluid composed of only one species of particle, but where the number N of particles is not constant. Now one additional variable is required to specify the state of such a system, i.e., we need three of the five variables T, S, P, V and N, so the fundamental thermodynamic relation is written:

$$dU = TdS - PdV + \mu dN, \tag{2.44}$$

whence $U = U(S,V,N)$. This can also be rewritten in terms of the Helmholtz free energy, Eq. (2.28), as

$$dF = -SdT - PdV + \mu dN, \tag{2.45}$$

whence $F = F(T,V,N)$. Alternatively, one may also write it in terms of the Gibbs free energy, Eq. (2.34), as

$$dG = -SdT + VdP + \mu dN, \tag{2.46}$$

whence $G = G(T,P,N)$. From Eqs. (2.44)–(2.46) we find the following equations of state:

$$\mu = \left(\frac{\partial U}{\partial N}\right)_{S,V} = \left(\frac{\partial F}{\partial N}\right)_{T,V} = \left(\frac{\partial G}{\partial N}\right)_{T,P}. \tag{2.47}$$

Comparison of the first of these equalities with Eq. (2.9) allows us to identify μ, the chemical potential, as the generalised force conjugate to N. The other two equations of state are analogous to those obtained before, except that we now need to keep two variables constant when differentiating.

2.2.3 Thermodynamic response functions

When in an infinitesimal process a system absorbs an amount of heat $đQ$, and as a result its temperature changes by dT, the *heat capacity* of the system is defined as the ratio

$$C_x = \left(\frac{đQ}{dT}\right)_x, \tag{2.48}$$

where x denotes all parameters kept constant during the process. C_x is an extensive quantity: heat capacities per unit mass, or per mole, are called *specific heats*, and are intensive quantities. The heat capacity, or the specific heat, are examples of *thermodynamic response functions*: they measure how a system responds to a given external stimulus – in this case, the uptake of heat from its surroundings.

We shall now derive expressions for the heat capacity at constant volume or at constant pressure. From the first law, Eq. (2.4), and from Eq. (2.1), we get, taking (V,T) as the independent variables:

$$đQ = \left(\frac{\partial U}{\partial T}\right)_V dT + \left[\left(\frac{\partial U}{\partial V}\right)_T + P\right] dV. \tag{2.49}$$

If, on the other hand, we take (P,T) as the independent variables:

$$đQ = \left[\left(\frac{\partial U}{\partial T}\right)_P + P\left(\frac{\partial V}{\partial T}\right)_P\right] dT + \left[\left(\frac{\partial U}{\partial P}\right)_T + P\left(\frac{\partial V}{\partial P}\right)_T\right] dP. \tag{2.50}$$

Setting $dV = 0$ in Eq. (2.49) and $dP = 0$ in Eq. (2.50) then yields

$$C_V = \left(\frac{\partial U}{\partial T}\right)_V, \tag{2.51}$$

$$C_P = \left(\frac{\partial E^*}{\partial T}\right)_P, \tag{2.52}$$

where we have used the definition of enthalpy, Eq. (2.22).

The heat capacity may also be written in terms of the derivatives of other state functions. Using the second law (for reversible processes), Eq. (2.49) becomes

$$dS = \frac{C_V}{T}dT + \frac{1}{T}\left[\left(\frac{\partial U}{\partial V}\right)_T + P\right]dV. \tag{2.53}$$

Because dS is an exact differential, we must have

$$\left(\frac{\partial}{\partial V}\right)_T \left(\frac{C_V}{T}\right) = \left(\frac{\partial}{\partial T}\right)_V \left[\frac{1}{T}\left(\frac{\partial U}{\partial V}\right)_T + \frac{P}{T}\right]. \tag{2.54}$$

Using Eq. (2.51) and performing the derivatives on the right-hand side of Eq. (2.54), we obtain

$$\left(\frac{\partial U}{\partial V}\right)_T = T\left(\frac{\partial P}{\partial T}\right)_V - P. \tag{2.55}$$

Substituting Eq. (2.55) into Eq. (2.53) then yields

$$TdS = C_V dT + T\left(\frac{\partial P}{\partial T}\right)_V dV, \tag{2.56}$$

whence

$$C_V = T\left(\frac{\partial S}{\partial T}\right)_V = -T\left(\frac{\partial^2 F}{\partial T^2}\right)_V, \tag{2.57}$$

where the second equality follows from Eq. (2.30). It can likewise be shown that

$$TdS = C_P dT - T\left(\frac{\partial V}{\partial T}\right)_P dP, \tag{2.58}$$

leading to

$$C_P = T\left(\frac{\partial S}{\partial T}\right)_P = -T\left(\frac{\partial^2 G}{\partial T^2}\right)_P, \tag{2.59}$$

where the second equality follows from Eq. (2.36).

One consequence of the third law of thermodynamics is that the heat capacity of any thermodynamic system must vanish at zero absolute temperature. Let us show this for a constant-volume process: from Eq. (2.57) we have

$$S(T_2, V) = S(T_1, V) + \int_{T_1}^{T_2} \frac{C_V}{T}dT. \tag{2.60}$$

Taking the limit $T_1 \to 0$, by Eq. (2.16) the integral on the right-hand side must remain finite, which can only happen if $C_V \to 0$ as $T \to 0$.

One further response function is the *compressibility*, which measures how the volume changes on varying the pressure. The *isothermal compressibility* is defined as

$$K_T = -\frac{1}{V}\left(\frac{\partial V}{\partial P}\right)_T = -\frac{1}{V}\left(\frac{\partial^2 G}{\partial P^2}\right)_T, \tag{2.61}$$

where the second equality follows from Eq. (2.37), and the *adiabatic compressibility* is defined as

$$K_S = -\frac{1}{V}\left(\frac{\partial V}{\partial P}\right)_S = -\frac{1}{V}\left(\frac{\partial^2 E^*}{\partial P^2}\right)_S, \tag{2.62}$$

where the second equality follows from Eq. (2.25).

Finally, the *coefficient of thermal expansion* measures how the volume changes in response to varying the temperature while keeping the pressure constant. It is defined as:

$$\alpha = \frac{1}{V}\left(\frac{\partial V}{\partial T}\right)_P. \tag{2.63}$$

The foregoing response functions are not all independent. Using Eq. (2.63), Eq. (2.58) may be rewritten as

$$TdS = C_P dT - \alpha TV dP. \tag{2.64}$$

Likewise, using Eq. (2.63) and the identity

$$\left(\frac{\partial x}{\partial y}\right)_z \left(\frac{\partial y}{\partial z}\right)_x \left(\frac{\partial z}{\partial x}\right)_y = -1,$$

Eq. (2.56) may be rewritten as

$$TdS = C_V dT + \frac{\alpha T}{K_T}dV. \tag{2.65}$$

Eliminating TdS between Eqs. (2.64) and (2.65) it can be shown, after some algebra, that

$$K_T(C_P - C_V) = TV\alpha^2. \tag{2.66}$$

It can also be shown (Huang, 1987, chap. 1) that

$$C_P(K_T - K_S) = TV\alpha^2. \tag{2.67}$$

Finally, it can be shown that thermal and mechanical stability require, respectively, that the heat capacities and the compressibilities should be positive. As we shall see in chapter 3, it follows from the so-called fluctuation-dissipation relations that the response functions must be positive; hence the two preceding equations imply that $C_P \geq C_V$ and $K_T \geq K_S$, and, furthermore, that

$$\frac{C_P}{C_V} = \frac{K_T}{K_S}. \tag{2.68}$$

2.2.4 Magnetic systems

We shall now apply the thermodynamic formalism developed in the preceding sections to magnetisable systems. Consider a sample of material placed inside a solenoid connected to a generator. An electric current passes through the solenoid, so the sample is subjected to a spatially uniform magnetic field. If there is no hysteresis,[5] it can be shown (see Reif (1985, chap. 11.1) or Callen (1985, Appendix B)), that the work performed by the generator to set up a magnetic field of strength H and increasing the total magnetic moment of the sample by an amount dM is

$$dW = \mu_0 H dM, \tag{2.69}$$

where $\mu_0 = 4\pi \times 10^{-7}$ H m^{-1} (henry per metre) is the magnetic permeability of vacuum, and M is the component of the total magnetic moment of the sample along the applied magnetic field. The fundamental thermodynamic relation is now written:[6]

$$dU = TdS + \mu_0 H dM. \tag{2.70}$$

Comparing this with Eq. (2.17), the equations of state and Maxwell's relations for magnetic systems follow from those obtained earlier for fluids by making the substitutions

$$V \to -M, \quad P \to \mu_o H. \tag{2.71}$$

The minus sign is a consequence of the fact that the magnetic moment increases when the applied magnetic field increases, whereas the volume decreases on increasing the pressure. H is the strength of the applied magnetic field, which does not depend on any sample properties. M is an extensive quantity; the intensive quantity conjugate to it is the applied magnetic induction, $B_0 = \mu_0 H$. We shall follow convention and define partial derivatives at constant H, rather than constant B_0; indeed, it is H that is controlled experimentally, by selecting the current that flows through the solenoid. It is important to note, however, that the partial derivatives will be taken with respect to $B_0 = \mu_0 H$, not H. Using relations (2.71) and our earlier results for fluids, we find, for the heat capacities at constant magnetic moment and constant magnetic field:

$$C_M = \left(\frac{\partial U}{\partial T} \right)_M = T \left(\frac{\partial S}{\partial T} \right)_M = -T \left(\frac{\partial^2 F}{\partial T^2} \right)_M, \tag{2.72}$$

$$C_H = \left(\frac{\partial E^*}{\partial T} \right)_H = T \left(\frac{\partial S}{\partial T} \right)_H = -T \left(\frac{\partial^2 G}{\partial T^2} \right)_H, \tag{2.73}$$

---

[5] For a *ferromagnetic* sample, these results are valid within a single *magnetic domain* (see Chapter 7), as hysteresis is a consequence of the polydomain nature of most magnetic materials. In other words, these results are also valid for a *monodomain* ferromagnet.

[6] Note that U thus defined includes the energy of the magnetic field in the absence of any magnetisable material. This is what is meant by '...the work performed to set up a magnetic field...'.

where F is the magnetic analogue of the Helmholtz free energy, given as before by Eq. (2.28); and E^* and G are, respectively, the magnetic analogues of the enthalpy and Gibbs free energy, Eqs. (2.22) and (2.34) with the substitutions (2.71):

$$E^* = U - \mu_0 MH, \tag{2.74}$$

$$G = U - TS - \mu_0 MH. \tag{2.75}$$

Turning now to the remaining response functions, the compressiblity is replaced by the *magnetic susceptibility*, as follows: if we define the *magnetisation* $\mathcal{M}$ of the sample as the magnetic moment per unit volume, the isothermal and adiabatic susceptibilities are

$$\chi_T = \frac{1}{\mu_0} \left(\frac{\partial \mathcal{M}}{\partial H} \right)_T = \frac{1}{V\mu_0} \left(\frac{\partial M}{\partial H} \right)_T = -\frac{1}{V\mu_0^2} \left(\frac{\partial^2 G}{\partial H^2} \right)_T, \tag{2.76}$$

$$\chi_S = \frac{1}{\mu_0} \left(\frac{\partial \mathcal{M}}{\partial H} \right)_S = \frac{1}{V\mu_0} \left(\frac{\partial M}{\partial H} \right)_S = -\frac{1}{V\mu_0^2} \left(\frac{\partial^2 E^*}{\partial H^2} \right)_S. \tag{2.77}$$

It is useful to define an alternative form of magnetic work that includes only the work performed to magnetise the sample, i.e., to set up a magnetic moment of strength M in the sample along the direction of the applied field, but which does not include the energy of the magnetic field in the absence of any sample. This is (Reif, 1985)

$$d\!\!\!{}^{-}\!W = -\mu_0 M dH, \tag{2.78}$$

leading to a different form of the fundamental thermodynamic relation:

$$dU = TdS - \mu_0 M dH. \tag{2.79}$$

The appropriate mapping between fluids and magnetic systems is now

$$V \to \mu_0 H, \quad P \to M. \tag{2.80}$$

It is left as an exercise for readers to show that, if Eq. (2.79) is used instead of Eq. (2.70), in Eqs. (2.72), (2.73), (2.76) and (2.77) the internal energy will switch roles with the enthalpy, and the Helmholtz free energy will switch roles with the Gibbs free energy. Thus the two mappings between fluids and magnetic systems, Eqs. (2.71) and (2.80), are thermodynamically equivalent.

Is is also left as an exercise to show, using Eqs. (2.72) and (2.73) for the heat capacities, and Eq. (2.76) and (2.77) for the magnetic susceptibilities, that the magnetic analogues of Eqs. (2.63), (2.66) and (2.67) are

$$\alpha_H = \left(\frac{\partial \mathcal{M}}{\partial T} \right)_H, \tag{2.81}$$

$$\chi_T (C_H - C_M) = T\alpha_H^2, \tag{2.82}$$

$$C_H (\chi_T - \chi_S) = T\alpha_H^2. \tag{2.83}$$

2.2.5 Electric systems*

One further application of thermodynamics is to electrical systems, in particular dielectrics. Conductors are less interesting: because the electric field inside a conductor vanishes to a good approximation, results for conductors may be formally obtained from those for dielectrics by taking the limit of an infinite dielectric constant, $\epsilon_r \to \infty$. For these reasons, and also because electric systems are less frequently discussed in the literature, we present here a brief treatment of the thermodynamics of dielectrics.

Consider a dielectric sample of relative permittivity ϵ_r, placed between the plates of a charged parallel-plate capacitor. The sample is thus subjected to a spatially uniform electric field. It can be shown (Landau and Lifshitz, 1960, chap. 2, §10) that the work performed by the capacitor on setting up a field of strength E and raising the total dipole moment $\mathcal{P}$ of the sample by $d\mathcal{P}$ is given by

$$\text{d}W = EdD, \tag{2.84}$$

where $D = \epsilon E + \mathcal{P}$ is the component of the electric displacement vector along the direction of the applied field, with ϵ the absolute permittivity of the sample. If we assume that the sample is inserted into the capacitor at constant voltage, i.e., that the capacitor plates are connected to a voltage source, then $E = $ const. and $dD = d\mathcal{P}$, whence

$$\text{d}W = Ed\mathcal{P}. \tag{2.85}$$

The fundamental thermodynamic relation is now written:[7]

$$dU = TdS + Ed\mathcal{P}, \tag{2.86}$$

and the equations of state and Maxwell's relations for electric systems follow from those obtained earlier for fluids by making the substitutions

$$V \to -\mathcal{P}, \quad P \to E. \tag{2.87}$$

As in the case of magnetic systems, the minus sign is a consequence of the fact that the dipole moment increases when the applied electric field increases, whereas the volume decreases on increasing the pressure. E is the strength of the applied electric field, which does not depend on any sample properties. $\mathcal{P}$ is an extensive variable; the conjugate intensive variable is the electric field E. In what follows we shall define partial derivatives at constant E, rather than constant D; indeed, it is E that is controlled experimentally, by selecting the voltage across the capacitor.

Making the substitutions (2.87), the heat capacities at constant dipole moment and at constant electric field follow:

$$C_{\mathcal{P}} = \left(\frac{\partial U}{\partial T} \right)_{\mathcal{P}} = T \left(\frac{\partial S}{\partial T} \right)_{\mathcal{P}} = -T \left(\frac{\partial^2 F}{\partial T^2} \right)_{\mathcal{P}}, \tag{2.88}$$

$$C_{E} = \left(\frac{\partial E^*}{\partial T} \right)_{E} = T \left(\frac{\partial S}{\partial T} \right)_{E} = -T \left(\frac{\partial^2 G}{\partial T^2} \right)_{E}, \tag{2.89}$$

---

[7] Note that U thus defined includes the energy of the electric field in the absence of any sample (Landau and Lifshitz, 1960). This is what is meant by '…the work performed by the capacitor to set up a field of strength E…'

where F is the electric analogue of the Helmholtz free energy, given as before by Eq. (2.28); and E^* and G are, respectively, the electric analogues of the enthalpy and Gibbs free energy, Eqs. (2.22) and (2.34) with the substitutions (2.87):

$$E^* = U - \mathcal{P}E, \tag{2.90}$$

$$G = U - TS - \mathcal{P}E. \tag{2.91}$$

Turning now to the remaining response functions, the compressibility is, as in the case of magnetic systems, replaced by the *electric susceptibility*, as follows: if we define the *polarisation* $\mathcal{P}$ of the sample as the dipole moment per unit volume, the isothermal and adiabatic susceptibilities are

$$\chi_T = \left(\frac{\partial \mathcal{P}}{\partial E} \right)_T = \frac{1}{V} \left(\frac{\partial P}{\partial E} \right)_T = -\frac{1}{V} \left(\frac{\partial^2 G}{\partial E^2} \right)_T, \tag{2.92}$$

$$\chi_S = \left(\frac{\partial \mathcal{P}}{\partial E} \right)_S = \frac{1}{V} \left(\frac{\partial P}{\partial E} \right)_S = -\frac{1}{V} \left(\frac{\partial^2 E^*}{\partial E^2} \right)_S, \tag{2.93}$$

As in the case of magnetic systems, it is useful to define an alternative form of electric work that includes only the work performed to polarise the sample, i.e., to set up a dipole moment of strength $\mathcal{P}$ in the sample along the direction of the applied field when the field strength increases by dE, but which does not include the energy of the electric field in the absence of any sample. This is (see Landau and Lifshitz (1960, chap. 2, §11) or Hill (1986, chap. 12.1)):

$$đW = -\mathcal{P}dE, \tag{2.94}$$

leading to a different form of the fundamental thermodynamic relation:

$$dU = TdS - \mathcal{P}dE. \tag{2.95}$$

The appropriate mapping between fluids and electric systems is now

$$V \to E \quad P \to \mathcal{P}. \tag{2.96}$$

It is left as an exercise for readers to show that, if Eq. (2.95) is used instead of Eq. (2.86), in Eqs. (2.88), (2.89), (2.92) and (2.93) the internal energy will switch roles with the enthalpy, and the Helmholtz free energy will switch roles with the Gibbs free energy. Thus the two mappings between fluids and electric systems, Eqs. (2.87) and (2.96), are thermodynamically equivalent.

Problems

2.1 Show that, if Eq. (2.79) is used for the fundamental thermodynamic relation of magnetic systems instead of Eq. (2.70), in Eqs. (2.72), (2.73), (2.76) and (2.77) the internal energy will switch roles with the enthalpy, and the Helmholtz free energy will switch roles with the Gibbs free energy

2.2 Redo the preceding problem for electric systems.

2.3 Using the definitions of heat capacities, Eqs. (2.72) and (2.73), and of susceptibilities, Eqs. (2.76) and (2.77), derive Eqs. (2.81)–(2.83) for magnetic systems.

References

Callen, H. 1985. *Thermodynamics and an Introduction to Thermostatistics*, 2nd edition. Wiley.

Hill, T. L. 1986. *An Introduction to Statistical Thermodynamics*. Dover Publications.

Huang, K. 1987. *Statistical Mechanics*, 2nd edition. Wiley.

Landau, L. D., and Lifshitz, E. 1960. *Electrodynamics of Continuous Media*. Pergamon Press.

Reif, F. 1985. *Fundamentals of Statistical and Thermal Physics*. McGraw-Hill.

3 The postulates of statistical physics. Thermodynamic equilibrium

3.1 Introduction

We saw in the preceding chapter that a given physical system can be described at either the macroscopic or the microscopic level. Of course, both macroscopic and microscopic quantities correspond to the same physical reality, so they must be related. Consider, for example, the pressure exerted by a volume of gas on the walls of a container. Macroscopically, this can be measured, e.g., with a manometer or some other pressure gauge. Microscopically, we know from the kinetic theory of gases that the pressure is related to the rate of momentum transfer per unit area when the gas molecules collide with the pressure gauge. We may likewise interpret the temperature of a gas as the mean translational kinetic energy of its molecules. Therefore, if we are able to interpret macroscopic quantities in microscopic terms, then we must also be able to cast the laws of thermodynamics in the language of statistical mechanics. This is what is known traditionally as *statistical thermodynamics*. More modernly we speak of *statistical physics*, to emphasise the greater generality of its methods and their applicability to problems such as magnetism and blackbody radiation, which we shall encounter in later chapters.

The laws of classical mechanics can, under certain conditions, be applied to the motion of atoms and molecules in a gas. Suppose we wish to calculate some properties of a monatomic gas kept at room temperature and atmospheric pressure. To a good approximation, classical mechanics may be used to describe the present state of the system and its time evolution, once the present state of each particle (in this case classically distinguishable atoms) is known. On the microscopic level, the state of such a gas is specified by the position coordinates, x, y and z, and the components of the velocity, v_x, v_y and v_z (or, equivalently, the components of the linear momentum p_x, p_y and p_z) of each constituent atom at every instant in time. In other words, the state of each atom is specified by six parameters, hence it can be mapped onto a point in a six-dimensional space known as the *phase space* of that particular atom. This point is said to *represent* the system (in this case a single atom). Because a macroscopic sample of gas typically contains a number of atoms of the order of Avogadro's number, $N_A = 6.02 \times 10^{23} \text{ mol}^{-1}$, then the microscopic state of such a gas is specified by $\approx 6 \times 6 \times 10^{23}$ parameters at any given time: this is the dimensionalty of the phase space of the system. Provided the initial values of all these parameters are known, it is in principle possible to predict the future behaviour of the gas, taking into account collisions between atoms, and between atoms and the container walls. However, this task is obviously beyond the computing power we have available. Moreover, analysing the result of such a calculation would be an extremely laborious task, and therefore of little practical interest. The very large number of particles in a sample of macroscopic dimensions effectively renders any exact treatment impossible, while at the same time allowing us to extract very accurate predictions using statistics. Statistical physics is concerned with computing mean values, and deviations from the mean, of thermodynamically relevant quantities. For macroscopic systems, the deviations from the mean are quite negligible in most cases. We owe this 'statistical leap' mostly to Boltzmann (1877), with major contributions by Maxwell (1869), Gibbs (1902) and Einstein (1905) ('all of Einstein's contributions to quantum theory are of a statistical nature' (Pais, 1982, chap. 4)), among others. This statistical approach was not well received by the physics community of the late nineteenth century, for whom the idea of physics as an exact science was incompatible with probability considerations, or with the then not yet well-established corpuscular nature of matter.

Although statistical mechanics was developed by Boltzmann, Maxwell and Gibbs within the framework of classical mechanics, we shall formulate the principles of statistical physics in the language of quantum mechanics. This allows us to use simpler and more general arguments, which are therefore better suited to an introductory treatment of the subject. Statistical physics as developed by Einstein and other twentieth-century physicists is actually underpinned by quantum mechanics, but for the most part requires only elementary quantum concepts such as quantum states and energy levels to be understood. Only the quantum gases and ferromagnetism, which we shall encounter in later chapters, will require a (slightly) higher level of sophistication. Finally, we shall concentrate on the statistical physics of equilibrium systems, as non-equilibrium phenomena are not readily amenable to the (quasi) elementary approach we have adopted in this book.

3.2 The postulates of statistical physics

We shall formulate statistical physics in terms of *statistical ensembles*, as originally proposed by Maxwell (1879), Boltzmann (1884, 1887) and Gibbs (1902) (Sklar, 1993, chap. 2). This relies on a number of postulates, the validity of which can only be verified *a posteriori* – i.e., it is inferred from the agreement between theory and experiment. Although many physical phenomena have been explained in this way, the validity of the postulates is still an open question, which will not be addressed here (Sklar, 1993).

Gibbs' work provides a recipe for computing the equilibrium value of a thermodynamic quantity under the set of constraints imposed on the system: first, select the appropriate statistical ensemble; second, compute the ensemble average of the statistical variable corresponding to the thermodynamic quantity we wish to find; finally, identify the ensemble average with the equilibrium value of the thermodynamic quantity. This is the programme we shall follow in the chapters dealing with statistical thermodynamics.

The ensemble average was defined in Chapter 1. There remains to introduce the concept of *ensemble, statistical ensemble* or *thermodynamic ensemble*: this is a set of $\mathcal{N} \gg 1$ virtual copies of the system under study, called elements of the ensemble, which are all identical from the point of view of thermodynamics. Consider, for example, a closed, isolated system of fixed volume: its thermodynamic state is fully specified by the variables U, V and N, whose values are the external constraints imposed on the system. All ensemble elements have the same U, V and N, but they are not identical at the microscopic level. In fact there is a very large number of microscopic states, or *microstates* (quantum or classical, whatever the case might be), that correspond to the same thermodynamic state, or *macrostate*. At any time, the elements of an ensemble characterised by a given set of constraints are distributed over all possible microstates compatible with the specified macrostate. Properties such as the pressure or the magnetic moment of the macrostate can in principle be calculated by averaging these quantities over all ensemble elements, i.e., over all microstates compatible with the given macrostate. This leads us to the

First Postulate: *The time average, taken over a sufficiently long time, of a physical quantity of a thermodynamic system equals the ensemble average of that same quantity, in the limit $\mathcal{N} \to \infty$.*

This postulate – known as the *ergodic hypothesis* – is important because it allows us to replace the time average of a physical quantity by its instantaneous average over a collection of virtual copies of the system that all correspond to the same thermodynamic state, under a given set of constraints (Hill, 1985, chap. 1.2) (see Figure 3.1). Unlike the time average, the ensemble average does not require us to solve the time evolution equations for the constituent particles of the system; instead, it can be found by the methods we shall describe later. Conversely, at equilibrium the ensemble average of the quantity under study must be time-independent in the limit of an infinitely large ensemble, $\mathcal{N} \to \infty$. The statistical description adopted thus leads us to a definition of equilibrium: *A macroscopic system is in equilibrium only if the statistical ensemble representing it has a time-independent probability distribution* (see Figure 3.2).

Consider an isolated system of volume V, internal energy U, which contains N particles. If the system is in thermodynamic equilibrium, its macrostate is completely specified by (U, V, N). If, however, the system is not in equilibrium, then additional variables are required to specify its state, such as its temperature gradient, for example. Denoting any additional variables collectively by y (this may be a set of variables $y_1, y_2, \ldots$), a non-equilibrium macrostate will be specified by (U, V, N, y). A complete mechanical description of a microstate is extremely complicated: it requires knowledge of, for example, the positions and momenta of $\approx 10^{23}$ particles, within the limitations imposed by the Heisenberg uncertainty relation.[1] In statistical mechanics we do not need to know which microstate the system is in: all we need to know is *how many microstates are compatible with a given macrostate*. We shall assume that this is a well-defined number, and define the *statistical weight* $\Omega(U, V, N, y)$ of a macrostate as the number of microstates compatible with the macrostate specified by U, V, N and y. Counting microstates is easier in quantum than in classical mechanics, as we can always force a system to have discrete energy levels by confining it in a box of arbitrary (but finite) dimensions (Schiff, 1968, chap. 4, §15).

---

[1] More rigorously, it requires knowledge of the wavefunction representing the microstate.

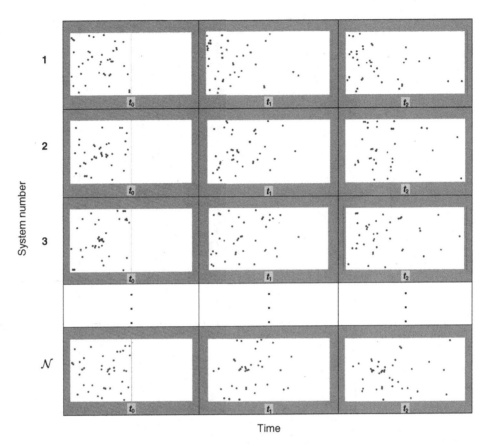

Time

Figure 3.1 Snapshots of a computer simulation of the time evolution of an ensemble representing a closed isolated system, i.e., at constant *(U,V,N)*. The system is composed of 40 particles confined in a box. At time zero, all the particles are located in the left half of the box, but the position and velocity of each individual particle is not known. Each row in the figure shows the time evolution of the *n*th virtual copy of the system, with $n = 1, 2, \ldots, \mathcal{N}$. The initial position and velocity of each particle are random variables drawn from an appropriate probability distribution. The state of the system at some time t_i is found by averaging over all copies in column $t = t_i$. At t_0 the probability of finding a particle in the left half of the box is $p = 1$, whereas the probability of finding a particle in the right half of the box is $q = 0$. (From http://chemconnections.org/Java/molecules/index.html.)

Note that a thermodynamic system with a well-defined energy and a well-defined number of states need not have an 'exact' energy or an 'exact' number of states. When we say that a macrostate has energy U, what we mean is that its energy lies within δU of U, where $\delta U \ll U$. This is a consequence of the Heisenberg uncertainty principle, according to which the energy may not be determined with infinite precision and we must have $\delta U \delta t \geq h/2$, where δt is the time it takes to perform a measurement of U. This complication is, however, inconsequential from the thermodynamic point of view, and so we shall ignore it in what follows. In quantum mechanics, the number of microstates, or quantum states, associated with a given macrostate, i.e., the statistical weight of the macrostate, is called the *degeneracy* of the macrostate.

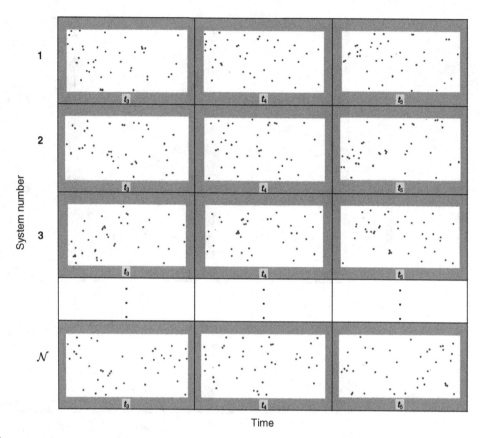

Figure 3.2 After a sufficiently long time, the particles are uniformly distributed throughout the box. Now $p = q = 1/2$ and the probabilies do not change in time. The ensemble is now time-independent, i.e., equilibrium has been reached. (From http://chemconnections.org/Java/molecules/index.html.)

Second Postulate: *The $\mathcal{N}$ ($\mathcal{N} \to \infty$) elements of an ensemble representing an isolated system at equilibrium are uniformly distributed over all possible microstates.*

This postulate is of paramount importance: basically it says that, because we do not know which microstate the system is actually in, we may boldly assume all microstates compatible with a given macrostate to be equally probable. This is the *postulate of equal* a priori *probabilities* (PEPP).

Let us assume that the elements of the ensemble are initially distributed over the accessible microstates in some arbitrary way, e.g., that they only occupy some particular subset of all accessible microstates. This corresponds, for example, to all particles in a gas occupying only the left half of the container (see Figure 3.1). As the system evolves, however, the elements of the ensemble will transition between different microstates (in the gas example, this happens because particles collide with one another and with the container walls). Hence the system will eventually occupy all accessible microstates. It is to be expected that, after a sufficiently long time, all elements in the ensemble representing the system will be uniformly distributed over all accessible microstates, in accordance with the PEPP, and this distribution will no longer change. In other words, *it is expected that any isolated system will, whatever its initial conditions, eventually reach an equilibrium state, in which all its accessible microstates are equally probable*[2] (Reif, 1985, chap. 3.4). The probability of any given microstate is therefore a constant, C; it follows from the normalisation condition that $C = 1/\Omega(U, V, N)$.

A consequence of the first two postulates is that, if we wait long enough, the system will spend approximately the same time in each microstate. This is the *quantum ergodic hypothesis* (Hill, 1985, chap. 1.2), whence we conclude that *the time the system spends in a given microstate is proportional to its probability*.

3.3 Isolated systems

3.3.1 Microcanonical ensemble

As we saw in the preceding chapter, the isolated system is particularly useful. Indeed, if the system under study, call it $\mathcal{A}$, is not isolated, but in contact with another system, $\mathcal{A}'$, we can view the combined system $\mathcal{A}_0 = \mathcal{A} + \mathcal{A}'$ as an isolated system and use the results derived earlier.

For simplicity we shall assume, as in the preceding section, that the volume is the only external parameter. Our combined, isolated system is then specified by the internal energy U (i.e., its internal energy takes a value between U and $U + \delta U$, where $\delta U \ll U$), volume V and particle number N. Calculations will be performed in what we shall call the *microcanonical ensemble*: this is *a very large number of virtual copies of the system, all isolated and in equilibrium, all with the same internal energy, volume and number of particles, and which are uniformly distributed over all accessible microstates compatible with the external constraints imposed on the system, in accordance with the PEPP*.

Let $\Omega(U)$ be the total number of microstates of the above system, where for notational simplicity we have omitted explicit reference to V and N. Now suppose that there is a subset of $\Omega(U, y_k)$ microstates for which the quantity y takes the value y_k (y could be, e.g., the pressure, the magnetic moment, etc., or some internal partition of the system, as we shall see later). By the PEPP, the probability $p(y_k)$ that $y = y_k$ equals the ratio of the number of microstates for which $y = y_k$ to the total number of microstates:

$$p(y_k) = \frac{\Omega(U, y_k)}{\Omega(U)} \propto \Omega(U, y_k), \qquad (3.1)$$

and the microcanonical ensemble average of y will be

$$\bar{y} = \sum_k p_k y_k = \frac{\sum_k \Omega(U, y_k) y_k}{\Omega(U)}, \qquad (3.2)$$

---

[2] Notice that we have said nothing about how the system reaches equilibrium, or how long it takes to get there. Boltzmann made a key contribution to the statistical study of this problem, but we shall not discuss it here (see, e.g., Reif (1985, chaps 12–15) or Sklar (1993)).

where the sum over k is over all permissible values of y. This ensemble average must be close to the most likely value of y, i.e., that which is realised in most elements of the ensemble. It can be shown on the basis of very general arguments that this is indeed the case for macroscopic systems (Reif, 1985, chap. 2). In practice, it is extremely difficult to find $\Omega(U,V,N)$ except for very simple systems, because of the restriction of constant internal energy. Thus it is in general not possible to use the microcanonical equations. This difficulty can be circumvented by performing calculations in an alternative ensemble, the *canonical ensemble*, where the restriction of constant energy does not apply. However, the fundamental importance of the isolated system, as represented by the microcanonical ensemble, is apparent from the general results derived in this section, and which we shall use in the next few sections.

3.3.2 Connection with thermodynamics. Entropy

Consider an isolated system $\mathcal{A}_0$, specified by (U_0, V_0, N_0). The system is partitioned into two subsystems, $\mathcal{A}$ and $\mathcal{A}'$, by an internal wall that allows the exchange of heat, work (i.e., it is movable) and particles between the two subsystems (see Figure 3.3). $\mathcal{A}$ and $\mathcal{A}'$ are specified, respectively, by (U,V,N) and (U',V',N'), such that

$$U_0 = U + U', \quad V_0 = V + V', \quad N_0 = N + N'. \tag{3.3}$$

In writing Eq. (3.3) we are assuming that the total internal energy is the sum of the energies of each of the subsystems, i.e., we are neglecting the interaction energy between the two subsystems. This does not mean that the two subsystems do not interact; they certainly do – indeed, if they did not, then the global system would never reach equilibrium. However, the interaction energy between the two subsystems usually scales with the area of the dividing surface, whereas the bulk internal energy of either subsystem scales with its volume. For systems of macroscopic dimensions the volume terms dominate over the surface terms and it is licit to take the total internal energy as just the sum of the internal energies of the individual subsystems.[3]

The constraint that the total energy, volume and number of particles are fixed reduces the number of independent variables, which we can take to be those pertaining to subsystem $\mathcal{A}$. In terms of these, the statistical weight of the full system $\mathcal{A}_0$ specified by Eqs. (3.3)

$$\Omega_0(U_0, V_0, N_0; U, V, N) = \Omega(U, V, N)\Omega'(U_0 - U, V_0 - V, N_0 - N), \tag{3.4}$$

where Ω and Ω' are the statistical weights of subsystems $\mathcal{A}$ and $\mathcal{A}'$, respectively. What this equation is telling us is that any microstate of $\mathcal{A}$ can be combined with any microstate of $\mathcal{A}'$ to give a microstate of $\mathcal{A}_0$. In other words, Eq. (3.4) expresses the assumption that the microstates of the two subsystems are independent.

The connection between the microcanonical ensemble and thermodynamics is established by introducing the statistical definition of *entropy*:

$$S(U, V, N) = k_B \ln \Omega(U, V, N), \tag{3.5}$$

---

[3] This is true only if the interparticle forces are short-ranged, which is the case in most systems. One notable exception is (unscreened) Coulombic forces.

where $k_B = 1.38 \times 10^{-23}$ J K^{-1} is known as *Boltzmann's constant*. Using Eqs. (3.4) and (3.5), the entropy of the full system $\mathcal{A}_0$ is given by

$$S_0(U_0, V_0, N_0; U, V, N) = S(U, V, N) + S'(U_0 - U, V_0 - V, N_0 - N), \qquad (3.6)$$

i.e., the entropy is an extensive quantity, as required by thermodynamics. This follows from its definition as the logarithm of the statistical weight.

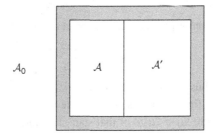

Figure 3.3   Two non-isolated systems, $\mathcal{A}$ and $\mathcal{A}'$, making up isolated system $\mathcal{A}_0$.

Equation (3.5), known as *Boltzmann's formula*, is one of the most fundamental equations of the whole of physics. Its significance lies in the fact that it relates a macroscopic property of a system – its entropy – to its microscopic properties, embodied in the statistical weight Ω – the number of microstates compatible with the given U, V and N. Earlier, Clausius had formulated a purely thermodynamic definition of entropy. Boltzmann then related this to microscopic behaviour using classical mechanics arguments, although he did not actually write down Eq. (3.5) in its present form – this was only done later, by Planck. In fact, it was Planck who first introduced Boltzmann's constant k_B in 1900, when writing down his blackbody radiation law, which we shall derive in Chapter 6 on the quantum ideal gas. Until about 1911 k_B was known as Planck's constant (Pais, 1982, chap. 4). It is the inclusion of k_B in Eq. (3.5) that makes statistical physics consistent with thermodynamics: equations for the same physical quantities, derived via either the statistical or the thermodynamic routes, are the same, with the same constants and the same numerical prefactors.

3.3.3 Equilibrium conditions for an isolated system

Consider a system as depicted in Figure 3.4 and described by Eqs. (3.3). The system is specified by the values of $(U_0, V_0, N_0; U, V, N)$ and its statistical weight is $\Omega_0(U_0, V_0, N_0; U, V, N)$. U, V and N are a particular realisation of the additional variable y used to define statistical weight in Section 3.2. If the system has reached its equilibrium state and all internal constraints are fixed, then in accordance with the PEPP all accessible microstates are equally probable and the probability of finding the system in any given state is proportional to the statistical weight Ω_0 of the state.

Let us now assume that some internal constraint is removed, e.g., that the wall partitioning the system into two subsystems is initially fixed, but is then allowed to move. In this case y is just the volume V: the two subsystems will change their volumes so as to keep the pressure uniform throughout the whole system, but the total volume V_0 must remain

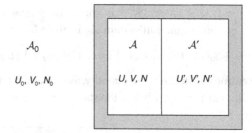

Figure 3.4   An isolated system composed of two subsystems, $\mathcal{A}$ and $\mathcal{A}'$, which can exchange heat, work and particles between each other.

constant. Thus y changes from its initial value y_i to some final value y_f. The probability $p_f \equiv p(y_f)$ of finding the system in a state with $y = y_f$ is, by Eq. (3.1), proportional to the statistical weight $\Omega_f \equiv \Omega(y_f)$ of that state. It is intuitively obvious that removing a constraint can only increase the number of accessible microstates, or at best keep it unchanged, so $\Omega_f \geq \Omega_0$ (in our specific example, removing the dividing wall clearly increases the number of microstates compatible with the external constraints on the system).

In general, $\Omega(y)$ has a very sharp maximum at some $y = \bar{y}$. This means that the probability that the elements of the ensemble will be distributed only over the original Ω_0 states is much less than $p(\bar{y})$. Consequently, at equilibrium, most elements of the ensemble will have $y \approx \bar{y}$, As discussed earlier, *equilibrium has been reached as the most probable state*, where all elements of the ensemble are uniformly distributed over the Ω_f accessible microstates. The foregoing argument may be summarised as follows:

When a constraint on an isolated system is removed, the variable y associated with that constraint will take such a value as maximises the statistical weight $\Omega(U, V, N, y)$. This value of y will then specify the equilibrium state at fixed U, V and N (Reif, 1985, chap. 3.1).

The above statement can be rephrased using Boltzmann's formula, Eq. (3.5): *The entropy of an isolated system never decreases (indeed in general it increases) and is maximal at equilibrium. This is the statistical interpretation of the second law of thermodynamics: the state of maximum entropy (i.e., the equilibrium state of a system, according to classical thermodynamics) is the most probable state.* We shall next derive the equilibrium condition for an isolated system by maximising its entropy.

Thermal equilibrium. Absolute temperature. Distribution function for the energy

We start by assuming that the internal wall in Figure 3.4 allows heat exchange between the two subsystems, but is fixed and impermeable to particle fluxes. We shall find the equilibrium condition for this system by maximising the entropy of $\mathcal{A}_0$, while taking the energy of system $\mathcal{A}$ as the independent variable. From Eq. (3.6):

$$\left(\frac{\partial S_0}{\partial U}\right)_{V,N} = \left(\frac{\partial S}{\partial U}\right)_{V,N} + \left(\frac{\partial S'}{\partial U}\right)_{V,N} = \left(\frac{\partial S}{\partial U}\right)_{V,N} + \left(\frac{\partial S'}{\partial U'}\right)_{V',N'}\frac{dU'}{dU} = 0. \qquad (3.7)$$

Because $U_0 = $ constant, it follows from Eq. (3.3) that $dU'/dU = -1$, hence the *condition for thermal equilibrium* of the global system $\mathcal{A}_0$ is

$$\left(\frac{\partial S}{\partial U}\right)_{V,N} = \left(\frac{\partial S'}{\partial U'}\right)_{V',N'}. \tag{3.8}$$

From this, and from the definition of thermal equilibrium in classical thermodynamics, we conclude that the quantity $(\partial S/\partial U)_{V,N}$ must be related to the system's temperature. Using Boltzmann's formula to define a statistical mechanical temperature variable as

$$\beta = \left(\frac{\partial \ln \Omega}{\partial U}\right)_{V,N}, \tag{3.9}$$

the equilibrium condition (3.8) becomes

$$\beta = \beta'. \tag{3.10}$$

β is a measure of a fundamental property of the system: how its statistical weight depends on its energy. If we now define the *absolute temperature T* of a system as

$$\frac{1}{T} = k_B \beta = \left(\frac{\partial S}{\partial U}\right)_{V,N}, \tag{3.11}$$

the condition of thermal equilibrium simplifies to

$$T = T', \tag{3.12}$$

i.e., the two subsystems must have the same temperature. $1/T = k_B \beta$ is the intensive variable conjugate to the energy. As we shall show in Chapter 5, definition (3.11) with Boltzmann's constant k_B is consistent with the absolute temperature of a classical ideal gas and with the entropy S as defined in thermodynamics. The statistical weight of most macroscopic systems is a rapidly increasing function of the energy, of the form $\Omega \propto U^\nu$, where ν is the number of degrees of freedom (Reif, 1985, chap. 2.5). From Eq. (3.11) we thus conclude that *the absolute temperature is positive*. Negative absolute temperatures are, however, possible if Ω is a decreasing function of U, which is the case whenever there is an upper bound to the energy of the quantum states of a system, as we shall see when studying the ideal paramagnetic solid.

We now reformulate the foregoing derivation in terms of the probability that the wall in system $\mathcal{A}_0$ is positioned such that subsystem $\mathcal{A}$ has energy U and subsystem $\mathcal{A}'$ has energy $U' = U_0 - U$, where for simplicity we do not refer explicitly to V and N. This is:

$$p(U) = \frac{\Omega_0(U_0, U)}{\Omega_0(U_0)} \propto \Omega_0(U_0, U), \tag{3.13}$$

where U is a realisation of the generic variable y introduced in Eq. (3.1). From Eq. (3.4),

$$p(U) \propto \Omega(U)\Omega'(U_0 - U). \tag{3.14}$$

Taking logarithms of both sides of this equation yields $\ln p(U) = \text{constant} + \ln \Omega(U) + \ln \Omega'(U_0 - U)$. Then using the definition of entropy, Eq. (3.5), equilibrium condition (3.8) becomes

$$\frac{\partial \ln p(U)}{\partial U} = 0, \tag{3.15}$$

i.e., *p(U) is maximised at equilibrium.*

Advanced topic 3.1   Gaussian distribution of the energy

From the requirement that the probability should be maximised at equilibrium it can be shown that the probability that system $\mathcal{A}_0$ is partitioned in such a way that subsystem $\mathcal{A}$ has energy U and subsystem U' has energy U' is, to a very good approximation, Gaussian. It follows that *the mean value of U coincides with its most probable value.* The proof which we now give follows Reif (1985, chap. 3.7).

As noted above, at equilibrium $p(U)$ has a maximum for some value of U, which we denote $\tilde{U}$. Because $p(U)$ is usually very sharply peaked at $U = \tilde{U}$, we shall work instead with $\ln p(U)$, which is more slowly varying. This will be expanded in powers of the energy difference $\eta = U - \tilde{U}$. Start by expanding $\ln \Omega$:

$$\ln \Omega(U) = \ln \Omega(\tilde{U}) + \eta \left(\frac{\partial \ln \Omega}{\partial U} \right)_{U = \tilde{U}} + \frac{1}{2} \eta^2 \left(\frac{\partial^2 \ln \Omega}{\partial U^2} \right)_{U = \tilde{U}} + \cdots$$

Introducing β as defined in Eq. (3.9) and further defining

$$\lambda = - \left(\frac{\partial^2 \ln \Omega}{\partial U^2} \right)_{U = \tilde{U}} = - \left(\frac{\partial \beta}{\partial U} \right)_{U = \tilde{U}},$$

the above expansion can be rewritten as

$$\ln \Omega(U) = \ln \Omega(\tilde{U}) + \beta \eta - \frac{1}{2} \lambda \eta^2 + \cdots$$

We may likewise expand $\ln \Omega'$ in powers of $U' - \tilde{U}' = -(U - \tilde{U}) = -\eta$, with the result

$$\ln \Omega'(U') = \ln \Omega(\tilde{U}') - \beta' \eta - \frac{1}{2} \lambda' \eta^2 + \cdots$$

Adding the two expansions yields

$$\ln \left[\Omega(U) \Omega'(U') \right] = \ln \left[\Omega(\tilde{U}) \Omega'(\tilde{U}') \right] + (\beta - \beta') \eta - \frac{1}{2} (\lambda + \lambda') \eta^2 + \cdots$$

At equilibrium $\beta = \beta'$ and the term linear in η vanishes. If U is not too far from $\tilde{U}$, the expansion may be truncated at second order, and from Eq. (3.14) we find

$$\ln p(U) = \ln p(\tilde{U}) - \frac{1}{2} \lambda_0 \eta^2, \quad \lambda_0 = \lambda + \lambda',$$

whence

$$p(U) = p(\tilde{U}) \exp \left[-\frac{1}{2} \lambda_0 \left(U - \tilde{U} \right)^2 \right].$$

Note, however, that λ_0 must not be negative, otherwise $p(U)$ would have a minimum, rather than a maximum, at $U = \tilde{U}$, which is unstable: $p(U)$ would grow as U moved away from $\tilde{U}$ and no equilibrium state would ever be reached, which is inconsistent with our starting hypothesis. On the other hand, either λ or λ' may be negative, provided their sum is always positive. We have thus proved that $p(U)$ is Gaussian and we may therefore identify $\tilde{U}$ with the mean value of U (see Eq. (1.47)):

$$\overline{U} = \tilde{U} \qquad\qquad q.e.d.$$

We emphasise that this result is only valid if the systems are not too far from equilibrium, so that truncating the series expansions at second order in $U - \tilde{U}$ does not introduce any large errors.

Let us now assume that system $\mathcal{A}_0$ is partitioned in such a way that $\lambda' \simeq 0$, i.e., the temperature of $\mathcal{A}'$ is nearly constant. In this case we must have $\mathcal{A}' \gg \mathcal{A}$ and $\mathcal{A}'$ is known as a *temperature reservoir* or *heat reservoir*; we shall study this in more detail in the canonical ensemble. Now $\lambda_0 \simeq \lambda > 0$ and

$$\lambda = -\left(\frac{\partial \beta}{\partial U}\right)_V = \frac{1}{k_B T^2}\left(\frac{\partial T}{\partial U}\right)_V = \frac{1}{k_B T^2 C_V} \geq 0, \qquad (3.16)$$

whence we conclude that *the temperature of a system is an increasing function of its energy, and (thermal) stability requires that the heat capacity should be positive.* $\partial T / \partial U \geq 0$ implies that *heat flows from the hotter subsystem to the cooler subsystem.*[4]

From the expression for $p(U)$ we can identify λ as the inverse variance of the energy, which yields the following *fluctuation-dissipation relation*:

$$\sigma^2(U) = k_B T^2 C_V, \qquad (3.17)$$

thus named because it relates a statistical quantity, namely the variance of the energy, to a macroscopic quantity, the heat capacity at constant volume. Equation (3.17) is a particular instance of a more general result known as the *fluctuation-dissipation theorem* (Landau *et al.*, 1980, chap. XII), which relates the variance of a microscopic quantity to the corresponding thermodynamic response function. We shall encounter further examples of this theorem later in the book. Because both the mean energy and the heat capacity are proportional to the number of particles N, the relative fluctuation of the energy, Eq. (1.20), equals

$$\delta(U) = \frac{\sigma(U)}{\overline{U}} \propto \frac{1}{\sqrt{N}},$$

whence $\lim_{N\to\infty} \delta(U) = 0$. This means that *the probability distribution for the internal energy of a macroscopic, non-isolated system is very sharply peaked at its mean value, which may therefore be identified with the internal energy as defined in thermodynamics.* In other words, the state of a non-isolated system may be specified by its mean energy (see Section 4.6).

---

[4] This is only true if the absolute temperature is positive, as is the case for most systems. If negative absolute temperatures are permissible then T is a non-monotomic function of the energy, as we shall see when studying spin systems.

If subsystems $\mathcal{A}$ and $\mathcal{A}'$ are initially at different temperatures, then system $\mathcal{A}_0$ will evolve towards thermal equilibrium according to

$$\frac{dS_0}{dt} = \left(\frac{\partial S_0}{\partial U}\right)_{V,N} \frac{dU}{dt} = \left(\frac{1}{T} - \frac{1}{T'}\right)\frac{dU}{dt} > 0. \tag{3.18}$$

So if, e.g., $T < T'$, then $dU/dt > 0$ and energy will flow from the hotter to the cooler subsystem. This is Planck's formulation of the second law of thermodynamics.

Mechanical equilibrium. Pressure

We now assume that the wall is movable and allows heat, but not particles, to be exchanged between the two subsystems. Taking U and V as the independent variables, Eq. (3.7) generalises to

$$dS_0 = \left[\left(\frac{\partial S}{\partial U}\right)_{V,N} + \left(\frac{\partial S'}{\partial U'}\right)_{V',N'} \frac{dU'}{dU}\right] dU$$

$$+ \left[\left(\frac{\partial S}{\partial V}\right)_{U,N} + \left(\frac{\partial S'}{\partial V'}\right)_{U',N'} \frac{dV'}{dV}\right] dV = 0. \tag{3.19}$$

Because this equation must be satisfied for arbitrary dU and dV, both terms in square brackets must vanish. Equations (3.3) then yield the condition for thermal equilibrium, Eq. (3.8), and the *condition for mechanical equilibrium*:

$$\left(\frac{\partial S}{\partial V}\right)_{U,N} = \left(\frac{\partial S'}{\partial V'}\right)_{U',N'}, \tag{3.20}$$

whence we conclude that the quantity $(\partial S/\partial V)_{U,N}$ must be related to the system's pressure. If we define the *pressure* as[5]

$$P = T\left(\frac{\partial S}{\partial V}\right)_{U,N} = \frac{1}{\beta}\left(\frac{\partial \ln\Omega}{\partial V}\right)_{U,N}, \tag{3.21}$$

and use the condition for thermal equilibrium, the condition for mechanical equilibrium can be written in a very simple form:

$$P = P', \tag{3.22}$$

i.e., the two subsystems must have the same pressure. Equation (3.21) relating P, T and V is the *equation of state* of the system in the microcanonical ensemble. Here P/T is the intensive variable conjugate to the volume.

If subsystems $\mathcal{A}$ and $\mathcal{A}'$ are initially at different pressures, then system $\mathcal{A}_0$ will evolve towards mechanical equilibrium according to

$$\frac{dS_0}{dt} = \left(\frac{\partial S_0}{\partial V}\right)_{U,N} \frac{dV}{dt} = \frac{1}{T}(P - P')\frac{dV}{dt} > 0. \tag{3.23}$$

So if, e.g., $P > P'$, then $dV/dt > 0$ and *the subsystem with higher pressure will expand.*

---

[5] The prefactor of T is included so that the pressure has the correct dimensions.

Equilibrium with respect to particle exchange. Chemical potential

Finally, we consider a movable wall that is permeable to both heat and particles. In this case the equilibrium condition is written:

$$dS_0 = \left[\left(\frac{\partial S}{\partial U} \right)_{V,N} + \left(\frac{\partial S'}{\partial U'} \right)_{V',N'} \frac{dU'}{dU} \right] dU$$

$$+ \left[\left(\frac{\partial S}{\partial V} \right)_{U,N} + \left(\frac{\partial S'}{\partial V'} \right)_{U',N'} \frac{dV'}{dV} \right] dV$$

$$+ \left[\left(\frac{\partial S}{\partial N} \right)_{U,V} + \left(\frac{\partial S'}{\partial N'} \right)_{U',V'} \frac{dN'}{dN} \right] dN = 0. \qquad (3.24)$$

Proceeding as before, we again find the conditions for thermal and mechanical equilibrium, Eqs. (3.8) and (3.20), and the *condition for equilibrium under particle exchange* (or *condition for diffusive equilibrium*):

$$\left(\frac{\partial S}{\partial N} \right)_{U,V} = \left(\frac{\partial S'}{\partial N'} \right)_{U',V'}. \qquad (3.25)$$

If $\mathcal{A}_0$ contains only one particle species and two phases (e.g., liquid and vapour), then Eq. (3.25) is the equilibrium condition for particle exchange between the two phases. Together with the conditions for thermal and mechanical equilibrium, it describes *phase equilibrium*. If we now define the *chemical potential* as

$$\mu = -T \left(\frac{\partial S}{\partial N} \right)_{U,V} = -\frac{1}{\beta} \left(\frac{\partial \ln \Omega}{\partial N} \right)_{U,V}, \qquad (3.26)$$

where the prefactor T is included so that μ has dimensions of energy, then the condition for equilibrium under particle exchange is

$$\mu = \mu', \qquad (3.27)$$

i.e., at equilibrium the two phases must have the same chemical potential. Here $-\mu/T$ is the intensive variable conjugate to the number of particles.

If subsystems $\mathcal{A}$ and $\mathcal{A}'$ initially have different chemical potentials, then system $\mathcal{A}_0$ will evolve towards equilibrium under particle exchange according to

$$\frac{dS_0}{dt} = \left(\frac{\partial S_0}{\partial N} \right)_{U,V} \frac{dN}{dt} = \frac{1}{T} \left(\mu' - \mu \right) \frac{dN}{dt} > 0. \qquad (3.28)$$

So if, e.g., $\mu > \mu'$, then $dN/dt < 0$ and *there will be a net particle flux from the subsystem at higher chemical potential to that at lower chemical potential.*

3.3.4 Infinitesimal quasi-static processes

Consider an infinitesimal quasi-static process in which subsystem $\mathcal{A}$, owing to its interaction with system $\mathcal{A}'$, occupies a succession of equilibrium states, from an equilibrium state specified by (U, V, N) to another equilibrium state specified by $(U + dU, V + dV, N + dN)$. The entropy change associated with this infinitesimal process is

$$dS(U, V, N) = \left(\frac{\partial S}{\partial U} \right)_{V,N} dU + \left(\frac{\partial S}{\partial V} \right)_{U,N} dV + \left(\frac{\partial S}{\partial N} \right)_{U,V} dN. \tag{3.29}$$

Using the definitions of temperature, Eq. (3.11), pressure, Eq. (3.21), and chemical potential, Eq. (3.26), this yields the *fundamental thermodynamic relation*:

$$dU = TdS - PdV + \mu dN. \tag{3.30}$$

The above result can be generalised to other forms of work than mechanical. If $x_1, x_2, \ldots, x_n$ are external variables, e.g., magnetic or electric fields, etc., we define the *generalised force* X_i, conjugate to x_i as

$$X_i = T \left(\frac{\partial S}{\partial x_i} \right)_{T, x_{j \neq i}} = \frac{1}{\beta} \left(\frac{\partial \ln \Omega}{\partial x_i} \right)_{T, x_{j \neq i}}. \tag{3.31}$$

Equation (3.31) is the general form of the *equation of state* in the microcanonical ensemble,[6] leading to the *most general form of the fundamental thermodynamic relation*:

$$dU = TdS - \sum_i X_i dx_i + \mu dN. \tag{3.32}$$

Once the fundamental relation has been established, we can use thermodynamics to study the system. In the case of a *reversible* infinitesimal quasi-static process, we can identify $\sum_i X_i dx_i$ as the *work performed by the system* upon an infinitesimal variation of the external variables $x_1, x_2, \ldots, x_n$. The interaction between subsystem $\mathcal{A}$ and subsystem $\mathcal{A}'$ described by this term is therefore associated with a variation of the external variables, leading to changes in the quantum energy levels of subsystem $\mathcal{A}$. If, for example, the interaction between subsystems is purely mechanical, then the varying external variable is the volume, and as we know from quantum mechanics, the spacing between energy levels will change.

Likewise, if the interaction is magnetic, then on increasing the magnitude of the applied magnetic field, the spacing between energy levels will also increase, as we shall see in the next subsection on the isolated paramagnetic solid. For sufficiently slow variations of the external variables, the only effect will be to change the spacing between energy levels: the energy of the system will change, but the system will remain on the same energy level and the process will be reversible. If, however, the variation is not sufficiently slow, the system may transition between energy levels and the process will be *irreversible*. In the latter case the work performed modifies the energy levels enough to induce transitions between them, and we will have $đW_{\text{irrev}} > -\sum_i X_i dx_i$. In the foregoing discussion we have disregarded the term μdN, as there is no way to change N arbitrarily slowly: adding or removing a single particle will change the quantum state of the system, as a system with

---

[6] It can be shown using Eq. (3.2) that $\beta \overline{X_i} = (\partial \ln \Omega / \partial x_i)_{\beta, x_{j \neq i}}$ (Reif, 1985, chap. 3.8). The mean value of X_i is then identified with its thermodynamic value.

a different number of particles is described by a different wavefunction, with unpredictable consequences.

The most general interaction between the two subsystems corresponds to not fixing any of the external variables and allowing heat exchange between the subsystems. The change in the system's energy not associated with variation of the external variables is, by definition, the heat exchanged between the two subsystems. In a reversible process we may identify this heat with TdS, whereas in an irreversible process it follows from the foregoing discussion and from the fundamental relation that $đQ_{\text{irrev}} < TdS$, i.e., in an irreversible process some of the work performed generates entropy. For a general quasi-static infinitesimal process we may thus write that $đQ \leq TdS$, where the equality holds if the process is reversible.

If heat exchange is allowed between subsystems but the external variables (volume, magnetic field, etc) are kept constant, the change in energy is due exclusively to a thermal interaction. Because the energy levels are not affected, the change in energy can only result from a change in the distribution of ensemble elements over the different energy levels, given by Eq. (3.13). This is the basis for the *statistical interpretation of the interaction we call heat*.

3.3.5 The isolated paramagnetic solid

A simple model for a paramagnetic solid in a magnetic field $\mathbf{H}$ is a set of N atoms or ions, each with magnetic moment μ, located at the sites of a crystal lattice. As a first approximation we neglect the vibrational motion of the atoms or ions about their equilibrium positions as well as the interactions between the magnetic moments and the lattice, and between each pair of magnetic moments. This is justified because in a paramagnetic solid, all these interactions are much weaker than that of the magnetic moments with the applied field.[7] In our model we shall therefore retain only this latter interaction, the energy of which is known as the *Zeeman energy*; for the ith magnetic moment μ_i this equals

$$\varepsilon_i = -\boldsymbol{\mu_i} \cdot \mathbf{B}, \quad i = 1, \ldots, N, \tag{3.33}$$

where $\mathbf{B}$ is the *local field* or *magnetic induction* acting on the ith magnetic moment. The Zeeman energy is quantised according to the orientation of the magnetic moment relative to the field, as we shall now show. The magnetic moment μ of each atom is given by:

$$\boldsymbol{\mu} = -g\mu_B\mathbf{J}, \tag{3.34}$$

where the total angular momentum $\hbar\mathbf{J} = \hbar\mathbf{L} + \hbar\mathbf{S}$ is the sum of the orbital angular momentum $\hbar\mathbf{L}$ and the spin angular momentum $\hbar\mathbf{S}$ of the atom. The proportionality constant is the product of the *Bohr magneton* $\mu_B = e\hbar/2m = 9.27 \times 10^{-24}$ J T^{-1} (with e the proton charge, $\hbar = h/2\pi$ the reduced Planck's constant, and m the electron mass), which is a common unit of atomic magnetic moment; and g, a numerical factor that depends

---

[7] It is, however, important to note that in reality the magnetic moments must interact among themselves, otherwise the system would never reach equilibrium in zero field.

on the magnitude of $\hbar J$ via the Landé equation (Kittel, 1995, chap. 14). If, as usual, we take the local field to lie along OZ, then the Zeeman energy ε_i, given by Eq. (3.33), will be proportional to the allowed values ot J_z, the component, or projection, of $\mathbf{J}$ along the field direction: $J_z = -J, -J+1, \ldots, J-1, J$. There are, therefore, $2J+1$ such projections, which we shall study in Chapter 4 as an exemplary application of the canonical ensemble formalism. Here we shall restrict ourselves to the case $\mathbf{L} = \mathbf{0}$, i.e., there is no orbital angular momentum and the only contribution to the total atomic angular momentum is due to unpaired electron spins, so that each atom has spin 1/2. In this case $J_z = \pm 1/2$ and $g \simeq 2$ and the projection of the magnetic moment along the local field can take only two values:

$$\mu_i = g\mu_B J_z = \pm \mu_B. \tag{3.35}$$

It follows that the Zeeman energy ε_i can itself take only two values, corresponding to magnetic moment orientations parallel (minimum energy) and antiparallel (maximum energy) to the field:

$$\varepsilon_i = \mp \mu_B B. \tag{3.36}$$

The spacing between these two energy levels is, consequently, $\Delta\varepsilon = 2\mu_B B$.

We are now in a position to apply the microcanonical formalism derived earlier to our model of an ideal paramagnetic solid. We take the N spins on the lattice to be N non-interacting, isolated subsystems with just two possible energy states.[8] We shall start by finding the energy E of the N spins when n of them are parallel to the field and $N-n$ are antiparallel:

$$U \equiv E(n) = \sum_{i=1}^{N} \varepsilon_i = (N-2n)\mu_B B. \tag{3.37}$$

As can be seen from Figure 3.5, n sets the energy but does not uniquely specify the microstate.

The number of microstates with energy $E(n)$, i.e., the statistical weight of the macrostate with energy $E(n)$, equals the number of different ways in which n spins can be parallel to the field while $N-n$ spins are antiparallel:

$$\Omega(n) = \frac{N!}{n!(N-n)!}. \tag{3.38}$$

This quantity is a measure of how ordered the system is: if $n = N$ (all spins parallel to field), then $\Omega = 1$ is a *minimum* and the state is *maximally ordered*. By Boltzmann's formula *this state has zero entropy* and its energy is $E(N) = -N\mu_B B$, which is the *lowest* or *ground-state energy of the system*. The statistical weight Ω is *maximised* for $n = N/2$: this corresponds to a state of *maximum disorder*, with *maximum entropy* and *zero energy*. If $n < N/2$, the statistical weight Ω will decrease until it reaches its minimum value $\Omega = 1$ for $n = 0$, which

---

[8] The system is ideal because the spins do not interact with each other, and isolated because the spins do not interact with the lattice where they are located

is a state of zero entropy, but now of *maximum energy*, $E_0 = N\mu_B B$. Inserting Eq. (3.38) into the Boltzmann formula, we obtain the general expression for the entropy:

$$S(n) = k_B \ln \Omega(n) = k_B \ln \frac{N!}{n!(N-n)!}. \tag{3.39}$$

In a macroscopic system, N and n are sufficiently large that we can use Stirling's formula for their logarithms, $\ln x! \approx x\ln x - x$ (see Appendix 1.B), whence

$$\frac{1}{k_B}S(n) = \ln N! - [\ln n! + \ln(N-n)!] \simeq N\ln N - n\ln n - (N-n)\ln(N-n). \tag{3.40}$$

The temperature is found from Eqs. (3.11) and (3.37) as

$$\frac{1}{T} = \left(\frac{\partial S}{\partial E}\right)_{H,N} = \frac{\partial S}{\partial n}\left(\frac{\partial n}{\partial E}\right)_{H,N} = k_B \ln \frac{N-n}{n}\left(-\frac{1}{2\mu_B B}\right)$$

$$= \frac{k_B}{2\mu_B B} \ln \frac{n}{N-n}, \tag{3.41}$$

with the following limiting behaviours:

- If $n > N/2$, $\partial S/\partial E > 0 \Rightarrow T > 0$;
- If $n < N/2$, $\partial S/\partial E < 0 \Rightarrow T < 0$;
- If $n = N/2$, $\partial S/\partial E = 0 \Rightarrow T = \infty$.

This implies that states for which $0 \le n \le N/2$, i.e., states with more spins antiparallel than parallel to the field, are *negative absolute temperature states*: as can be seen from Figure 3.6, their energies are higher than those of positive absolute temperature states.

The question now arises of whether a spin system can ever be found in a negative absolute temperature state. As discussed earlier, the energy of most macroscopic systems is unbounded from above (e.g., the kinetic energy of an n-particle system can be arbitrarily large), so the statistical weight, and consequently the entropy, are increasing functions of the energy. A different situation may result in some part of a system that interacts only

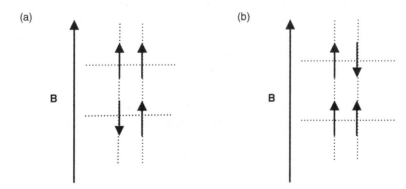

Figure 3.5 Spin configurations (a) and (b) have the same energy but correspond to two different microstates: although all spins are *identical*, those that are antiparallel to the field are *located* at different lattice sites in (a) and (b). Exchanging two parallel spins in either configuration does not generate a new microstate.

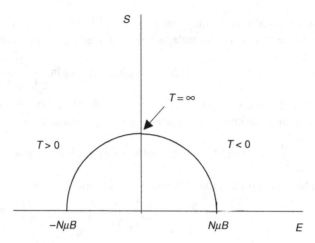

Figure 3.6   The entropy S as a function of energy E has a maximum for $E = 0$, at which the absolute temperature is infinite.

very weakly with all other parts of the same system: now this particular part may reach equilibrium, but not be in equilibrium with all the other parts: i.e., one part of the system is in equilibrium, but the system as a whole is not. A temperature may then be defined for that part of the system which need not be the same as that (those) of the other part(s). If the energy of that part of the system is bounded from above, then its absolute temperature may be negative. This is indeed the case with the spin system studied, where the spin–lattice interaction is much weaker than the spin–spin interaction (which in turn is weak in a paramagnet). The spin–lattice relaxation time (i.e., the time it takes for spins to reach equilibrium with the lattice) is much larger than the spin–spin relaxation time. We can therefore define a specific temperature, the *spin temperature*, associated with the spin system.

Negative absolute temperatures can be realised in a nuclear magnetic resonance (NMR) experiment: nuclear magnetic moments can be aligned using a sufficiently strong magnetic field. If the field direction is then quickly inverted, not all magnetic moments will follow the field and many will be temporarily antiparallel to the field. Through spin–lattice interactions, this unstable state will eventually evolve to the (lower-energy) global thermodynamic equilibrium state in which most spins are aligned parallel to the field (the excess spin energy is absorbed by the lattice, thereby increasing the amplitude of its vibrational motion, i.e., its temperature). Finally, if a system at positive absolute temperature is brought into contact with another system at negative absolute temperature, then heat will flow from the system at negative temperature to the system at positive temperature, i.e., negative temperatures are 'hotter'.

To conclude this example, we shall now calculate the projection along OZ of the total magnetic moment:

$$\mathbf{M} = \sum_{i=1}^{N} \mu_i. \tag{3.42}$$

Because each magnetic moment has only two possible states, its projection along OZ is simply

$$M = \sum_{i=1}^{N} \mu_i = N\mu_B + (N-n)(-\mu_B) = -(N-2n)\mu_B, \qquad (3.43)$$

which is maximal for $n = N$ and vanishes for $n = N/2$, i.e., it is maximal when all spins are aligned parallel to the field and the entropy is minimal, and minimal in the fully disordered state of maximal entropy. Note that for $n > N/2$ we have $M(n) > 0$ and $T > 0$, which is the most commonly encountered situation.

Finally, from Eqs. (3.37) and (3.43) we find

$$U = -\mathbf{M} \cdot \mathbf{B}. \qquad (3.44)$$

Eq. (3.41) can be solved with respect to n, which allows us to write n in terms of the temperature T, the magnetic induction B, and the total number of spins N. By so doing, we switch from describing an isolated system specified by its energy E (which is in turn determined by the number of parallel spins given by Eq. (3.37)), B and N, to describing a (magnetic) system in contact with a temperature reservoir, i.e., specified by T, B, and N. As we shall see later, such a system is represented by the canonical ensemble, but the two ensembles give the same results in the limit $N \to \infty$ (see section 4.6). This limit has actually already been invoked, when using Stirling's formula to find the entropy from the statistical weight. We thus obtain, for the number of parallel spins,

$$n = N\frac{e^x}{e^x + e^{-x}}, \quad x = \frac{\mu_B B}{k_B T}, \qquad (3.45)$$

and for the number of antiparallel spins,

$$N - n = N\frac{e^{-x}}{e^x + e^{-x}}, \quad x = \frac{\mu_B B}{k_B T}. \qquad (3.46)$$

Substituting Eq. (3.45) into Eq. (3.43), we find the total magnetic moment in terms of the magnetic induction and the temperature. We defer analysis of this result to our study of the paramagnetic solid in the canonical ensemble in the next chapter. For now, we merely note that n/N is the probability p_+ to find a spin aligned parallel to the field, and $(N-n)/n$ is the probability p_- to find a spin aligned antiparallel to the field. In summary,

$$p_\pm = \frac{e^{\pm x}}{e^x + e^{-x}}, \qquad (3.47)$$

which result will arise most naturally in the context of the canonical ensemble, where its magnetic induction and temperature dependences will be investigated.

3.4 General conditions for thermodynamic equilibrium. Thermodynamic potentials revisited

3.4.1 Isolated system: the second law

Consider an isolated system. If some external variable of the system (such as the volume, magnetic field, etc.) is varied sufficiently slowly, the system will undergo a quasi-static process where it occupies a sequence of equilibrium states, all represented by the microcanonical ensemble.

The elements of a microcanonical ensemble (virtual copies of the real system) are uniformly distributed over all Ω_i accessible microstates. If in the process the number of microstates does not change, the entropy will also not change. Upon (sufficiently slow) restoration of the initial conditions, all microstates will still be equally probable, and the process will be reversible. On the other hand, if we vary some external variable (e.g., remove a constraint in such a way as to initiate a spontaneous process in the thermodynamic sense), then the number of available microstates will increase; let it be Ω_f in the final equilibrium state. The entropy will thus vary from $S_i = k_B \ln \Omega_i$ in the initial equilibrium state, to $S_f = k_B \ln \Omega_f$ in the final equilibrium state. Because $\Omega_f > \Omega_i$, the change in entropy in a spontaneous process is

$$\Delta S = S_f - S_i = k_B \ln \frac{\Omega_f}{\Omega_i} > 0. \tag{3.48}$$

If now the constraint is restored sufficiently slowly, the ensemble elements will remain uniformly distributed over the Ω_f microstates. Because $\Omega_f > \Omega_i$, restoring the constraint will not return the system to its initial state; hence the process is irreversible. In summary, for an isolated system we have

$$\Delta S = S_f - S_i \geq 0, \tag{3.49}$$

where the equality holds for reversible processes. Equation (3.49) is a re-statement of the statistical interpretation of the second law of thermodynamics: it sets a criterion for the direction of naturally occurring processes and for the equilibrium state of an isolated system.

3.4.2 General criterion for equilibrium

We now wish to derive a general criterion for the equilibrium of systems that are not isolated. This will be done by treating our system as a subsystem of a large, isolated, system to which criterion (3.49) applies. The procedure to be followed is the same as in Section 3.3.2: the isolated system, denoted $\mathcal{A}_0$, is specified by (U_0, V_0, N_0) and partitioned by an internal wall into two subsystems, $\mathcal{A}$ and $\mathcal{A}'$, which in the most general case can exchange heat between themselves, as well as perform work on each other. Our purpose is to find the equilibrium condition for system $\mathcal{A}$ in terms of its own thermodynamic quantities, so

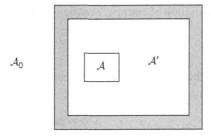

Figure 3.7 An isolated system $\mathcal{A}_0$ composed of $\mathcal{A}$, the system under study, and $\mathcal{A}'$, the reservoir. $\mathcal{A}$ and $\mathcal{A}'$ are not isolated; they can exchange heat and work with each other.

we partition system $\mathcal{A}_0$ in such a way that $\mathcal{A}' \gg \mathcal{A}$, and denote system $\mathcal{A}'$ *the reservoir* (Figure 3.7).

System $\mathcal{A}'$ is chosen to be large enough that its heat capacity is much larger than that of system $\mathcal{A}$, so its temperature will remain constant in spite of energy being exchanged between the two systems. We call such a system a *temperature reservoir* or *thermostat* (in fact it is an energy reservoir). At equilibrium system $\mathcal{A}$, which is embedded in the reservoir (e.g., a water bottle placed inside a pool) will have the same temperature as the reservoir. Likewise, although the wall between $\mathcal{A}$ and $\mathcal{A}'$ is free to move, because $\mathcal{A}'$ is much bigger than $\mathcal{A}$ its pressure does not change, so at equilibrium $\mathcal{A}$ will have the same pressure as $\mathcal{A}'$. In this case, $\mathcal{A}'$ is a *pressure reservoir*. We start by assuming that the system and its reservoir are not in (mechanical or thermal) equilibrium with each other. Let S_0 be the entropy of $\mathcal{A}_0$; from Eq. (3.49), $S_0 = S + S'$. Because $\mathcal{A}_0$ is isolated, for any finite process we must have, by Eq. (3.6), that

$$\Delta S_0 = \Delta S + \Delta S' \geq 0. \tag{3.50}$$

Let us further assume that in this process $\mathcal{A}$ absorbs an amount of heat Q from the reservoir, and performs work $W = P' \Delta V$ against the constant pressure of the reservoir. Now clearly reservoir A' has absorbed an amount of heat $Q' = -Q$, at constant temperature T', so the number of microstates accessible to it, which was initially $\Omega'(U')$, is now $\Omega'(U' + Q')$. Now $\mathcal{A}' \gg \mathcal{A}$, so the change in the number of states will be small and we can expand the logarithm of $\Omega'(U' + Q')$ about its initial value:

$$\ln \Omega'(U' + Q') = \ln \Omega'(U') + \left(\frac{\partial \ln \Omega'}{\partial U'}\right)_V Q' + \frac{1}{2}\left(\frac{\partial^2 \ln \Omega'}{\partial U'^2}\right)_V Q'^2 + \cdots$$

$$= \ln \Omega'(U') + \beta' Q' + \frac{1}{2}\left(\frac{\partial \beta'}{\partial U'}\right)_V Q'^2 + \cdots \tag{3.51}$$

where we have used the definition of β, Eq. (3.9). Because $\mathcal{A}'$ is a temperature reservoir, β' does not change appreciably and we can neglect its derivatives, leading to

$$\ln \Omega'(U' + Q') - \ln \Omega'(U') = \beta' Q'. \tag{3.52}$$

From the definition of entropy, equqtion (3.5), it then follows that

$$\Delta S' = \frac{Q'}{T'} = -\frac{Q}{T'}.$$ (3.53)

Combining this with inequality (3.50) and the first law of thermodynamics, $Q = \Delta U + W$, we then obtain

$$\Delta S_0 = \Delta S - \frac{\Delta U + W}{T'} \geq 0,$$ (3.54)

which is a criterion for the direction of natural processes, as well as for the equilibrium of a non-isolated system, in terms of quantities pertaining to that system and to its surroundings. Only in very special cases is this criterion formulated exclusively in terms of quantities pertaining to the system alone, as we shall see next. The equality only holds for reversible processes. Let us now suppose that the wall between $\mathcal{A}$ and $\mathcal{A}'$ is adiabatic. System $\mathcal{A}$ is thus thermally insulated. If in addition the wall is fixed, then the state of system $\mathcal{A}$ is specified by

$$U = \text{constant}, \quad V = \text{constant}, \quad \Delta S \geq 0.$$

We can use inequality (3.54) to replace the constraint of constant energy by a constraint of constant entropy, leading to

$$S = \text{constant}, \quad V = \text{constant}, \quad \Delta U \leq 0,$$

i.e., *the energy is minimised at equilibrium*, which allows us to formulate the *minimum energy principle: the equilibrium value of any non-fixed internal variable of a system is such that it minimises the energy, at fixed entropy.*

We conclude that *the maximum entropy principle (at fixed energy) is equivalent to the minimum energy principle (at fixed entropy)* (Callen, 1985, chap. 5.1). Callen draws a very illuminating analogy with the definition of a circle: it is the closed curve of minimum length that encloses a given area; or, of all curves of given length, that which encloses the largest area.

3.4.3　System in contact with a temperature reservoir

Let us first consider a system undergoing an isothermal process, such that its temperature is the same as that of the reservoir: $T = T'$. If we now *define* the state function of the system as $F = U - TS$, we can rewrite condition (3.54) as

$$\frac{\Delta(TS - U) - W}{T} = \frac{-\Delta F - W}{T} \geq 0.$$ (3.55)

In all thermodynamically relevant situations the absolute temperature is positive; hence we can rewrite the above as

$$-\Delta F \geq W,$$ (3.56)

which implies that the *maximum amount of work* that can be performed by a system in contact with a temperature reservoir is $(-\Delta F)$, where the equality holds for reversible processes. If the external variables are kept constant, then no work is performed and

$$\Delta F \leq 0. \tag{3.57}$$

This tells us which way a system will naturally evolve when in contact with a temperature reservoir. It is written exclusively in terms of quantities pertaining to the system, and as such it replaces the requirement that the entropy of an isolated system must increase, condition (3.49). We thus arrive at the following formulation: *if the external variables of a system are fixed and the system is in contact with a temperature reservoir, the function $F = U - TS$ will decrease and the condition for stable equilibrium is*

$$F = \text{minimum}. \tag{3.58}$$

3.4.4 System in contact with a temperature and pressure reservoir

Let us now suppose that the wall between $\mathcal{A}$ and $\mathcal{A}'$ is free to move. We write the first law of thermodynamics in the form

$$Q = \Delta U + P' \Delta V + W^*, \tag{3.59}$$

where we have singled out $P' \Delta V$, the work performed by $\mathcal{A}$ against the (constant) reservoir pressure P', and W^* denotes any other work performed by $\mathcal{A}$ in this process. Proceeding as before, we find

$$\Delta S_0 = \Delta S - \frac{\Delta U + P' \Delta V + W^*}{T'} \geq 0. \tag{3.60}$$

If the system is in both thermal and mechanical equilibrium with the reservoir, then $T = T'$ and $P = P'$. Defining the state function $G = U - TS + PV$, condition (3.60) can be rewritten in the form

$$\frac{\Delta(TS - U - PV) - W^*}{T} = \frac{-\Delta G - W^*}{T} \geq 0, \tag{3.61}$$

which implies

$$-\Delta G \geq W^*, \tag{3.62}$$

i.e., that the maximum amount of work that system $\mathcal{A}$ can perform on the pressure reservoir is $(-\Delta G)$. As before, the equality holds for reversible processes. If all external variables except the volume are fixed, then $W^* = 0$ and

$$\Delta G \leq 0. \tag{3.63}$$

Thus we conclude that *if all the external variables of a system except its volume are fixed and the system is in contact with a temperature reservoir (i.e., both its pressure and its temperature are fixed), the function $G = U - TS + PV$ will decrease and the condition for stable equilibrium is*

$$G = \text{minimum}. \tag{3.64}$$

3.4.5 Isolated system in contact with a pressure reservoir

Next, suppose that the wall between $\mathcal{A}$ and $\mathcal{A}'$ is adiabatic and free to move. The first law of thermodynamics now yields

$$Q = \Delta U + P' \Delta V = 0. \tag{3.65}$$

Because the system is in mechanical equilibrium with the reservoir, $P = P'$. Defining the state function $E^* = U + PV$, condition (3.65) can be rewritten in the form

$$Q = \Delta E^* = 0, \tag{3.66}$$

which implies that

$$E^* = \text{constant}, \tag{3.67}$$

$$\Delta S \geq 0. \tag{3.68}$$

We conclude that *in an isolated system at constant $E^* = U + PV$ and fixed P, the entropy is maximised at equilibrium, just as in a system at constant energy and fixed V.*

Combining condition (3.50) with Eqs. (3.53) and (3.65) yields, for $P = P'$,

$$\Delta S_0 = \Delta S - \frac{\Delta E^*}{T'} \geq 0, \tag{3.69}$$

whence *at constant entropy, E^* is minimised at equilibrium.*

From $Q = \Delta E^*$ it follows that E^* is the 'natural heat function' at constant pressure. Likewise, at constant volume we have $Q = \Delta U$, so U is the natural heat function in that case.

If we now identify the state functions F, G and E^* with, respectively, the Helmholtz energy, Eq. (2.28); the Gibbs energy, Eq. (2.34); and the enthalpy,[9] Eq. (2.22); it becomes apparent why, in view of the equilibrium conditions just derived, it is appropriate to name these functions *thermodynamic potentials*.

3.4.6 Legendre transformations. General formulation of the conditions for thermodynamic equilibrium*

Equilibrium conditions appropriate to a given thermodynamic system may be derived most generally using *Legendre transforms*. We shall introduce these by considering first a simple example: let $f(x, y)$ be a function of variables x and y; then

$$f(x,y) \Rightarrow df = u\,dx + v\,dy, \quad u = \frac{\partial f}{dx}, \quad v = \frac{\partial f}{dy}.$$

Suppose we want to switch variables form (x, y) to (u, y). Now an infinitesimal change in f must be expressed in terms of du and dy. This is accomplished by defining a new function

---

[9] The enthalpy is sometimes denoted H. Here we have opted for the alternative symbol E^*, as we are using H for the magnetic field.

$g(u,y)$ as $g = f - ux$:

$$dg = df - xdu - udx = -xdu + vdy \Rightarrow x = -\frac{\partial g}{\partial u}, \quad v = \frac{\partial g}{\partial y},$$

Function $g(u,y)$ is the *partial Legendre transform of f(x,y) with respect to x*. We may likewise define the *total Legendre transform* as

$$h = f - ux - vy \Rightarrow dh = -xdu - ydv.$$

We now apply this to thermodynamics. The internal energy in terms of its variables is written

$$U \equiv U(x_0, \ldots, x_n, x_{n+1}, \ldots, x_t), \tag{3.70}$$

where $x_0 = S$ is the entropy. In this notation, the fundamental thermodynamic relation, Eq. (2.8), becomes

$$dU = \sum_{k=0}^{t} X_k dx_k, \tag{3.71}$$

where we have used the definitions of conjugate force, Eq. (2.9), and of temperature, Eq. (2.10). The partial Legendre transform of U with respect to $x_0, \ldots, x_n$ is now obtained by replacing $x_0, \ldots, x_n$ by their conjugate forces $X_0, \ldots, X_n$, as follows:

$$\underset{x_0,\ldots,x_n}{\mathcal{L}} U = U - \sum_{k=0}^{n} X_k x_k = \mathcal{W}(X_0, \ldots, X_n, x_{n+1}, \ldots, x_t), \tag{3.72}$$

where $\mathcal{W}$ must not be confused with work, W. The partial derivatives of $\mathcal{W}$ are easily found to be

$$\left(\frac{\partial \mathcal{W}}{\partial X_k} \right)_{X_{i \neq k}} = -x_k, \quad k = 0, \ldots, n, \tag{3.73}$$

$$\left(\frac{\partial \mathcal{W}}{\partial x_k} \right)_{x_{i \neq k}} = X_k, \quad k = n+1, \ldots, t, \tag{3.74}$$

whence we obtain the fundamental thermodynamic relation in terms of $\mathcal{W}$ as

$$d\mathcal{W} = \sum_{k=0}^{n} (-x_k) dX_k + \sum_{k=n+1}^{t} X_k dx_k. \tag{3.75}$$

The thermodynamic potentials are Legendre transforms of the internal energy with respect to their natural variables. This allows the fundamental thermodynamic relation to be written in terms of more convenient variables (e.g., temperature rather than entropy, pressure rather than volume). It is left as an exercise for readers to check that

$$F(T,V,N) = \underset{S}{\mathcal{L}} U(S,V,N), \tag{3.76}$$

$$E^*(S,P,N) = \underset{V}{\mathcal{L}} U(S,V,N), \tag{3.77}$$

$$G(T,P,N) = \underset{S,V}{\mathcal{L}} U(S,V,N). \tag{3.78}$$

In the next chapter we shall use the thermodynamic potentials to make a connection between statistical mechanics and thermodynamics. As an aside, we note that, in classical mechanics, the Hamiltonian is the Legendre transform of the Lagrangean with respect to the generalised velocities (Goldstein, 1980, chap. 8.1). The Legendre transform thus enables us to switch from a formulation of classical mechanics in terms of generalised coordinates and velocities (Lagrangean) to one in terms of generalised coordinates and momenta (Hamiltonian).

The conditions for thermodynamic equilibrium can be formulated very generally in terms of partial Legendre transforms of the energy: *The equilibrium value of any internal variable of a system in contact with a reservoir of constant intensive variables* $X_i', X_j', \ldots$ *will be such as to minimise* W *at constant* $X_i = X_i', X_j = X_j', \ldots$.

The interested reader is refereed to Callen (1985, chaps. 5.2, 5.3 and 6.1) for more details. In the next section we apply the thermodynamic potential formalism to study phase equilibria (or 'changes between states of matter') in a thermodynamic system.

3.4.7 Phase equilibria

Here we shall use the results derived in earlier sections to establish the conditions for phase equilibrium in systems at constant temperature and pressure, which is a situation of great experimental interest. If we further assume that the number of particles is also constant, such a system will be specified by T, P and N. It then follows from the fundamental thermodynamic relation in the form of Eq. (2.46) that the appropriate thermodynamic potential is the Gibbs energy, $G(T, P, N)$. This is an extensive quantity, and therefore proportional to the number of particles. If our system is homogeneous and composed of only one particle species, we may write

$$G(T, P, N) = Ng(T, P), \tag{3.79}$$

where $g(T, P)$ is the Gibbs energy per particle. Equation (2.46) then yields

$$\mu = \left(\frac{\partial G}{\partial N} \right)_{T,P} = g(T, P), \tag{3.80}$$

i.e., the chemical potential equals the Gibbs free energy per particle. Combining the two preceding equations, we find

$$G(T, P, N) = N\mu. \tag{3.81}$$

Inserting this into the definition of G, Eq. (2.34), yields

$$U = TS - PV + \mu N. \tag{3.82}$$

Differentiating this and taking into account the fundamental thermodynamic relation, Eq. (2.44), we arrive at

$$SdT - VdP + Nd\mu = 0, \tag{3.83}$$

which is known as the *Gibbs–Duhem relation*: this tells us that the intensive variables T, P and μ cannot all vary independently. As such it complements Eq. (2.44), which relates the infinitesimal variations of the extensive variables U, S, V and N. Equilibrium conditions

(3.64) together with Eq. (3.81) allow us to write that, in a system of fixed particle number at equilibrium,

$$\mu = \text{minimum}. \tag{3.84}$$

Consider now a system composed of two phases, 1 and 2 (e.g., 1 = liquid and 2 = vapour). The Gibbs energy of this system is

$$G(T,P,N_1,N_2) = N_1 g_1(T,P) + N_2 g_2(T,P). \tag{3.85}$$

From Eq. (3.64), equilibrium requires that

$$dG(T,P,N_1,N_2) = \left[N_1 \left(\frac{\partial g_1}{\partial T} \right)_{P,N_1,N_2} + N_2 \left(\frac{\partial g_2}{\partial T} \right)_{P,N_1,N_2} \right] dT$$
$$+ \left[N_1 \left(\frac{\partial g_1}{\partial P} \right)_{T,N_1,N_2} + N_2 \left(\frac{\partial g_2}{\partial P} \right)_{T,N_1,N_2} \right] dP$$
$$+ g_1(T,P)dN_1 + g_2(T,P)dN_2 = 0. \tag{3.86}$$

Since $T = \text{constant}$, $P = \text{constant}$ and $N = N_1 + N_2 = \text{constant}$, we obtain

$$g_1(T,P) = g_2(T,P). \tag{3.87}$$

Now each phase is homogeneous and composed of just one particle species, hence Eq. (3.81) allows us to write $G_i = \mu_i N_i \Rightarrow \mu_i = g_i(T,P)$ $(i = 1,2)$, whereupon Eq. (3.87) becomes

$$\mu_1 = \mu_2. \tag{3.88}$$

We thus conclude that the condition for equilibrium of a two-phase system at constant temperature and pressure is the same as that for an isolated system, Eq. (3.27).

An isotropic fluid such as water may exhibit not just two-phase equilibria (solid–liquid, liquid–vapour and solid–vapour) but also three-phase equilibria: solid–liquid–vapour. Complex fluids, e.g., polymers or liquid crystals, will have even more complicated phase diagrams. Because three-phase equilibrium presupposes equilibrium between any pair of phases involved, we must have

$$g_1(T,P) = g_2(T,P) = g_3(T,P). \tag{3.89}$$

Whereas Eq. (3.87) defines a curve as the locus of two-phase equilibria in the $(T - P)$ plane, Eq. (3.89) expresses the intersection of two such curves, for $g_1 = g_2$ and $g_2 = g_3$. The point of intersection of these curves in the phase diagram is called the *triple point* (see Figure 3.8). As an example, the triple point of water is at $P = 4.58$ mmHg and $T = 0.01°C$ (in 'everyday' units). Because ice, liquid water and water vapour coexist only at this one temperature, the triple point of water is used as the fixed point of the absolute temperature scale: by definition (1954), the triple point of water occurs at $T_t = 273.16$ K, and one kelvin is defined as $1/273.16$ of the temperature interval between $T = 0$ and $T = T_t$. The absolute temperature scale thus defined is almost identical to the 'historical' temperature scale based on dividing the temperature interval between the freezing and boiling points of water into 100 degrees.

The Gibbs phase rule

The topology of the phase diagrams of systems with an arbitrary number of components – i.e., how many phases can coexist in three-, two-, or one-dimensional regions of thermodynamic space – is governed by the *Gibbs phase rule*. Conditions for phase equilibrium in multicomponent systems are derived by generalising Eq. (3.88); here we shall follow Callen (1985, chap. 9).

Consider a system with r components and M phases. The Gibbs energy of each of these phases is a function of $T, P, N_1, N_2, \ldots, N_r$. We shall view each phase as a separate (sub)system: the actual system is the sum of all these parts. The walls partitioning the system into subsystems are, therefore, purely conceptual and do not restrict the flow of heat, work, or particles.

For definiteness, take a container filled with a binary mixture, and kept at such constant temperature and pressure that a liquid phase L coexists with a solid phase S. Our purpose is to find the compositions of the coexisting phases. Let x_1^ϕ be the mole fraction of component 1 in phase ϕ ($\phi = L, S$); its chemical potential in the liquid phase is thus $\mu_1^L(T, P, x_1^L)$, and in the solid phase $\mu_1^S(T, P, x_1^S)$. Note that, unlike in homogeneous systems, the subscript now denotes the component and not the phase. For component 1, coexistence then requires

$$\mu_1^L(T, P, x_1^L) = \mu_1^S(T, P, x_1^S). \tag{3.90}$$

Because in each phase ϕ we must have $x_1^\phi + x_2^\phi = 1$, the chemical potential of component 2 can be written in terms of x_1^ϕ, so for component 2 we likewise have

$$\mu_2^L(T, P, x_1^L) = \mu_2^S(T, P, x_1^S). \tag{3.91}$$

If T and P are known, Eqs. (3.90) and (3.91) can be solved for x_1^L and x_1^S.

Now suppose three phases coexist in the system, call them *I*, *II* and *III*. For component 1 we have

$$\mu_1^I(T, P, x_1^I) = \mu_1^{II}(T, P, x_1^{II}) = \mu_1^{III}(T, P, x_1^{III}), \tag{3.92}$$

which are actually two equations. Another pair of equations is obtained by equating the chemical potentials of component 2 in the three phases, giving a total of four equations for five unknowns: T, P, x_1^I, x_1^{II} and x_1^{III}. Now only T or P may be fixed freely: the three composition variables and either T or P are given by the solution of the four equations. Consequently, although in a two-component system two phases may coexist for any pair of values of the temperature and pressure, three phases may only coexist at one special value of the pressure, if the temperature has been fixed, or at one special value of the temperature, if the pressure has been fixed.

We next investigate whether four phases may coexist in the same system. Proceeding as before, equality of the chemical potential of each component in all four phases gives three equations per component, i.e., a total of six equations, for the six unknowns $T, P, x_1^I, x_1^{II}, x_1^{III}$ and x_1^{IV}, implying that the four phases can only coexist at one special value of the temperature and one special value of the pressure. On the other hand, five phases cannot coexist in a two-component system. because the eight equations expressing equality of the chemical potentials of the two components overdetermine the seven unknowns.

Let us now formulate a general rule for a system with r components and M coexisting phases. The variables describing the system are T and P, which are the same for all phases, plus $x_1^I, x_2^I, \ldots, x_{r-1}^I$ mole fractions of phase I, $x_1^{II}, x_2^{II}, \ldots, x_{r-1}^{II}$ mole fractions of phase II, etc., adding up to a total of $2 + M(r - 1)$ variables. On the other hand, there are r different chemical potentials, one for each component; the equality of chemical potentials of each component in M phases generates $M - 1$ equations per component, hence the total number of equations is $r(M - 1)$. Therefore, the number of variables f that can be assigned arbitrary values is just the number of variables minus the number of equations:

$$f = 2 + M(r - 1) - r(M - 1) = r - M + 2 \Leftrightarrow f + M = r + 2, \qquad (3.93)$$

which relation is known as the *Gibbs phase rule*. Quantity f can be interpreted as the number of *thermodynamic degrees of freedom* (Callen, 1985): this is the number of intensive variables that may be varied independently without changing the number of phases in the system. To see that this is so, we shall use the Gibbs–Duhem relation, (3.83). In a one-component, two-phase system, there are three intensive variables, T, P and μ (because at equilibrium the chemical potentials in both phases are the same) and two Gibbs–Duhem relations, one for each phase. This leaves one degree of freedom. In an r-component system, degrees of freedom may be counted likewise: there are $2 + r$ intensive variables: $T, P, \mu_1, \mu_2, \ldots, \mu_r$, each of which is the same in all phases. Now for each phase there is one Gibbs–Duhem relation, The number of independent variables, i.e., of degrees of freedom, is therefore $2 + r - M$, as given by the Gibbs phase rule, Eq. (3.93). The Gibbs phase rule may then be formulated as follows: *In an r-component system in which M phases coexist, one may freely assign values to $r - M + 2$ variables from the set $\{T, P, \mu_1, \mu_2, \ldots, \mu_r\}$ (or, alternatively, from the set composed of T, P and r independent mole fractions).*

As an example, consider a one-component, three-phase system: $r = 1$ and $M = 3$: hence $f = 0$, as we saw in the case of water, for which three phases may coexist only at a single point in the phase diagram – the triple point of water.

The Clausius–Clapeyron equation

The line along which two phases coexist, defined by Eq. (3.87), is the solution of a differential equation, which we shall now establish. Consider the P–T phase diagram of a one-component system shown in Figure 3.8a. By Eq. (3.87), at each point of coordinates (T, P) located *on* the coexistence line, the Gibbs free energies per particle of the two phases must be equal:

$$g_1(T, P) = g_2(T, P). \qquad (3.94)$$

Differentiation yields

$$dg_1 = dg_2, \qquad (3.95)$$

where

$$dg_i = \left(\frac{\partial g_i}{\partial T}\right)_P dT + \left(\frac{\partial g_i}{\partial P}\right)_T dP = -s_i dT + v_i dP. \qquad (3.96)$$

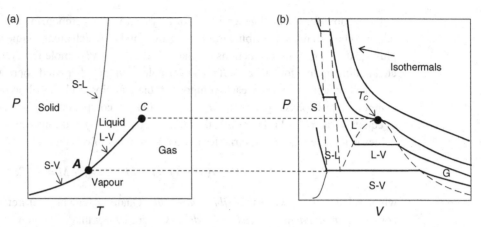

Figure 3.8 (a) Schematic $P-T$ phase diagram of a one-component system exhibiting a *triple point*. The lines are loci of two-phase coexistence, according to Eq. (3.88). C is the *critical point*, at which the liquid–vapour coexistence line terminates. (b) Schematic $P-V$ phase diagram of the same system as in (a). The solid lines are isotherms; the dashed line is the envelope of two-phase coexistence, called the *binodal line*.

Equating $dg_1 = dg_2$, we obtain

$$(s_1 - s_2)dT = (v_1 - v_2)dP \Rightarrow \frac{dP}{dT} = \frac{\Delta s}{\Delta v}. \qquad (3.97)$$

Recall that Δs and Δv are the entropy and volume changes *per mole*. Multiplying both sides of the preceding equation by the number of moles in the system, we arrive at the *Clausius–Clapeyron equation*:

$$\frac{dP}{dT} = \frac{\Delta S}{\Delta V} \qquad (3.98)$$

The Clausius–Clapeyron equation tells us how the pressure changes with temperature at a phase transition, i.e., the slope of the P vs T lines in a (P, T) phase diagram, in terms of the changes in entropy and volume. Applications of this equation are discussed, e.g., in Reif (1985, 8.5) or Mandl (1988, chap. 8.5). It is interesting to note that, for substances such as water, which contract on freezing, $\Delta V_{S-L} < 0$. Because the liquid phase is more disordered than the solid, $\Delta S_{S-L} > 0$. From Eq. (3.98) it then follows that the solid–liquid transition line has negative slope.

The liquid–vapour coexistence line terminates at the *critical point*, marked C in Figure 3.8a: beyond this point there is no discontinuous transition between the liquid and the vapour. Note that the liquid–solid coexistence line does not terminate at a critical point. It is apparent from the $P-V$ phase diagram of Figure 3.8b that the critical point is a point of inflection, where $(\partial P/\partial V)_{T=T_c} = (\partial^2 P/\partial V^2)_{T=T_c} = 0$. If the vapour is compressed at constant $T < T_c$, condensation will occur at constant pressure, along one of the horizontal lines in the coexistence region (marked L–V). On transitioning from the vapour to the liquid, the volume decreases by ΔV. When $T \to T_c^-$, ΔV goes to zero. Beyond the critical point, i.e., for $T > T_c$, there is no abrupt change in the systems properties upon isothermal compression (see Chapter 10 and Reif (1985, chap. 8.5) for more details).

A phase transition is an isothermal (constant temperature) and isobaric (constant pressure) reversible process (if the volume is not constrained). It also involves the exchange of heat between the system and its surroundings. We define the *latent heat ℓ* of a transition as *the amount of heat exchanged between a substance undergoing a phase transition and its surroundings, per unit mass (alternatively, per mole) of the substance.* The temperature at which a given phase transition occurs depends on the ambient pressure, and the latent heat in turn depends on the temperature. The entropy change at the transition may be written in terms of the latent heat as

$$S_2 - S_1 = \int_1^2 \frac{d\,Q_{\mathrm{rev}}}{T} = m\frac{\ell}{T} \Rightarrow \Delta s = \frac{\ell}{T}, \tag{3.99}$$

where we have defined the *specific entropy* as the entropy per unit mass, $s = S/m$. If we further define the *specific volume* as the volume per unit mass, $v = V/m$, the Clausius–Clapeyron Eq. (3.98) can be rewritten as

$$\frac{dP}{dT} = \frac{\ell}{T\Delta v}. \tag{3.100}$$

This is a very important thermodynamic relation: it allows the latent heat of a transition to be found by measuring the slope of the coexistence line and the change in specific volume across the transition. We shall come back to this in more detail in Chapter 10.

Problems

3.1 Consider an isolated monatomic crystalline solid, where atoms initially occupy all N lattice sites. A point defect is created when an atom migrates to another location within the solid. In particular, if the atom migrates from a lattice site to an interstitial site, leaving behind a vacancy, the resulting point defect is called a *Frenkel defect*. Let ε be the energy cost of creating a Frenkel defect. Assuming that the number of interstitial sites is the same as the number of lattice sites, which in turn is the same as the number N of atoms in the lattice, and that the number of defects is sufficiently small that interactions between defects may be neglected, find:
   (a) The entropy of creation of n defects, $S(n)$. What is the entropy of the crystal when $N/2$ atoms reside at interstitial sites?
   (b) The temperature $T(n)$ of the crystal at thermal equilibrium. What is its temperature when $N/2$ atoms reside at interstitial sites?
   (c) The temperature dependence of the number of defects when $n \ll N$. Check that the same result can be obtained by minimising the Helmholtz energy.

3.2 Consider an ideal isolated system composed of N spins $1/2$ at fixed positions, in the presence of an applied magnetic field H. Let μ_i be the component along the direction of the magnetic field of the magnetic moment associated with the ith spin. The internal energy is fixed to be equal to that of n spins aligned with the field. The system is then partitioned into two subsystems that can exchange heat between themselves, such that subsystem 1 contains N_1 spins, of which n_1 are aligned with the field, and subsystem 2 contains N_2 spins, of which n_2 are aligned with the field.

(a) Find n_1 and n_2 when the global system (subsystem 1 + subsystem 2) is at equilibrium (take n_1 as the independent variable).

(b) Check that at equilibrium the two subsystems have the same temperature, which is the same as that of the global system.

3.3 A simple model of an ideal, isolated macromolecule consists of $N(\gg 1)$ segments of length a, all aligned along OX. Each segment is represented by a vector, $\mathbf{a}_i$ ($i = 1,\ldots,N$), which may point in either the positive or negative direction of OX: $\mathbf{a} = \pm a\hat{\mathbf{x}}$. Let

$$\mathbf{L} = \sum_{i=1}^{N} \mathbf{a}_i$$

be the end-to-end vector of the chain. Recall what you learnt about random walks in one dimension and denote by n the number of segments pointing in the positive direction of OX. Assume that each segment pointing in the positive direction has energy ε_+, and each segment pointing in the negative direction has energy ε_-.

(a) Find the mean and variance of $L = \sqrt{\mathbf{L} \cdot \mathbf{L}}$. (Hint: write down the probabilities of adding a segment in the positive or negative direction, p_+ and p_- respectively, in terms of the energies ε_+ and ε_-. Further assume that p_+/p_- is inversely proportional to $\varepsilon_+/\varepsilon_-$.)

(b) Find the chain entropy and the Helmholtz energy, using the approximation $\ln y! \approx y \ln y - y$.

(c) Find the chain temperature at equilibrium, expressed in terms of $\Delta\varepsilon = \varepsilon_+ - \varepsilon_-$, N and n.

(d) In the equation of state (3.31) for this model chain, the generalised coordinate x_i is L and the generalised force X_i is the tension τ. Show that

$$\tau = -\frac{k_B T}{2na^2}L,$$

where the approximation $\ln(1+y) \approx y$ with $y = L/n \ll 1$ has been used. This result implies that the chain will contract when subjected to an extensional tension, *because its elasticity is of entropic origin*. Check that you obtain the same result from

$$\tau = -\left(\frac{\partial F}{\partial L}\right)_T.$$

3.4 Show that, for an isolated system at equilibrium, the relative probability of a fluctuation of some quantity y resulting in a change of entropy ΔS is $e^{\Delta S/k_B}$. Estimate this probability for macroscopic systems when $\Delta S < 0$, for which experimentally ΔS is of order Nk_B.

3.5 Compute the total amount of energy required to convert one litre of water, initially at $20°C$, into vapour at $100°C$. (Data: specific heat of water at 1 atm =

4.187 kJ kg^{-1} K^{-1}, latent heat of water at 1 atm = 2270 kJ kg^{-1}. Assume the density of liquid water to be constant and equal to 10^3 kg m^{-3}.)

References

Callen, H. 1985. *Thermodynamics and an Introduction to Thermostatistics*, 2nd edition. Wiley.

Goldstein, H. 1980. *Classical Mechanics*, 2nd edition. Addison-Wesley.

Hill, T. L. 1985. *An Introduction to Statistical Thermodynamics*. Dover Publications.

Huang, K. 1987. *Statistical Mechanics*, 2nd edition. Wiley.

Kittel, C. 1995. *Introduction to Solid State Physics*, 7th edition. Wiley.

Landau, L. D., Lifshitz, E. M., and Pitaevskii, L. P. 1980. *Statistical Physics, Part 1*, 3rd edition. Pergamon Press.

Mandl, F. 1988. *Statistical Physics*, 2nd edition. Wiley.

Pais, A. 1982. *Subtle is the Lord*. Oxford University Press.

Reif, F. 1985. *Fundamentals of Statistical and Thermal Physics*. McGraw-Hill.

Schiff, L. 1968. *Quantum Mechanics*, 3rd edition. McGraw-Hill.

Sklar, L. 1993. *Physics and Chance*. Cambridge University Press.

4 Statistical thermodynamics: developments and applications

4.1 Introduction

To study the statistics of a thermodynamic system that is not isolated, and consequently whose internal energy is not specified, we shall follow a procedure used in the previous chapter: we consider an isolated global system $\mathcal{A}_0$, partitioned by an internal wall into two subsystems $\mathcal{A}$ and $\mathcal{A}'$, such that $\mathcal{A}' \gg \mathcal{A}$, where $\mathcal{A}$ is the system under study. Subsystem $\mathcal{A}$ is embedded in subsystem $\mathcal{A}'$, which shall be referred to as the reservoir (see Figure 4.1). What kind of a reservoir this is will depend on the interactions allowed between the two subsystems. We shall consider a heat reservoir, a heat and particle reservoir, and a heat and pressure reservoir: these are the three most important cases from the point of view of applications, and also for the purpose of establishing the statistical basis of thermodynamics.

We shall use the following notation for the energy of the system: U is the internal energy in the thermodynamic sense, while E is the energy computed by applying the methods of statistical physics to microscopic (classical or quantum) models. For systems without a specified internal energy, E should be regarded as a random variable. For macroscopic systems we shall assume $\overline{E} = U$, as discussed in Chapter 3.

4.2 System in equilibrium with a heat reservoir

4.2.1 Canonical ensemble

We assume that the wall separating the two subsystems is immovable and allows the two subsystems to exchange energy[1] but not particles. Then system $\mathcal{A}'$ acts as a heat reservoir for system $\mathcal{A}$. We assume that system $\mathcal{A}_0$ is isolated and specified by (U_0, V_0, N_0), and that the partition is such that $V_0 = V + V'$ and $N_0 = N + N'$.

System $\mathcal{A}$ will be in one of its available quantum states, or microstates, labelled $1, 2, \ldots, r, \ldots$ These microstates have energies $E_1 \leq E_2 \leq \ldots \leq E_r \leq \ldots$ From quantum mechanics we know that $E_r = E_r(V, N)$ if no other constraints or interactions are to be

---

[1] The energy exchanged between the two subsystems cannot be mechanical work, because the wall is immovable. It can be other types of work, or heat (i.e., kinetic energy) transferred through the wall, which is then not perfectly rigid. In this case, the volumes V and V' of both systems will undergo small fluctuations around their mean values, which we assume to be fixed.

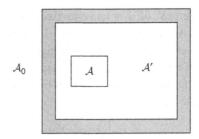

Figure 4.1   An isolated system, $\mathcal{A}_0$, partitioned by an internal wall into two subsystems: $\mathcal{A}$, the system under study; and $\mathcal{A}'$, the reservoir.

taken into account. On the other hand, different microstates can have the same energy: these are known as degenerate microstates. Therefore, when counting the microstates, in the above sequence of energies a given energy may appear more than once. The number of states with the same energy is called the *degeneracy* of that energy level, which is its statistical weight $\Omega(E_r)$ as defined in the preceding chapter. We assume

$$U_0 = E_r + E'. \tag{4.1}$$

Recall the discussion following this equation in Section 3.3.2. Furthermore, according to the discussion in Section 3.3.3, the probability of finding system $\mathcal{A}_0$ partitioned in such a way that system $\mathcal{A}$ has energy E_r and system $\mathcal{A}'$ has energy $E' = U_0 - E_r$ is given by Eq. (3.14):

$$p(E_r) \propto \Omega(E_r)\Omega'(U_0 - E_r), \tag{4.2}$$

where $\Omega(E_r)$ is the degeneracy of microstate r. According to the Boltzmann formula, we can write down the probability of finding the system in microstate r as a function of the entropy of $\mathcal{A}'$:

$$p_r \propto \exp\left[S'\left(U_0 - E_r\right)/k_B\right]. \tag{4.3}$$

We shall now use the fact that $\mathcal{A}'$ is a heat reservoir for $\mathcal{A}$, and therefore the energy and heat capacity of $\mathcal{A}'$, which are extensive thermodynamic quantities, are much greater then the energy and heat capacity of $\mathcal{A}$. If follows that, for microstates near equilibrium, $E_r \ll U_0$ and therefore $E' \simeq U_0$: although heat is exchanged, the temperature of the reservoir will remain almost constant. Under such conditions, we can expand the entropy of the reservoir in a Taylor series around U_0 , i.e., in powers of $\eta = E' - U_0 = -E_r$:[2]

$$S'\left(U_0 - E_r\right) = S'(U_0) - E_r\left(\frac{\partial S'}{\partial E'}\right)_{E'=U_0} + \frac{E_r^2}{2}\left(\frac{\partial^2 S'}{\partial E'^2}\right)_{E'=U_0} + \dots \tag{4.4}$$

---

[2] Recall that the Taylor series expansion of $f(x)$ in powers of $\eta = x - x_0$ is

$$f(x) = f(x_0) + \eta\left(\frac{df(x)}{dx}\right)_{x=x_0} + \frac{1}{2}\eta^2\left(\frac{d^2f(x)}{dx^2}\right)_{x=x_0} + \dots$$

The derivative $(\partial S'/\partial E')$ is taken at fixed energy $E' = U_0$, and consequently does not depend on E_r. According to definition (3.11), this derivative equals the inverse temperature $1/T$ of the reservoir. The second- and higher-order terms then involve derivatives of the temperature of the reservoir, which is assumed (almost) constant; these terms can therefore be neglected. Dividing the above result by Boltzmann's constant and using the definition of β, Eq. (3.9), we get:

$$\frac{1}{k_B} S'(U_0 - E_r) = \frac{1}{k_B} S'(U_0) - \beta E_r. \tag{4.5}$$

Because $S'(U_0)$ is a constant, by Eq. (4.3) the probability of finding system $\mathcal{A}$ with energy E_r is

$$p_r \propto e^{-\beta E_r}. \tag{4.6}$$

The term $e^{-\beta E_r}$ is called the *Boltzmann factor*. The inverse of the normalisation constant in Eq. (4.6) is known as the *partition function* of the system:

$$Z = \sum_r e^{-\beta E_r} \equiv Z(T, V, N), \tag{4.7}$$

which is a sum over the microstates available to the system (the symbol Z comes from the German *Zustandssumme*, 'sum over states'). This function plays a major role in the formalism of statistical physics, as we shall see. Probability p_r is then given by

$$p_r = \frac{e^{-\beta E_r}}{Z}. \tag{4.8}$$

We thus arrive at the following conclusion: The probability that a system in thermal equilibrium with a heat reservoir is in a given microstate r depends only on the temperature T of the reservoir and on the energy E_r of the state. The probability distribution (4.8) is called the *canonical distribution*.

A system in thermal equilibrium with a heat reservoir is statistically described by a *canonical ensemble*, which is a statistical ensemble of $\mathcal{N} \to \infty$ copies of the system distributed over the available microstates according to the canonical distribution.

When the microstates are degenerate, by Eq. (4.2) the probability that the system is in a state of energy E_r is

$$p_r = \frac{\Omega(E_r) e^{-\beta E_r}}{Z}, \tag{4.9}$$

since there are $\Omega(E_r)$ different states with the same energy, all of them equally probable. The partition function is now calculated as a sum over energies:

$$Z = \sum_{E_r} \Omega(E_r) e^{-\beta E_r}. \tag{4.10}$$

As we shall see later, it is sometimes convenient to use the continuum-states approximation. This is the case when the energy separation between consecutive states is small compared with the thermal energy:

$$\Delta E = E_{r+1} - E_r \ll k_B T. \tag{4.11}$$

Indeed, the ratio of the probabilities of two consecutive states is

$$\frac{p_{r+1}}{p_r} = e^{-\beta \Delta E}, \tag{4.12}$$

so for $\Delta E \approx k_B T$, $p_{r+1}/p_r = e^{-1} \approx 0.37$, whereas for $\Delta E \approx 0.001 k_B T$, $p_{r+1}/p_r = e^{-0.001} \approx 0.9990$, Hence in the latter case, the energy varies almost continuously.[3] Under these conditions, when calculating the partition function we can replace the sum over E_r by an integral over the continuous variable E:

$$Z = \sum_{E_r} \Omega(E_r) e^{-\beta E_r} \simeq \int g(E) e^{-\beta E} dE, \tag{4.13}$$

where

$$g(E) dE \tag{4.14}$$

is the number of microstates with energies in the interval $(E, E + dE)$, and $g(E)$ is the *density of states*. Correspondingly, the canonical probability is written

$$p(E) dE = \frac{1}{Z} g(E) e^{-\beta E} dE, \tag{4.15}$$

where $p(E)$ is a probability density.

4.2.2 Mean value and variance of the energy

From Eqs. (4.8) or (4.9) for the canonical distribution, the mean energy of the system is easily obtained:

$$\overline{E} = \sum_r p_r E_r = -\left(\frac{\partial \ln Z}{\partial \beta} \right)_{V,N}. \tag{4.16}$$

For the variance of the energy, from

$$\frac{\partial^2 \ln Z}{\partial \beta^2} = \frac{\partial}{\partial \beta} \left[\frac{1}{Z} \sum_r (-E_r) e^{-\beta E_r} \right] = \sum_r E_r^2 \frac{e^{-\beta E_r}}{Z} - \left(\sum_r E_r \frac{e^{-\beta E_r}}{Z} \right)^2 = \overline{E^2} - \overline{E}^2,$$

we find

$$\sigma^2(E) = \overline{E^2} - \overline{E}^2 = \left(\frac{\partial^2 \ln Z}{\partial \beta^2} \right)_{V,N}. \tag{4.17}$$

From Eq. (4.16) and the definiton of β, Eq. (3.9), we can write

$$\sigma^2(E) = -\left(\frac{\partial \overline{E}}{\partial \beta} \right)_{V,N} = k_B T^2 \left(\frac{\partial \overline{E}}{\partial T} \right)_{V,N}. \tag{4.18}$$

This implies that the variance of the energy vanishes at the absolute zero of temperature, i.e., that the internal energy of the system equals its mean energy in this limit. Hence *at*

---

[3] Note that the inequality $p_{r+1} < p_r$ holds when the absolute temperature of the system is positive. States with $p_{r+1} > p_r$ would correspond to negative absolute temperatures. These will be discussed in the chapter on magnetic systems.

absolute zero a canonical ensemble reduces to a microcanonical ensemble specified by $(\overline{E}, V, N)$

For macroscopic systems, in general we can equate the mean energy with the internal energy, from which there results the following relation between the variance of the energy and the heat capacity at constant volume:

$$\sigma^2(E) = k_B T^2 C_V, \tag{4.19}$$

whereby we recover the fluctuation-dissipation relation, Eq. (3.17). Because the variance is positive, this relation implies $C_V \geq 0$. It further implies that the variance of the energy vanishes at zero absolute temperature, where the heat capacity also vanishes in accordance with the third law (see the discussion in Section 2.2.3). Consistently with the above, the relative fluctuation of the energy is

$$\delta(E) = \frac{\sigma(E)}{\overline{E}} \propto \frac{\sqrt{C_V}}{\overline{E}} \propto \frac{\sqrt{N}}{N} = \frac{1}{\sqrt{N}}. \tag{4.20}$$

which shows that, when the thermodynamic energy U is equated with the mean value $\overline{E}$, evaluated by the methods of statistical physics, the relative error is clearly negligible for macroscopic systems.

4.2.3 Infinitesimal quasi-static processes. Connection with thermodynamics

Consider a system in equilibrium with a heat reservoir. Suppose it is carried to a new equilibrium state by an infinitesimal quasi-static process such that the external parameters and the temperature vary slowly enough that the system always stays very close to equilibrium and is therefore described by a canonical ensemble. The infinitesimal variation of the mean energy in such a process is

$$\overline{E} = \sum_r p_r E_r \Rightarrow d\overline{E} = \sum_r p_r dE_r + \sum_r E_r dp_r. \tag{4.21}$$

Let us examine the first term in the right-hand side of Eq. (4.21) in the case where the volume is the only external parameter:

$$\sum_r p_r dE_r = \sum_r p_r \left(\frac{\partial E_r}{\partial V}\right) dV = \sum_r p_r(-P_r) dV,$$

where P_r is the pressure of the system in microstate r, and we have used the definition of pressure, Eq. (2.19). If the volume change is slow enough that no transitions are induced between quantum states, and the system thus remains in the same microstate r while its volume changes by dV, then

$$\sum_r p_r dE_r = -\overline{P} dV, \tag{4.22}$$

where we have defined the mean pressure of the system as

$$\overline{P} = \sum_r p_r P_r. \tag{4.23}$$

According to the discussion of Section 3.3.4, under these conditions the process is reversible and we can identify $\overline{P}dV$ as the work performed by the system:

$$d\!\!\!\!^{-} W_{\mathrm{rev}} = \overline{P}dV. \tag{4.24}$$

From Eq. (4.23), the definition of pressure, Eq. (2.19, and Eq. (4.8), we obtain the equation of state of the system:

$$P = k_B T \left(\frac{\partial \ln Z}{\partial V} \right)_T. \tag{4.25}$$

Let us now examine the second term on the right-hand side of Eq. (4.21). Noting that Eq. (4.8) can be written as

$$E_r = -\frac{1}{\beta} \left(\ln Z + \ln p_r \right),$$

and that

$$\sum_r dp_r = d \left(\sum_r p_r \right) = 0,$$
$$\underbrace{}_{=1}$$

we have

$$\sum_r E_r dp_r = -\frac{1}{\beta} \sum_r \left(\ln Z + \ln p_r \right) dp_r = -\frac{1}{\beta} \sum_r dp_r \ln p_r. \tag{4.26}$$

We now introduce a *general definition* of *entropy*

$$S = -k_B \sum_r p_r \ln p_r, \tag{4.27}$$

i.e., the quantity $-S/k_B$ is the mean value of the logarithm of the microstate probabilities. We shall discuss later the relation between this general definition of entropy and the Boltzmann formula. From Eq. (4.27) and the normalisation requirement we have

$$dS = -k_B \sum_r dp_r \ln p_r. \tag{4.28}$$

Substituting this into Eq. (4.26) we get

$$\sum_r E_r dp_r = TdS. \tag{4.29}$$

In that part of the (reversible) process that leads to Eq. (4.29), the energy levels of the system did not change but their probabilities did change, i.e., the entropy changed. Then, according to the discussion in Section 3.3.4, the corresponding change in energy is due to the heat exchanged with the heat reservoir:

$$TdS = d\!\!\!\!^{-} Q_{\mathrm{rev}}. \tag{4.30}$$

To conclude our analysis of Eq. (4.21), we substitute Eqs. (4.22) and (4.29) into Eq. (4.21). If we now equate the mean energy and the mean pressure, respectively, with the internal energy and the pressure of the system, we arrive at the fundamental thermodynamic relation for the case where the volume is the single external parameter and the number of particles is kept constant:

$$dU = TdS - PdV. \tag{4.31}$$

This holds for reversible as well as irreversible infinitesimal processes, as discussed in Section 2.2.1.

We now note that from Eqs. (4.22)–(4.25) we have

$$d\!\!\!/\, W_{\mathrm{rev}} = \frac{1}{\beta} \left(\frac{\partial \ln Z}{\partial V} \right)_T dV. \tag{4.32}$$

Thus both the work and the mean energy, the latter given by Eq. (4.16), are functions of $\ln Z$, whence we can conclude that there should be a close relationship between $d\overline{E}$ and dW, as we know from thermodynamics. We can therefore expect thermodynamic quantities to be expressed as functions of $\ln Z$. Let us start with the entropy. From definition (4.27) and probability distribution (4.8) we find

$$S = k_B \ln Z + k_B \beta \sum_r p_r E_r,$$

which can be rewritten as

$$S = k_B \left(\ln Z + \beta \overline{E} \right), \tag{4.33}$$

with $\overline{E}$ given by Eq. (4.16). This immediately yields

$$TS = k_B T \ln Z + \overline{E},$$

or, equivalently,

$$F \equiv \overline{E} - TS = -k_B T \ln Z, \tag{4.34}$$

which we identify as the *Helmholtz free energy*, Eq. (2.28). In the canonical ensemble, the connection with thermodynamics is made via the Helmholtz free energy:

$$F(T,V,N) = -k_B T \ln Z(T,V,N). \tag{4.35}$$

It follows that F is a function of its natural variables (T,V,N), as expressed in the fundamental thermodynamic relation (2.45). It is easy to show that the function defined by Eq. (4.35) is (correctly) extensive: we consider two distinguishable systems $\mathcal{A}$ and $\mathcal{B}$, coupled with each other and in equilibrium with a reservoir at temperature T. Recalling the discussion in Section 3.3.2, we assume that the interaction energy between the two systems is negligible compared with the total energy. The partition functions for the two systems are

$$Z_A = \sum_i e^{-\beta \epsilon_i} \quad , \quad Z_B = \sum_j e^{-\beta \epsilon_j},$$

and we have

$$Z_{A+B} = \sum_{i,j} e^{-\beta(\epsilon_i + \epsilon_j)} = Z_A Z_B.$$

Hence, the total Helmholtz energy is

$$F_{A+B} = -k_B T \ln Z_{A+B} = -k_B T \ln Z_A - k_B T \ln Z_B = F_A + F_B.$$

Furthermore, $F(T, V, N)$ defined by Eq. (4.35) obeys thermodynamic relation (2.43):

$$-T^2 \left[\frac{\partial}{\partial T} \left(\frac{F}{T} \right) \right]_{V,N} = -T^2 \left[\frac{\partial}{\partial T} (-k_B \ln Z) \right]_{V,N}$$

$$= k_B T^2 \left(\frac{\partial \ln Z}{\partial \beta} \right)_{V,N} \frac{d\beta}{dT} = -\left(\frac{\partial \ln Z}{\partial \beta} \right)_{V,N} = \overline{E}.$$

Using definition (4.35) and Eq. (4.25), we recover the thermodynamic equation of state:

$$P = -\left(\frac{\partial F}{\partial V} \right)_V. \qquad (4.36)$$

We now discuss the relation between the definition of entropy, Eq. (4.27) and Boltzmann's formula. Consider a system where the only accessible microstates are $\Omega(E, V, N)$. Now, the postulate of equal *a priori* probabilities implies that $p_r = 1/\Omega(E, V, N)$ for all r. If we substitute this into Eq. (4.27) and take into account that probabilities are normalised to unity, we find Boltzmann's formula:

$$S = -k_B \ln(1/\Omega) = k_B \ln \Omega.$$

The *microcanonical ensemble* can thus be seen as a *degenerate canonical ensemble*, i.e., where all accessible states have the same energy.

The preceding results can be generalised to an arbitrary number of external parameters x_i. The first term on the right-hand side of Eq. (4.21) can be written

$$\sum_r p_r dE_r = \sum_r p_r \left(\sum_i \frac{\partial E_r}{\partial x_i} dx_i \right) = \sum_i \left(\sum_r p_r \frac{\partial E_r}{\partial x_i} \right) dx_i$$

$$= \sum_i \left[\sum_r p_r (-X_{ir}) \right] dx_i = -\sum_i \overline{X}_i dx_i, \qquad (4.37)$$

where the generalised force X_{ir} conjugate to the generalised coordinate x_i in state r is defined as

$$X_{ir} = -\left(\frac{\partial E_r}{\partial x_i} \right)_{x_{j \neq i}}.$$

The mean value of X_{ir} is given by

$$\overline{X}_i = \sum_r p_r X_{ir} = \frac{1}{\beta} \left(\frac{\partial \ln Z}{\partial x_i} \right)_{\beta, x_{j \neq i}} = -\left(\frac{\partial F}{\partial x_i} \right)_{\beta, x_{j \neq i}}, \qquad (4.38)$$

where we have used the definitions of Z, Eq. (4.7), and of F, Eq. (4.35). This is the general form of the equation of state in the canonical ensemble. As stated earlier, the partition function plays a key role in this formalism, as all quantities describing the system can be expressed in terms of the logarithm of Z. Again, we can only equate $\sum_i \overline{X}_i dx_i$ with the work performed by the system if the process is reversible.

It would be tempting to write a general relation for the variance of the generalised force X_i, conjugate with the variable x_i, in terms of the partition function of the system:

$$\sigma^2(X_i) = \frac{1}{\beta^2}\left(\frac{\partial^2 \ln Z}{\partial x_i^2}\right)_{\beta, x_{j \neq i}}, \qquad (4.39)$$

and thence derive a corresponding fluctuation-dissipation relation;

$$\sigma^2(X_i) = \frac{1}{\beta}\left(\frac{\partial \overline{X}_i}{\partial x_i}\right)_{\beta, x_{j \neq i}} = k_B T \Theta_T, \qquad (4.40)$$

where Θ_T is a generalised isothermal susceptibility. However, these two equations are only valid if the energy E_r depends linearly on the parameter x, as we shall now show:

$$\frac{\partial^2 \ln Z}{\partial x^2} = \frac{\partial}{\partial x}\left(\frac{1}{Z}\frac{\partial Z}{\partial x}\right) = \frac{\partial}{\partial x}\left[\frac{1}{Z}\sum_r \underbrace{\left(-\beta\frac{\partial E_r}{\partial x}\right)}_{\beta X_r} e^{-\beta E_r}\right] = \beta\frac{\partial \overline{X}}{\partial x}$$

$$= \frac{1}{Z}\left[\sum_r\left(-\beta\frac{\partial^2 E_r}{\partial x^2}\right) + \sum_r\left(-\beta\frac{\partial E_r}{\partial x}\right)^2\right]e^{-\beta E_r}$$

$$- \frac{1}{Z^2}\left[\sum_r\left(-\beta\frac{\partial E_r}{\partial x}\right)e^{-\beta E_r}\right]^2.$$

Thus if E_r is linear in x or, recalling the definition of X_r, if X_r does not depend on x, then the first term inside the first square brackets vanishes and we get

$$\frac{\partial^2 \ln Z}{\partial x^2} = \beta^2\left(\overline{X^2} - \overline{X}^2\right) = \beta^2 \sigma^2(X).$$

This is not true, for example, of the pressure fluctuations in a gas (see Problem 4.3), because the energy depends non-linearly on the volume, as we know from the quantum mechanical study of a particle in a box.

Finding the partition function is often a formidable task except for very simple systems, such as the ideal gas and a few other examples to be discussed in Section 4.3. Some of these difficulties stem from the restriction $N = $ constant. In this case it is often easier to work in the grand canonical ensemble, where the number of particles is not fixed.

4.2.4 The third law

To discuss the third law of thermodynamics statistically, we consider the entropy of a system described by a canonical ensemble in the limit that the temperature goes to absolute zero. We write the energy levels of the system without repetition: $E_0 < E_1 < E_2 < \cdots < E_r < \cdots$, where the generic level E_r has degeneracy Ω_r. At low enough temperatures such that $E_1 - E_0 \gg k_B T$, we have, according to the canonical probability distribution (4.9)

$$\frac{p_1}{p_0} = \frac{\Omega_1}{\Omega_0} e^{-\beta(E_1 - E_0)} \simeq 0 \Rightarrow p_0 \gg p_1, \tag{4.41}$$

hence *the system tends to reside in its ground state when $T \to 0$ and, consequently,* $\lim_{T \to 0} \overline{E} = E_0$. Consider now the partition function as given by Eq. (4.10): as the system has a negligibly small probability of being found in any state other than its ground state, the only terms to be retained in the sum over states are those corresponding to the lowest energy level, i.e., the Ω_0 states of energy E_0, and consequently $\lim_{T \to 0} Z = \Omega_0 e^{-\beta E_0}$. Under these conditions, it follows from Eq. (4.33) for the entropy that

$$\lim_{T \to 0} S = k_B [(\ln \Omega_0 - \beta E_0) + \beta E_0] = k_B \ln \Omega_0. \tag{4.42}$$

Hence, *as the temperature of the system goes to zero, its entropy tends to a constant value that is independent of the thermodynamic parameters of the system.* This is a statement of the third law of thermodynamics. It follows from the assumption of discrete energy levels and is therefore *a consequence of the quantum description of the system.* If we further assume that the ground state is non-degenerate (i.e., $\Omega_0 = 1$), which is believed to be always the case, we can make a stronger statement: *the entropy of a thermodynamic system vanishes at the absolute zero of temperature*:

$$\lim_{T \to 0} S = 0. \tag{4.43}$$

4.3 The ideal solid

In this section we shall apply the canonical ensemble formalism to a few ideal systems consisting of identical and localised particles in crystals. Specifically, we shall evaluate the contributions of the paramagnetic orientational order, and of the thermal lattice vibrations, to the heat capacity of solids.

4.3.1 Paramagnetism

We model a paramagnetic solid as composed of N localised, non-interacting magnetic moments in equilibrium with a heat reservoir. In the preceding chapter we studied the contribution of lattice atoms or ions to the magnetism of crystals as an ideal isolated system of spins 1/2. Now we shall take into account the fact that the system is not isolated, but rather in thermal equilibrium with the rest of the solid, which acts as a heat reservoir. This interaction is usually weak enough that its contribution to the total energy can be treated independently of other features of the solid, in particular lattice vibrations, which will be studied later.

Ideal system of N spins 1/2

We recall Section 3.3.5: in a paramagnetic solid, each spin 1/2 in a local field $\mathbf{B}$ has energy $\varepsilon_\pm = \mp \mu_B B$, the minus sign corresponding to an orientation 'parallel' to the field (minimum energy) and the plus sign to an orientation 'antiparallel' to the field (maximum energy). In the case of a paramagnetic sample, we can neglect the influence of the material on the local field, and consequently take the local field to be the same as the applied field: $\mathbf{B} \simeq \mathbf{B}_0 = \mu_0 \mathbf{H}$ (i.e., the magnetic field and the magnetic induction are the same physical quantity, only their units are different), where $\mu_0 = 4\pi \times 10^{-7}$ H m^{-1} (henry metre^{-1}) is the magnetic permeability of free space. As discussed in Section 2.2.4, the external thermodynamic parameter is the applied field $\mathbf{B}_0 = \mu_0 \mathbf{H}$, hence differentiation will be with respect to this parameter.

To find the partition function of N localised and identical spins 1/2, we recall from Section 3.3.5 that the total energy when n spins are parallel and $N-n$ spins are antiparallel to the field is $E(n) = (N - 2n)\mu_B B$ and that the degeneracy of such a state is given by the statistical weight $\Omega(n) = N!/n!(N-n)!$, i.e., the number of different ways in which we can have spins parallel to the field at n lattice sites, and spins antiparallel to the field at $N-n$ lattice sites. The partition function is, from Eq. (4.10),

$$Z_N = \sum_{n=0}^{N} \Omega(n) e^{-\beta E(n)} = \sum_{n=0}^{N} \frac{N!}{n!(N-n)!} e^{(-\beta[n\varepsilon_- + (N-n)\varepsilon_+])}$$

$$= \sum_{n=0}^{N} \frac{N!}{n!(N-n)!} e^{-\beta n\varepsilon_-} e^{-\beta(N-n)\varepsilon_+}. \tag{4.44}$$

Defining $p = e^{-\beta\varepsilon_-}$ and $q = e^{-\beta\varepsilon_+}$, this can be rewritten as

$$Z_N = \sum_{n=0}^{N} \frac{N!}{n!(N-n)!} p^n q^{N-n},$$

which, using the binomial theorem, yields

$$Z_N = \left(e^{-\beta\varepsilon_-} + e^{-\beta\varepsilon_+} \right)^N. \tag{4.45}$$

This is actually a general result for a system of N identical, localised and independent subsystems each with two energy levels.

The individual partition function of a subsystem is

$$Z_1 \equiv Z(T, H, 1) = e^{-\beta\varepsilon_-} + e^{-\beta\varepsilon_+}. \tag{4.46}$$

Comparing this with the previous result, we find

$$Z_N = Z_1^N. \tag{4.47}$$

Specifically for the system under study,

$$Z_1 = e^{\beta \mu_B B} + e^{-\beta \mu_B B}. \tag{4.48}$$

If we now define

$$x = \frac{\mu_B B}{k_B T}, \tag{4.49}$$

we can write

$$Z_1 = 2\cosh x. \tag{4.50}$$

We can now apply the canonical ensemble formalism. From Eq. (4.16) we obtain the mean energy of the system:

$$\overline{E} = -N\left(\frac{\partial \ln Z_1}{\partial \beta}\right)_H = N\overline{\varepsilon} = -N\mu_B B \tanh x, \tag{4.51}$$

and from Eq. (4.38) the magnetic equation of state:

$$\overline{M} = \frac{1}{\beta} \frac{N}{\mu_0} \left(\frac{\partial \ln Z_1}{\partial H}\right)_\beta = N\overline{\mu} = N\mu_B \tanh x, \tag{4.52}$$

which gives the projection along the field direction of the mean magnetic moment of the system, Eq. (3.42), as a function of the temperature and field strength. Equations (4.51) and (4.52) allow us to derive the (classical) relation between the mean energy and the mean magnetic moment:

$$\overline{E} = -\overline{M}B = -\overline{\mathbf{M}} \cdot \mathbf{B}. \tag{4.53}$$

Note that we could have obtained this result by using the fact that, because we are dealing with an ideal system of identical particles,

$$E = \sum_{i=1}^{N} \varepsilon_i \Rightarrow \overline{E} = \sum_{i=1}^{N} \overline{\varepsilon_i} = N\overline{\varepsilon}. \tag{4.54}$$

Equation (4.51) then follows from

$$\overline{\varepsilon} = -\left(\frac{\partial \ln Z_1}{\partial \beta}\right)_H. \tag{4.55}$$

We may likewise derive Eq. (4.52) for the magnetic moment from

$$M = \sum_{i=1}^{N} \mu_i \Rightarrow \overline{M} = \sum_{i=1}^{N} \overline{\mu_i} = N\overline{\mu}, \tag{4.56}$$

and

$$\overline{\mu} = \frac{1}{\beta \mu_0} \left(\frac{\partial \ln Z_1}{\partial H}\right)_\beta. \tag{4.57}$$

From Eq. (4.17) we find the variance of the energy:

$$\sigma^2(E) = N\left(\frac{\partial^2 \ln Z_1}{\partial \beta^2}\right)_H = N\sigma^2(\varepsilon) = N\frac{4(\mu_B B)^2}{(e^x + e^{-x})^2} = N(\mu_B B \operatorname{sech} x)^2. \tag{4.58}$$

The heat capacity at constant magnetic field is given by

$$C_H = \left(\frac{\partial \overline{E}}{\partial T}\right)_H = Nk_B \left(x \operatorname{sech} x\right)^2. \tag{4.59}$$

Comparison of Eqs. (4.58) and (4.59) yields

$$\sigma^2(E) = k_B T^2 C_H, \tag{4.60}$$

which is a magnetic fluctuation-dissipation relation analogous to the fluid relation (4.19), which we could have easily obtained by making the substitution $V \to \mu_0 H$ in the latter equation.

To find the variance of the total magnetic moment of the sample, we use Eq. (1.74), which holds for N independent random variables with the same probability distribution, and Eq. (1.79) for the one-dimensional random walk, where we have replaced the step length l by the Bohr magneton μ_B:

$$\sigma^2(M) = N\sigma^2(\mu) = N\mu_B^2 4 p_+ p_-,$$

where p_+ and p_- are, respectively, the probabilities of finding one spin parallel or antiparallel to the field, given by the canonical distribution (4.8):

$$p_\pm = \frac{e^{\pm x}}{e^x + e^{-x}}, \tag{4.61}$$

with the result

$$\sigma^2(M) = N\mu_B^2 \operatorname{sech}^2 x. \tag{4.62}$$

Next, we define the magnetisation of the system, $\mathcal{M}$, as the mean total magnetic moment per unit volume and find the magnetic susceptibility via definition (2.76):

$$\mathcal{M} = \frac{\overline{M}}{V}, \tag{4.63}$$

$$\chi_T = \frac{1}{V}\frac{1}{\mu_0}\left(\frac{\partial \overline{M}}{\partial H}\right)_T = \frac{N\mu_B^2}{Vk_B T}\operatorname{sech}^2 x. \tag{4.64}$$

Comparison of Eqs. (4.62) and (4.64) yields

$$\sigma^2(M) = Vk_B T \chi_T, \tag{4.65}$$

which is another magnetic fluctuation-dissipation relation, where χ_T is the response function associated with the variance of M. As the variance is positive definite, this relation implies that $\chi_T > 0$.

Combining Eq. (4.65) with Eqs. (4.64) and (4.52), we obtain

$$\sigma^2(M) = \frac{k_B T}{\mu_0}\left(\frac{\partial \overline{M}}{\partial H}\right)_T = \frac{1}{(\mu_0 \beta)^2}\left(\frac{\partial^2 \ln Z}{\partial H^2}\right)_\beta, \tag{4.66}$$

which agrees with Eqs. (4.39) and (4.40), since the energy E is linear in the magnetic field.

Finally, the entropy is calculated from Eq. (4.33) together with Eqs. (4.47), (4.50) and (4.51):

$$S = Nk_B \left[\ln\left(2\cosh x\right) - x\tanh x\right]. \tag{4.67}$$

It is interesting to study the behaviour of the above thermodynamic quantities as functions of the dimensionless variable x, given by Eq. (4.49), which measures the ratio of magnetic to thermal energy. We shall study two limiting cases: (a) the high-temperature or weak-field limit; and (b) the low-temperature or strong-field limit. Let us first look at the mean energy and total mean magnetic moment. These two quantities have the same $\tanh x$ dependence:

(a) $\tanh x \approx x$ if $|x| \ll 1$,

(b) $\tanh x \approx 1$ if $|x| \gg 1$.

Hence in case (a) we have, for the mean energy,

$$\overline{E} \simeq -N\frac{(\mu_B B)^2}{k_B T} \propto \frac{B^2}{T}, \tag{4.68}$$

$$\lim_{x\to 0} \overline{E} = 0, \tag{4.69}$$

and for the mean total magnetic moment:

$$\overline{M} \simeq N\frac{\mu_B^2 B}{k_B T} \propto \frac{B}{T}, \tag{4.70}$$

$$\lim_{x\to 0} \overline{M} = 0. \tag{4.71}$$

In case (b) we have

$$\overline{E} \simeq -N\mu_B B = \lim_{x\to\infty} \overline{E}, \tag{4.72}$$

$$\overline{M} \simeq N\mu_B = \lim_{x\to\infty} \overline{M}. \tag{4.73}$$

These results imply that in case (a) there are as many parallel as there are antiparallel spins, whereas in case (b) alignment is mostly parallel to the field. Hence in the latter case the system tends to sit in its ground state, in which the mean energy equals the minimum energy and the mean magnetic moment reaches its maximum value.

It is instructive to rationalise these limiting results in terms of the entropy, given by Eq. (4.67).

(a) *High-temperature limit*

$$x\tanh x \simeq x^2 \quad \text{and} \quad \ln(2\cosh x) \simeq \ln 2.$$

Hence:

$$\lim_{x\to 0} S = Nk_B \ln 2 = k_B \ln 2^N. \tag{4.74}$$

This is the state of maximum entropy, which means that the system has maximum disorder. We shall come back to Eq. (4.74) later.

(b) *Low-temperature limit*

$$x\tanh x \simeq x \quad \text{and} \quad \ln(2\cosh x) \simeq x.$$

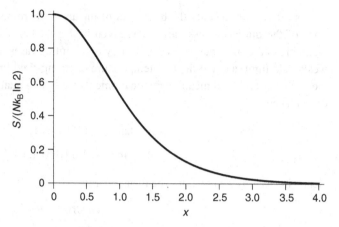

Figure 4.2 Entropy S given by Eq. (4.67) divided by its maximum value, Eq. (4.74), vs $x = \mu_B B / k_B T$. Note that S reaches a finite maximum at infinite temperature ($x = 0$).

Hence:

$$\lim_{x \to \infty} S = 0. \tag{4.75}$$

This is the state of minimum entropy, corresponding to maximum order. Figure 4.2 shows how the entropy changes as x is varied.

Once the system is in its ground state, with energy given by Eq. (4.72), we can understand Eq. (4.75) for the entropy on the basis of Boltzmann's formula:

$$S = 0 \Rightarrow \Omega = e^{S/k_B} = 1. \tag{4.76}$$

That is, there is only one microstate available (minimum Ω), in which all magnetic dipoles are aligned parallel to the field. On the other hand, because the system is macroscopic, we can identify its mean energy with its internal energy. Hence we can regard the system as specified by its mean energy (recall the discussion in Section 4.6) and understand Eq. (4.74) for the entropy in terms of the statistical weight:

$$S = k_B \ln 2^N \Rightarrow \Omega = 2^N. \tag{4.77}$$

That is, the number of accessible microstates is maximal (maximum Ω). Both limiting cases agree with the discussion of the isolated magnetic system in Section 3.3.5.

We should not forget that the canonical ensemble formalism is rooted in the canonical probability distribution (4.61); hence it is important to examine its limiting behaviour:

$$\text{High temperatures:} \quad \lim_{x \to 0} p_+ = \lim_{x \to 0} p_- = \frac{1}{2}, \tag{4.78}$$

$$\text{Low temperatures:} \quad \begin{cases} \lim_{x \to \infty} p_+ = 1, \\ \lim_{x \to \infty} p_- = 0. \end{cases} \tag{4.79}$$

We leave it as an exercise for readers to derive the limiting behaviours of the variance of the energy, Eq. (4.58); the heat capacity, Eq. (4.59); the variance of the magnetic moment, Eq. (4.62); and the magnetic susceptibility, Eq. (4.64).

Advanced topic 4.2   Remark on the thermodynamics of a paramagnetic system in equilibrium with a heat reservoir*

From the canonical definition of the Helmholtz free energy, Eq. (4.35), the magnetic equation of state (4.52) can be written as

$$\overline{M} = -\left(\frac{\partial F}{\partial (\mu_0 H)}\right)_T. \tag{4.80}$$

This equation can be derived using thermodynamic relations, starting from definition (2.78) of the magnetic work:

$$đ W = -M d(\mu_0 H) \Rightarrow dU = TdS - Md(\mu_0 H) \Rightarrow dF = -SdT - Md(\mu_0 H) \Rightarrow (4.80),$$

where we have used the definition of Helmholtz energy, $F = U - TS$. Hence definition (2.78) of the magnetic work is convenient for the study of paramagnetic systems (Landau *et al.*, 1980, chap. IV, §52). For a paramagnetic system in equilibrium with a heat reservoir, the appropriate thermodynamic potential is the Helmholtz free energy and the canonical ensemble is the appropriate statistical formalism.

Brillouin theory of paramagnetism*

Here we give a brief account of the quantum theory of paramagnetism. To this end we generalise our earlier study of paramagnetic systems to the case of (identical) atoms with total angular momentum **J** (in units of $\hbar$), as described in Section 3.3.5. If we take the OZ axis along the applied field, then the projection J_z of **J** along this axis may take $2J + 1$ values: $J_z = m$, with $m = -J, -J+1, \ldots, J-1, J$. Under these conditions, each dipole in a magnetic field has energy $\varepsilon = -\boldsymbol{\mu} \cdot \mathbf{B}$, where (for an atom in free space) $\boldsymbol{\mu} = -g\mu_B \mathbf{J}$; hence

$$\varepsilon = -\boldsymbol{\mu} \cdot \mathbf{B} = g\mu_B J_z B = g\mu_B m B = \mu_z B. \tag{4.81}$$

We now define a reduced variable[4]

$$y = g\mu_B B / k_B T, \tag{4.82}$$

in terms of which the single-atom partition function is:

$$Z_1 = \sum_{m=-J}^{J} e^{-ym} = \frac{e^{yJ} - e^{-y(J+1)}}{1 - e^{-y}} = \frac{e^{y(J+1/2)} - e^{-y(J+1/2)}}{e^{y/2} - e^{-y/2}}$$

$$= \frac{\sinh[y(J+1/2)]}{\sinh(y/2)}. \tag{4.83}$$

---

[4] From Eqs. (4.82), (4.49) and (3.35), $y = 2x$ for a system of spins 1/2, since $g \simeq 2$.

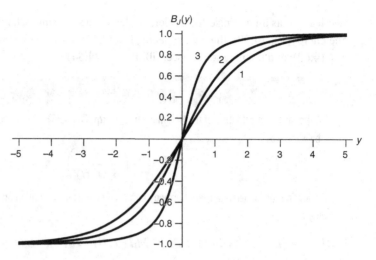

Figure 4.3 Dimensionless magnetisation $\mathcal{M}/[(N/V)g\mu_B J] = B_J(y)$ as given by Eq. 4.84), vs $y = g\mu_B B/k_B T$, for several values of the total angular momentum: (1) $J = 1/2$; (2) $J = 1$; (3) $J = 7/2$. Near the origin, the magnetisation increases linearly with y, in agreement with Eq. (4.88). Far from the origin, the magnetisation saturates at a constant value given by Eq. (4.91). Positive and negative magnetisations should be understood as characterising states where the dipoles point in opposite directions. Because the dipoles align with the field, the magnetisation and the field have the same sign, i.e., the same direction; one switches between positive and negative magnetisations by reversing the field direction.

The mean magnetic moment of an atom along the field direction is given by

$$\overline{\mu_z} = \frac{1}{\beta}\frac{1}{\mu_0}\left(\frac{\partial \ln Z_1}{\partial H}\right)_\beta = \frac{1}{\beta}\frac{d\ln Z_1}{dy}\left(\frac{\partial y}{\partial B}\right)_\beta = g\mu_B\frac{d\ln Z_1}{dy}.$$

Now from Eq. (4.83) we find

$$\frac{d\ln Z_1}{dy} = JB_J(y),$$

where we have defined the *Brillouin function* $B_J(y)$ as

$$B_J(y) = \frac{1}{J}\left\{\left(J+\frac{1}{2}\right)\coth\left[\left(J+\frac{1}{2}\right)y\right] - \frac{1}{2}\coth\left(\frac{y}{2}\right)\right\}, \tag{4.84}$$

which yields, for the mean magnetic moment of a single atom,

$$\overline{\mu_z} = g\mu_B J B_J(y). \tag{4.85}$$

The mean total magnetic moment along the field direction is

$$\overline{M} = N\overline{\mu_z}, \tag{4.86}$$

and the magnetisation is

$$\mathcal{M} = \frac{\overline{M}}{V}. \tag{4.87}$$

This is plotted in Figure 4.3.

We leave it as an exercise for readers to show that, at high temperatures ($y \ll 1$):

$$\mathcal{M} \simeq \frac{N}{V} \frac{g^2 \mu_B^2 B}{3k_B T} J(J+1),$$ (4.88)

and consequently, in this limit the magnetic susceptibility (2.76) equals

$$\chi_T \simeq \frac{N}{V} \frac{g^2 \mu_B^2}{3k_B T} J(J+1).$$ (4.89)

This is of the form of the *Curie law* of classical paramagnetism:

$$\chi_T = \frac{C}{T} \quad , \quad C = \frac{N}{V} \frac{g^2 \mu_B^2}{3k_B} J(J+1),$$ (4.90)

where C is the *Curie constant*. This constant can be found by fitting to experimental data the first of equations (4.90), or alternatively can be evaluated using the quantum mechanical result (4.90), modified so as to take into account the fact that the atoms are not in free space but at the sites of a crystal lattice (see, e.g., Kittel (1995, chap. 14)).

We leave it as an exercise to show that, in the low-temperature limit ($y \gg 1$), the maximum magnetisation is given by

$$\mathcal{M} \simeq \frac{N}{V} g \mu_B J,$$ (4.91)

which is temperature-independent.

From Eqs. (4.53), (4.85) and (4.86) we get the mean total energy:

$$\overline{E} = -MB = -N g \mu_B J B_J(y) B,$$ (4.92)

whence the heat capacity is

$$C_H = \left(\frac{\partial \overline{E}}{\partial T} \right)_H = N k_B y^2 J \frac{dB_J(y)}{dy}.$$ (4.93)

This is plotted in Figure 4.5.

4.3.2 Magnetic cooling

Magnetic cooling is a powerful method for obtaining low temperatures in a paramagnetic solid, and an interesting application of the results of the preceding section. The magnetic cooling of a solid sample is an example of adiabatic cooling, analogous to the cooling of a gas sample by adiabatic expansion. Here we follow Mandl (1988). Adiabatic cooling relies on the fact that the entropy of a system is a function of its temperature and some other external parameter, such as the volume or the applied magnetic field, which we shall call α. The magnetic entropy $S(T, \alpha)$ measures the disorder in the magnetic dipole alignment, and is a function of the temperature and of the applied field, of the form $S(B/T)$, as given by Eq. (4.67). If we decrease the magnetic field while keeping the entropy constant (in a reversible adiabatic process), then the temperature of the system will drop in order to keep the ratio B/T constant, i.e., to preserve the degree of magnetic order. To obtain significant temperature changes, one should work in a temperature range where the entropy varies

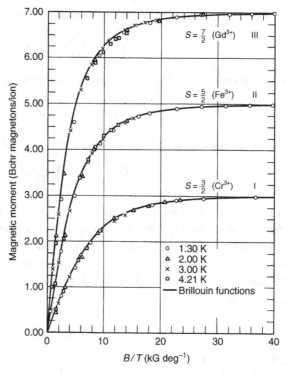

Figure 4.4 Experimental and theoretical magnetic moment (in units of μ_B, the Bohr magneton) vs B/T, for samples of several paramagnetic salts. The curves are Brillouin theory predictions (from Kittel (1995)). Original figure from Henry, W. E. 1952. Spin paramagnetism of Cr^{+++}, Fe^{+++}, and Gd^{+++} at liquid helium temperatures and in strong magnetic fields. *Phys. Rev.* **88**, 559–562. ©APS, https://journals.aps.org/pr/abstract/10.1103/PhysRev.88.559.] In all three cases, $J = S$ (the total electron spin of the ion) and $g = 2$. At $T = 1.3$ K and $B \approx 5 \times 10^4$ G, the magnetic moment is more than 99.5% of its saturation value (see Reif (1985, chap. 7.8) or Kittel (1995, chap. 14)).

strongly with the magnetic field, i.e., where the dipole alignment changes significantly. To simplify our analysis, we shall consider the case of spin-1/2 magnetic dipoles. From the behaviour of the entropy (Figure 4.2), the magnetisation (Figures 4.3 and 4.4) or the heat capacity (Figure 4.5), it can be inferred that it is convenient to work near $x \equiv \mu_B B/k_B T \simeq 1$, where $\mu_B = e\hbar/2m_e = 9.27 \times 10^{-24}$ A m^2. Let us start with a field $B \approx 1$ T,[5] for which $\mu_B B \approx 10^{-23}$ J. At room temperature, $T = 300$ K, the thermal energy is $k_B T \approx 4 \times 10^{-21}$ J, whence $x \approx 10^{-2}$. At $T = 1$ K, $k_B T \approx 1.4 \times 10^{-23}$ J and $x \approx 1$. We conclude that the adiabatic process should be started at $T \sim 1$ K (i.e., the temperature of liquid helium) in order to attain a significant degree of cooling.

Adiabatic demagnetisation, like adiabatic expansion, is a two-step process. In the first step, the paramagnetic sample is kept in contact with a heat reservoir at $T \approx 1$ K and is isothermally magnetised by the application of a field B_1. In the course of this step, the magnetic moments release heat ($\approx \mu_B B$ per spin) on aligning parallel to the field, which

---

[5] Note that this is a very strong field. For comparison, the Earth's magnetic induction is $\sim 10^{-4}$ T.

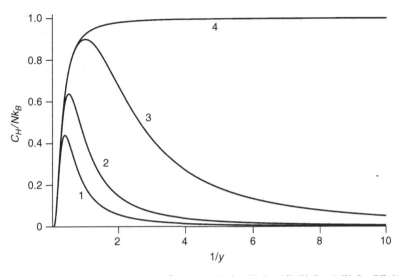

Figure 4.5   Dimensionless heat capacity C_H/Nk_B vs $1/y = k_B T/g\mu_B B$, for: (1) $J = 1/2$; (2) $J = 1$; (3) $J = 7/2$; (4) $J \to \infty$. The theory correctly predicts that $C_H \to 0$ when $T \to 0$, in agreement with the third law of thermodynamics. We note that, when the number of atomic magnetic states is finite, the heat capacity exhibits a maximum. This behaviour is characteristic of systems consisting of subsystems with a finite number of quantum states and is known as the *Schottky anomaly*. When the number of states is very large, $C_H \to Nk_B$ (the classical limit) when $T \to \infty$.

heat is transferred to the heat reservoir. In the second step, the sample is isolated from the reservoir and the field is decreased in a reversible and adiabatic process to a value much less than B_1. The field has to be decreased slowly enough for the process to be reversible (usually a few seconds are enough). The process is sketched in Figure 4.6.

The magnetic entropy is a function of the local field acting on each dipole. When no magnetic field is applied, there is a residual local field due to all the other dipoles, of order $B_2 \approx 10^{-2}$ T, much smaller than the applied field B_1. We can estimate the degree of cooling using these values. In the adiabatic step (paths b–c and d–e in Figure 4.6), the ratio B/T remains constant:

$$\frac{B_1}{T_1} = \frac{B_2}{T_2} \Rightarrow T_2 = T_1 \frac{B_2}{B_1} \simeq 1\,\text{K} \times \frac{1}{100} = 10^{-2}\,\text{K}.$$

We might ask ourselves whether a sequence of such processes could allow us to reach the absolute zero of temperature. The two curves plotted in Figure 4.6 converge to zero because from Eq. (4.67) $S(T = 0) = 0$, as required by the third law. Therefore, absolute zero can only be reached by an infinite number of steps. We thus conclude that it is not possible to reach the absolute zero of temperature by means of a finite number of adiabatic cooling processes. Actually, the weaker statement of the third law would suffice: 'As the temperature of the system goes to zero, its entropy tends to a constant that is independent of the thermodynamic parameters of the system.' In this case the two curves would tend to the same value on the vertical axis, and the conclusion would be the same as above. In any case, temperatures below 10^{-6} K have been attained by magnetic cooling.

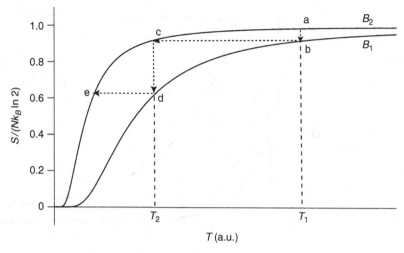

Figure 4.6 Sketch of the adiabatic demagnetisation of N spin-1/2 magnetic moments. The entropy of the system, Eq. (4.67), divided by its maximum value, Eq. (4.77), is plotted vs temperature for $B_1 \gg B_2$. Paths a–b and c–d are the isothermal magnetisation and paths b–c and d–e the reversible adiabatic demagnetisation. Note that at very low temperatures the entropy is so small as to be indistinguishable from zero on the scale of this plot.

4.3.3 Thermal vibrations of the crystal lattice. Einstein's theory

Consider a crystalline solid. At any finite temperature, the atoms or ions will oscillate around their equilibrium positions at the lattice sites. The heat capacity is a measure of how well the solid absorbs heat. In this section we shall study the contribution to the heat capacity of the thermal energy of vibration of the crystal lattice.

In insulator or semiconductor crystals, the heat capacity is dominated by the lattice thermal vibrations. In metals, the conduction electrons make a significant contribution as well, as we shall see when studying the free electron gas in Chapter 6. However, at room temperature this contribution is small compared with that of lattice vibrations.

In his pioneering work of 1907, Einstein posited that the atoms or ions of a crystalline monatomic solid should oscillate harmonically, and independently of each other, according to quantum mechanics. The energy of a quantum mechanical harmonic oscillator of angular frequency ω is

$$\varepsilon_n = \left(n + \frac{1}{2} \right) \hbar\omega, \quad n = 0, 1, 2, \dots \tag{4.94}$$

where n is the corresponding quantum number and $\hbar = h/2\pi$. The energy of a quantum oscillator therefore depends only on its frequency and quantum number. The *Einstein model* of the thermal vibrations of monatomic solids thus consists of a system of $3N$ identical and localised harmonic oscillators. The factor 3 takes into account that atoms or ions may oscillate in any of the three dimensions of space. Einstein further simplified his model by assuming that all atoms or ions vibrate with the same frequency ω_E (the Einstein frequency). As we shall see later, Debye theory improves on this by allowing a distribution of frequencies.

A system of oscillators in thermal equilibrium at temperature T can be described by the canonical ensemble. The partition function for a single oscillator is easily found:

$$Z_1 = \sum_{n=0}^{\infty} e^{-\beta \varepsilon_n} = \sum_{n=0}^{\infty} e^{-\left(n+\frac{1}{2}\right)\beta\hbar\omega} = e^{-\frac{x}{2}} \sum_{n=0}^{\infty} e^{-xn}, \quad x = \beta\hbar\omega.$$

This is a geometrical series of ratio $e^{-x} < 1$ that sums to $1/(1 - e^{-x})$ (exercise: verify), whence

$$Z_1 = \frac{e^{-\frac{x}{2}}}{1 - e^{-x}} = \frac{1}{2\sinh(x/2)}. \tag{4.95}$$

Next, we derive the mean energy of an oscillator using Eq. (4.16):

$$\bar{\varepsilon} = -\left(\frac{\partial \ln Z_1}{\partial \beta}\right)_{V,N} = -\left(\frac{\partial \ln Z_1}{\partial x}\right)_{V,N} \frac{dx}{d\beta} = \hbar\omega\left(\frac{1}{2} + \frac{e^{-x}}{1 - e^{-x}}\right)$$

$$\Rightarrow \bar{\varepsilon} = \hbar\omega\left(\frac{1}{2} + \frac{1}{e^{\beta\hbar\omega} - 1}\right), \tag{4.96}$$

where the first term is the *zero-point energy* of the oscillator:

$$\varepsilon_0 = \lim_{T\to 0} \bar{\varepsilon} = \frac{\hbar\omega}{2}. \tag{4.97}$$

On the other hand, the high-temperature limit is

$$\bar{\varepsilon} \simeq \hbar\omega\left(\frac{1}{2} + \frac{1}{x}\right) \simeq \frac{\hbar\omega}{x} = k_B T, \tag{4.98}$$

which is the classical limit (see Problem 4.2).

Because all oscillators are assumed to have the same frequency ω_E, we can write, for the mean energy of $3N$ independent oscillators,

$$\bar{E} = 3N\bar{\varepsilon}, \tag{4.99}$$

with $\bar{\varepsilon}$ given by Eq. (4.96). The heat capacity at constant volume then follows:

$$C_V = \left(\frac{\partial \bar{E}}{\partial T}\right)_V = 3N\left(\frac{\partial \bar{\varepsilon}}{\partial \beta}\right)_V \frac{d\beta}{dT} = \frac{3N\hbar\omega_E}{k_B T^2} \frac{\hbar\omega_E e^x}{(e^x - 1)^2},$$

which can be rewritten as

$$C_V = 3Nk_B \left(\frac{\theta_E}{T}\right)^2 \frac{e^{\theta_E/T}}{\left(e^{\theta_E/T} - 1\right)^2}, \quad \theta_E = \frac{\hbar\omega}{k_B}, \tag{4.100}$$

where θ_E is a material parameter called the *Einstein temperature*. This parameter can be extracted by fitting a curve to experimental results for the heat capacity. For most solids $\theta_E \approx 300$ K, hence the vibration frequency is $\nu_E = \omega_E/2\pi \approx 6 \times 10^{12}$ Hz, which lies in the infrared region of the electromagnetic spectrum.

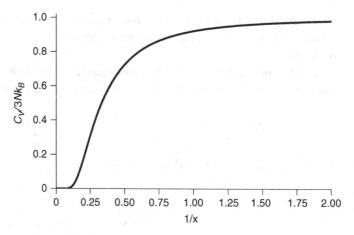

Reduced heat capacity at constant volume, Eq. (4.100), vs $1/x = T/\theta_E$. The classical result $C_V = 3\,Nk_B$ is recovered in the high-temperature limit, whereas at low temperatures $C_V \to 0$, as required by the third law. Note that at very low temperatures the heat capacity is so small as not to be distinguishable from zero on the scale of the plot.

The Einstein theory is now mostly of historical interest, as it gives an incomplete description of the thermal properties of solids, especially at low temperatures. We shall see later, when studying the phonon gas, that the Debye theory gives better quantitative results. Einstein's theory was important because it showed that lattice vibrations had to be quantised in order to explain why the heat capacity goes to zero at zero temperature, as found experimentally and as required by the third law of thermodynamics (see Figure 4.7). The theory also explains why $\lim_{T\to\infty} c_V = 3R$ (the Dulong and Petit law, where c_V is the specific heat) for monatomic solids. Indeed, at the time (1907), it was not understood why the specific heat of these solids at low temperatures dropped below the Dulong and Petit prediction at low temperatures. Following in the footsteps of the Planck theory of black-body radiation, Einstein's theory of the specific heat of monatomic solids played a pivotal role in the acceptance of quantum physics.

4.4 System in equilibrium with a heat and particle reservoir

4.4.1 Grand canonical ensemble

Before starting this section. it is useful to reread the introduction to the present chapter. Here we shall consider an isolated system partitioned into two subsystems by a wall that is fixed but allows heat and particle exchange. This could be a conceptual wall, in which case system $\mathcal{A}$ under study would be a small (but macroscopic and of fixed volume) part of isolated system $\mathcal{A}_0$; the remainder of the system being the reservoir $\mathcal{A}'$. We now generalise the procedure followed to derive the canonical distribution in Section 4.2.1, taking into account that now the system has no fixed energy and no fixed number of particles. System $\mathcal{A}$ is thus a heat and particle reservoir for system $\mathcal{A}$. As before, we assume that the isolated system, $\mathcal{A}_0$, is specified by (U_0, V_0, N_0), and partitioned in such a way that $V_0 = V + V'$, where V and V' are fixed. Now we need to label the accessible microstates of $\mathcal{A}$ with two indexes: the energy states with r and the particle number states with N. For each value of N, the system has a sequence of increasing energy levels:

$$E_{N1} \leq E_{N2} \leq \cdots \leq E_{Nr} \leq \cdots$$

With this notation, E_{Nr} is the energy of microstate r of system $\mathcal{A}$ with N particles. Correspondingly, the reservoir will have an energy E', a number of particles N' and a volume V' given by

$$E' = U_0 - E_{Nr}, \quad N' = N_0 - N, \quad V' = V_0 - V. \tag{4.101}$$

Proceeding as in Section 4.2.1, the probability to find the global system $\mathcal{A}_0$ in state (4.101) is

$$p_{Nr} \propto \Omega'(U_0 - E_{Nr}, N_0 - N, V_0 - V). \tag{4.102}$$

We can write this probability in terms of the entropy of the reservoir:

$$p_{Nr} \propto \exp\left[S'(U_0 - E_{Nr}, N_0 - N, V_0 - V)/k_B\right]. \tag{4.103}$$

Because $\mathcal{A}'$ is a reservoir, $E_{Nr} \ll U_0$, $N \ll N_0$, $V \ll V_0$, and we can expand S' in a Taylor series around U_0, N_0 and V_0:

$$S'(U_0 - E_{Nr}, N_0 - N, V_0 - V) = S'(U_0, N_0, V_0)$$
$$-E_{Nr}\left(\frac{\partial S'}{\partial E'}\right)_{E'=U_0} - N\left(\frac{\partial S'}{\partial N'}\right)_{N'=N_0} - V\left(\frac{\partial S'}{\partial V'}\right)_{V'=V_0}$$
$$+ \text{higher-order terms in } E_{Nr}, N \text{ and } V. \tag{4.104}$$

By definition (3.11), the derivative with respect to E' equals $\beta' = 1/k_B T'$, where T' is the temperature of the reservoir. Next, by definition (3.26), the derivative with respect to N' equals $-\mu'/T'$, where μ' is the chemical potential of the reservoir. Finally, by definition (3.21), the derivative with respect to V' equals P'/T', where P' is the pressure of the reservoir. As the volume V is fixed, this last term is a constant adding to $S'(U_0, N_0, V_0)$. All higher-order terms in the energy, the number of particles and the volume involve derivatives of the temperature, of the number of particles and of the pressure of the reservoir, respectively, and are therefore negligible. At equilibrium, the temperature and the chemical potential of $\mathcal{A}$ and of the reservoir $\mathcal{A}'$ are the same. Consequently, we can write for the entropy, Eq. (4.104):

$$S' = \text{constant} - \frac{E_{Nr}}{T} + \mu\frac{N}{T}, \tag{4.105}$$

and from Eq. (4.103) we get

$$p_{Nr} = \text{constant} \times e^{\beta(\mu N - E_{Nr})}. \tag{4.106}$$

The constant is determined by the requirement that p_{Nr} be normalised to unity. Its inverse is the *grand canonical partition function*:

$$Y = \sum_{N,r} e^{\beta(\mu N - E_{Nr})} \equiv Y(T, V, \mu). \tag{4.107}$$

Y is a sum over the microstates of the system, specified by N and r, and plays an analogous role to that of the canonical partition function Z, but now for a system in contact with a heat and particle reservoir. The probability to find the system in a state with N particles and energy E_{Nr} is then given by

$$p_{Nr} = \frac{e^{\beta(\mu N - E_{Nr})}}{Y}, \tag{4.108}$$

whence we conclude that *this probability depends only on the temperature and the chemical potential of the reservoir, and on the number of particles and energy of the system*, and is called the *grand canonical distribution*. A system in equilibrium with a heat and particle reservoir is described by a *grand canonical ensemble*.

The grand canonical partition function can be written as

$$Y(T, V, \mu) = \sum_{N} \left(\sum_{r} e^{-\beta E_{Nr}} \right) e^{\beta \mu N} = \sum_{N} Z(T, V, N) e^{\beta \mu N}, \tag{4.109}$$

where we have used definition (4.7) of the canonical partition function. Thus the grand canonical partition function can be seen as a sum over all possible values of N of the canonical partition functions of N-particle systems, each weighted by $e^{\beta \mu N}$.

Finally, the probability to find the system with N particles and any value of the energy, is obtained by summing Eq. (4.108) over the energy microstates:

$$p_N = \sum_{r} p_{Nr} = \frac{Z(T, V, N) e^{\beta \mu N}}{Y(T, V, \mu)}. \tag{4.110}$$

4.4.2 Connection with thermodynamics

From the definition of entropy, Eq. (4.27), combined with the grand canonical distribution, Eq. (4.108), we have

$$S = -k_B \sum_{N,r} p_{Nr} \ln p_{Nr} = -k_B \sum_{N,r} p_{Nr} (\beta \mu N - \beta E_{Nr} - \ln Y)$$

$$= \frac{\overline{E}}{T} - \mu \frac{\overline{N}}{T} + k_B \ln Y. \tag{4.111}$$

We now make the connection with thermodynamics via the following definition of the *grand potential* (or *grand canonical potential*):

$$J(T, V, \mu) = -k_B T \ln Y(T, V, \mu). \tag{4.112}$$

With this definition, Eq. (4.111) can be written as

$$J = \overline{E} - TS - \mu \overline{N}. \tag{4.113}$$

Upon equating the mean energy and the mean number of particles with their thermodynamic (i.e., macroscopic) values, Eq. (4.113) becomes a thermodynamic relation:

$$J = U - TS - \mu N. \tag{4.114}$$

Comparison with Eq. (3.82) yields

$$J = -PV, \tag{4.115}$$

which reveals the physical meaning of the grand canonical potential.

Advanced topic 4.3

We leave it as an exercise for readers to show that the potential J defined by Eq. (4.114) is the partial Legendre transform (3.72) of the energy with respect to S and N:

$$J(V,T,\mu) = \mathop{\mathcal{L}}_{S,N} U(S,V,N). \tag{4.116}$$

From the general formulation of equilibrium conditions in Section 3.4.6 we conclude that, in a system in contact with a heat and particle reservoir and with fixed external parameters, the grand potential will tend to decrease on approaching equilibrium, such that the (stable) equilibrium condition is

$$J = \text{minimum}. \tag{4.117}$$

We now derive the thermodynamics of the grand canonical ensemble. Start by writing the fundamental thermodynamic relation in terms of the grand canonical potential J, using Eqs. (4.114) and (2.44):

$$\left.\begin{aligned} dJ &= dU - TdS - SdT - \mu dN - Nd\mu \\ dU &= TdS - PdV + \mu dN \end{aligned}\right\} \Rightarrow dJ = -SdT - PdV - Nd\mu. \tag{4.118}$$

It follows that $J = J(V,T,\mu)$ and therefore the thermodynamic, Eq. (4.114), and statistical, Eq. (4.112), definitions of J are consistent. From Eq. (4.118) we obtain the following equations of state for a system specified by (T,V,μ) as the independent thermodynamic variables:

$$S = -\left(\frac{\partial J}{\partial T}\right)_{V,\mu}, \tag{4.119}$$

$$P = -\left(\frac{\partial J}{\partial V}\right)_{T,\mu}, \tag{4.120}$$

$$N = -\left(\frac{\partial J}{\partial \mu}\right)_{T,V}. \tag{4.121}$$

From Eq. (4.112) we can write the general thermodynamic and statistical forms of the equation of state, given respectively by

$$X_i = -\left(\frac{\partial J}{\partial x_i}\right)_{x_{j\neq i}} = \frac{1}{\beta}\left(\frac{\partial \ln Y}{\partial x_i}\right)_{x_{j\neq i}} = \overline{X}_i. \qquad (4.122)$$

Once the number of particles is known, we can find the mean energy:

$$\overline{E} - \mu\overline{N} = -\left(\frac{\partial \ln Y}{\partial \beta}\right)_{V,\mu}. \qquad (4.123)$$

Finally, by equating the second derivatives of the potential J, we obtain three Maxwell relations with (T, V, μ) as the independent thermodynamic variables:

$$\left(\frac{\partial S}{\partial \mu}\right)_{V,T} = \left(\frac{\partial N}{\partial T}\right)_{V,\mu}, \qquad (4.124)$$

$$\left(\frac{\partial P}{\partial \mu}\right)_{V,T} = \left(\frac{\partial N}{\partial V}\right)_{T,\mu}, \qquad (4.125)$$

$$\left(\frac{\partial P}{\partial T}\right)_{V,\mu} = \left(\frac{\partial S}{\partial V}\right)_{T,\mu}. \qquad (4.126)$$

4.4.3 Particle number fluctuations

The number of particles N is not fixed in a system described by a grand canonical ensemble, which is an open system. N is found using equation of state (4.121), where the equality of the mean value of N and its thermodynamic value are assumed, as they are in the more general Eq. (4.122). Likewise, we can express the variance of N in terms of the grand partition function:

$$\frac{\partial^2 \ln Y}{\partial \mu^2} = \frac{\partial}{\partial \mu}\left[\frac{1}{Y}\sum_{N,r}(\beta N)\,e^{\beta(\mu N - E_{Nr})}\right]$$

$$= \sum_{N,r}(\beta N)^2\frac{e^{\beta(\mu N - E_{Nr})}}{Y} - \left[\sum_{N,r}\beta N\frac{e^{\beta(\mu N - E_{Nr})}}{Y}\right]^2 = \beta^2\left(\overline{N^2} - \overline{N}^2\right),$$

whence

$$\sigma^2(N) = \frac{1}{\beta^2}\left(\frac{\partial^2 \ln Y}{\partial \mu^2}\right)_{T,V}. \qquad (4.127)$$

Using Eq. (4.121) for the mean value of N and Eq. (4.122), Eq. (4.127) becomes

$$\sigma^2(N) = \frac{1}{\beta}\left(\frac{\partial \overline{N}}{\partial \mu}\right)_{T,V}. \qquad (4.128)$$

Furthermore, using Eq. (4.125) we can write, following (Huang 1987):

$$\left(\frac{\partial \overline{N}}{\partial \mu}\right)_{T,V} = \left(\frac{\partial \overline{N}}{\partial P}\right)_{T,V}\left(\frac{\partial P}{\partial \mu}\right)_{T,V} = \left(\frac{\partial \overline{N}}{\partial P}\right)_{T,V}\left(\frac{\partial \overline{N}}{\partial V}\right)_{T,\mu}. \qquad (4.129)$$

The mean number of particles $\overline{N}$ is an extensive quantity, which when written as a function of V, T and μ has the form

$$\overline{N} = Vf(T,\mu) \Rightarrow \left(\frac{\partial \overline{N}}{\partial V}\right)_{T,\mu} = f(T,\mu) = \frac{\overline{N}}{V},$$

whence $n = \overline{N}/V$ is independent of $\overline{N}$ and of V. Substituting this into Eq. (4.129) we get

$$\left(\frac{\partial \overline{N}}{\partial \mu}\right)_{T,V} = \frac{\overline{N}}{V}\left(\frac{\partial \overline{N}}{\partial P}\right)_{T,V} = \overline{N}\left(\frac{\partial n}{\partial P}\right)_{T,V}.$$

Because n does not depend on $\overline{N}$ or V, we can replace the derivative at constant volume by a derivative at constant $\overline{N}$:

$$\left(\frac{\partial \overline{N}}{\partial \mu}\right)_{T,V} = \overline{N}\left(\frac{\partial n}{\partial P}\right)_{T,\overline{N}} = \overline{N}^2\left(\frac{\partial}{\partial P}\frac{1}{V}\right)_{T,\overline{N}}.$$

Upon substitution into Eq. (4.128), this gives

$$\sigma^2(N) = -k_B T \left(\frac{\overline{N}}{V}\right)^2 \left(\frac{\partial V}{\partial P}\right)_T, \tag{4.130}$$

which, combined with the definition of isothermal compressibility K_T, Eq. (2.61), leads to a fluctuation-dissipation relation:

$$\sigma^2(N) = k_B T \frac{\overline{N}^2}{V} K_T. \tag{4.131}$$

It follows that *the isothermal compressibility is the macroscopic measure of the variance of the number of particles*. As the variance is positive definite, this relation implies that $K_T \geq 0$, which is a stability condition with respect to fluctuations of the number of particles or the density. The relative fluctuation (1.20) of N is then

$$\delta^2(N) = \frac{\sigma^2(N)}{\overline{N}^2} = \frac{k_B T K_T}{V}. \tag{4.132}$$

This can be written in terms of the number density $n = N/V$:

$$\delta^2(n) = \frac{\sigma^2(N)}{V^2} \Rightarrow \delta^2(n) = \frac{k_B T K_T}{V}, \tag{4.133}$$

which implies that the density fluctuations in a fluid increase as the volume is decreased. We can further conclude that a very compressible gas (i.e., a low-density gas) is more susceptible to large density fluctuations than a denser fluid.

From the ideal gas equation of state $PV = Nk_B T$ we can estimate the order of magnitude of the fluctuations in the number of particles:

$$V = \frac{Nk_B T}{P} \Rightarrow \frac{\partial V}{\partial P} = -\frac{Nk_B T}{P^2}.$$

Hence from Eq. (4.132) we have

$$\delta(N) = \delta(n) \simeq \frac{1}{\sqrt{N}}. \tag{4.134}$$

This is an exact result for the ideal gas. We can then conclude that the density fluctuations are in general negligible for macroscopic systems. However, they play an important role in some physical phenomena, such as light scattering in the atmosphere due to local density fluctuations and the consequent fluctuations of the refractive index. According to Rayleigh theory, the mean intensity of the scattered light is proportional to the fourth power of the frequency of the incident radiation (Landau and Lifshitz, 1960, chap. XIV, §§93 and 94). Hence the (visible) light scattered away from the direction of incidence has a greater high-frequency content, which explains why the sky is blue, and the transmitted radiation has a greater low-frequency content, which explains the red colour of sunset (particularly on hot days). Another interesting phenomenon occurs near the critical point of a fluid (see Section 3.4.7): at the critical temperature, the first and second derivatives of the pressure with respect to the volume vanish (see Figure 3.8b), and therefore the isothermal compressibility K_T, Eq. (2.61), diverges, and large density fluctuations result. This is how Einstein's theory explains the phenomenon of critical opalescence, which occurs when a fluid near its critical point is illuminated [Landau and Lifshitz, 1960, chap. XIV, §95; Pais, 1982, chap. 5] (see also Figure 10.2).

4.5 System in equilibrium with a heat and pressure reservoir*

4.5.1 Isothermal–isobaric ensemble

Before starting this section, it is useful to reread the introduction to the present chapter. Here we shall consider an isolated system partitioned into two subsystems by a wall that is movable and allows heat exchange but not particle exchange. Hence system $\mathcal{A}'$ is a heat and pressure reservoir for system $\mathcal{A}$. When in equilibrium with the reservoir, system $\mathcal{A}$ will be specified by (T,P,N). The probability distribution to find system $\mathcal{A}$ in a given state with energy E and volume V is derived using the same method as for the grand canonical distribution, where now N is kept constant and V is allowed to change. Under these conditions, the entropy of the reservoir follows from Eqs. (4.104) and (4.103) upon substitution of E_{Nr} by E_{Vr}:

$$S' = \text{constant} - \frac{E_{Vr}}{T} - \frac{PV}{T}, \tag{4.135}$$

and the corresponding probability distribution is

$$p_{Vr} = \frac{e^{-\beta(E_{Vr}+PV)}}{Z^*} = \frac{e^{-\beta E_{Vr}^*}}{Z^*}, \tag{4.136}$$

where Z^* is the *isothermal–isobaric partition function*:

$$Z^* = \sum_{V,r} e^{-\beta(E_{Vr}+PV)} = \sum_{V,r} e^{-\beta E_{Vr}^*} \equiv Z^*(T,P,N) \tag{4.137}$$

and E_{Vr}^* is the *enthalpy* of the system in the microstate specified by V and r. A system in equilibrium with a heat and pressure reservoir is described by the isothermal–isobaric ensemble. *For a system specified by (T,P,N), the enthalpy plays the role of the energy in the canonical probability distribution and in the canonical partition function.*

Furthermore, we can write

$$Z^*(T,P,N) = \sum_V \left(\sum_r e^{-\beta E_{V_r}} \right) e^{-\beta PV} = \sum_V Z(T,V,N) e^{-\beta PV}, \tag{4.138}$$

where we have used definition (4.7) of the canonical partition function. Thus the isothermal–isobaric partition function can be seen as a sum over all possible values of the volume of canonical partition functions of systems of volume V, each weighted by $e^{-\beta PV}$.

Finally, the probability to find the system with volume V, and any value of the energy, is obtained by summing Eq. (4.136) over the energy microstates:

$$p_V = \sum_r p_{V_r} = \frac{Z(T,V,N) e^{-\beta PV}}{Z^*(T,P,N)}. \tag{4.139}$$

4.5.2 Connection with thermodynamics

Following the same procedure as for the grand canonical ensemble, from the definition of entropy, Eq. (4.27), combined with the isothermal–isobaric distribution, Eq. (4.136), we have:

$$S = \frac{\overline{E}}{T} + P\frac{\overline{V}}{T} + k_B \ln Z^*. \tag{4.140}$$

We now make the connection with thermodynamics via the following definition of the Gibbs energy:

$$G(T,P,N) = -k_B T \ln Z^*(T,P,N). \tag{4.141}$$

If we equate the mean energy and the mean volume with, respectively, the internal energy and the volume of the system, then substitution of Eq. (4.141) into Eq. (4.140) gives

$$G = U - TS + PV, \tag{4.142}$$

which is the thermodynamic definition of the Gibbs potential (see Section 2.2.2 for the formulation of thermodynamics in terms of G).

4.5.3 Volume fluctuations

From the equation of state (2.37) and the definition of G, Eq. (4.141), we get

$$V = \left(\frac{\partial G}{\partial P} \right)_{T,N} = \frac{1}{\beta} \left(\frac{\partial \ln Z^*}{\partial P} \right)_{T,N} = \overline{V}. \tag{4.143}$$

Following a similar procedure as in Section 4.4.3, it can be shown that the variance of the volume is

$$\sigma^2(V) = V k_B T K_T, \tag{4.144}$$

and the relative fluctuation is

$$\delta^2(V) = \frac{k_B T K_T}{V}. \tag{4.145}$$

Hence the relative fluctuation of the volume in a system specified by (T, P, N) is given by the same Eq. (4.133) as the relative fluctuation of the density in a system specified by (T, V, μ). The fluctuation-dissipation relation (4.144) implies that $K_T \geq 0$ is a necessary condition for the system to be stable with respect to volume fluctuations. Fluctuation-dissipation relations such as Eqs. (4.144) or (4.131) are very useful in numerical simulations of fluids, as they provide a method to obtain K_T from the variance of V or N. Furthermore, they allow us to compare statistical mechanical predictions with experimental results, thereby providing a test of the theoretical models or of the numerical methods employed.

4.6 Thermodynamic equivalence of the statistical ensembles

In this chapter we have reviewed some statistical ensembles describing single-component systems in thermodynamic equilibrium with a reservoir. Other ensembles can be defined, depending on which exchanges are allowed between the system and the reservoir (see, e.g., Hill (1986, chap. 1.7)).

Advanced topic 4.4

For readers who read through the Legendre transform in Section 3.4.6, we provide a general prescription to define the partition function appropriate for a particular system: if the system is in equilibrium with a reservoir through walls that are non-restrictive with respect to the extensive variables $x_0, \dots, x_n$ and restrictive with respect to the remaining extensive variables $x_{n+1}, \dots, x_t$, the partition function of the system is related to the appropriate thermodynamic potential via the Legendre transform:

$$\mathcal{Z}(X_0, \dots, X_n, x_{n+1}, \dots, x_t) = \exp\left(-\beta \underset{x_0, \dots, x_n}{\mathcal{L}} U\right).$$

Inverting this equation and applying it to the Helmholtz potential, Eq. (3.76), the Gibbs potential, Eq. (3.78), and the grand canonical potential, Eq. (4.116), yields their statistical definitions, Eqs. (4.35), (4.141) and (4.114) respectively.

In this section we show that the ensembles are equivalent for the purpose of evaluating thermodynamic quantities. We start with the canonical ensemble and consider the definition of entropy, Eq. (4.27):

$$S = -k_B \sum_r p_r \ln p_r.$$

From analysis of the energy fluctuations, Eq. (3.17), we conclude that the probability to find a system described by a canonical ensemble with an energy substantially different from its mean energy $\overline{E}$ is in practice negligible. We can then assume that the only probabilities p_r that contribute appreciably to statistical averages are those associated with $\overline{E}$. Now the number of states with energy $\overline{E}$ is the statistical weight $\Omega(\overline{E}, V, N)$; hence from the postulate of equal a priori probabilities we have, for all r,

$$p_r = \frac{1}{\Omega(\overline{E}, V, N)},$$

and from the expression for the canonical entropy we get

$$S = -k_B \ln \left[1 / \Omega(\overline{E}, V, N) \right] = k_B \ln \Omega(\overline{E}, V, N).$$

This method is similar to that used at the end of Section 4.2.3 to show that the microcanonical ensemble can be understood as a degenerate canonical ensemble, but with one important difference: 'a canonical ensemble (...) is, so to speak, *by its own choosing*, virtually a microcanonical ensemble' (Hill, 1986, chap. 2.2), specified by $(\overline{E}, V, N)$. The system of N spins 1/2 studied in Sections 3.3.5 and 4.3.2 is an example of this equivalence of ensembles when $N \to \infty$, as we saw.

We likewise conclude that in the grand canonical entropy

$$S = -k_B \sum_{N,r} p_{Nr} \ln p_{Nr},$$

the relevant p_{Nr} are those associated with $\overline{E}$ and $\overline{N}$. The corresponding number of states is $\Omega(\overline{E}, V, \overline{N})$, whence

$$S = k_B \ln \Omega(\overline{E}, V, \overline{N}),$$

and the grand canonical ensemble is, in practice, a microcanonical ensemble specified by $(\overline{E}, V, \overline{N})$. Also the isothermal–isobaric ensemble can be shown to be equivalent to a microcanonical ensemble specified by $(\overline{E}, \overline{V}, N)$. Indeed, the same argument can be applied to any other ensemble, thus leading to the conclusion that *although the Boltzmann entropy formula is in principle valid only for isolated systems, in practice it has general validity.*

Because the different ensembles are thermodynamically equivalent, we may select the working ensemble where calculations are easiest to perform. However, in numerical simulations, where system sizes are constrained by the available computer resources, different results will be obtained for the same quantity calculated in different ensembles.

Problems

4.1 A system consisting of N identical and distinguishable subsystems, each of which has two non-degenerate energy levels, is in equilibrium with a heat reservoir at temperature T.

(a) Find the energy of the system.

(b) Discuss the possibility of finding this system in a state of negative absolute temperature.

4.2 A classical one-dimensional harmonic oscillator of mass m has energy $\varepsilon = p^2/2m + Cx^2/2$, where p is the linear momentum and C is a constant. The angular frequency is given by $\omega = \sqrt{C/m}$.

   (a) Assuming that the oscillator is in thermal equilibrium at temperature T, show that its partition function is $z = 2\pi/h_0\beta\omega$, where h_0 is a constant with dimensions of angular momentum. Find the mean energy and discuss the result.

   (b) Consider now a system of N identical, localised and non-interacting oscillators. Find the energy, the heat capacity at constant volume and the Helmholtz free energy of this system.

4.3 *Show that the variance of the pressure of a system described by a canonical ensemble is

$$\sigma^2(P) = k_B T \left[\left(\frac{\partial \overline{P}}{\partial V} \right)_T - \overline{\left(\frac{\partial P}{\partial V} \right)_T} \right].$$

4.4 Derive the high-temperature and low-temperature limits of the variance of the energy, Eq. (4.58); of the heat capacity, Eq. (4.59); of the variance of the magnetic moment, Eq. (4.62); and of the magnetic susceptibility, Eq. (4.64), for a paramagnetic solid. Discuss the results.

4.5 *Show that the high-temperature and low-temperature limiting behaviours of the magnetisation in the Brillouin theory of paramagnetism are given by Eqs. (4.88) and (4.91) respectively.

4.6 Find the entropy and the heat capacity of an ideal system of N identical and localised quantum harmonic oscillators, all of the same angular frequency ω, and in equilibrium with a reservoir at temperature T. Derive the corresponding high- and low-temperature limits and discuss the results.

4.7 Consider a one-dimensional polymer chain composed of N monomers, under applied stress τ and in equilibrium with a reservoir at temperature T. Each monomer can be in one of two states: parallel to the chain, with energy ε_α and length a, or perpendicular to the chain, with energy ε_β and length b, where $\varepsilon_\alpha < \varepsilon_\beta$ and $a > b$ (both a and b are measured along the chain backbone).

   (a) Find the chain length L, taking the number N_α of monomers in state α as the independent variable.

   (b) Find the chain enthalpy $E^* = U - \tau L$.

   (c) By replacing the energy by the enthalpy in the partition function, find $\overline{L}$ using the equation of state:

$$\overline{L} = k_B T \left(\frac{\partial \ln Z^*}{\partial \tau} \right)_{T,N}.$$

   (d) *Discuss the procedure followed above in terms of the isothermal–isobaric ensemble.

4.8 *Using the isothermal–isobaric ensemble formalism, show that $\sigma^2(E^*) = k_B T^2 C_P$.

4.9 Consider a system consisting of N identical and distinguishable non-interacting sub-systems, each with two non-degenerate energy levels $\pm\varepsilon$, and in equilibrium with a heat and particle reservoir at temperature T and chemical potential μ.

(a) Use the grand canonical ensemble formalism to find the mean number of particles and the mean energy of the system.

(b) Check the thermodynamic equivalence of the grand canonical and canonical ensembles by comparing the preceding result with that of Problem 4.1.

References

Hill, T. L. 1986. *An Introduction to Statistical Thermodynamics*. Dover Publications.

Huang, K. 1987. *Statistical Mechanics*, 2nd edition. Wiley.

Kittel, C. 1995. *Introduction to Solid State Physics*, 7th edition. Wiley.

Landau, L. D., and Lifshitz, E. M. 1960. *Electrodynamics of Continuous Media*. Pergamon Press.

Landau, L. D., Lifshitz, E. M., and Pitaevskii, L.P. 1980. *Statistical Physics, Part 1*, 3rd edition. Pergamon Press.

Mandl, F. 1988. *Statistical Physics*, 2nd edition. Wiley.

Pais, A. 1982. *Subtle is the Lord*. Oxford University Press.

Reif, F. 1985. *Fundamentals of Statistical and Thermal Physics*. McGraw-Hill.

The classical ideal gas

5.1 Introduction

We can think of a gas as a fluid whose particles move fairly freely in space. The ideal or perfect gas is a limiting behaviour of a real gas, in which the interactions between particles are neglected. This approximation is valid when the potential energy of interaction between particles is negligible compared with their kinetic energy of motion. In some physical situations a real gas can be well described as an ideal gas. Fortunately, this is often the case with common gases under standard conditions for temperature and pressure; otherwise, evaluating the partition function of a real gas can become a formidable task. In the last section of this chapter we will address the statistical description of a real gas. This turns out to be necessary when we want to study phase transitions. In this chapter we shall restrict ourselves to the one-component gas.

As we saw from solid-state physics examples in the preceding chapter, evaluating statistical quantities is relatively easy for ideal systems consisting of localised, non-interacting particles. By non-interacting particles or subsystems we implicitly mean 'weakly interacting', so that thermal equilibrium is maintained by energy exchange, as discussed before. An ideal gas, however, is a system of non-interacting and *non-localised* particles. In a quantum mechanical description of the one-component gas, the particles are identical and *indistinguishable*, as we shall see in the next chapter. However, in a classical description the particles are considered identical and *distinguishable*: think of the molecules of a gas being labelled $1, 2, \ldots, N$. Thus the classical and the quantum statistical descriptions of the gas are different, and important consequences arise, namely that some classical thermodynamics results are inconsistent. In this chapter we shall give a 'corrected' classical description of the ideal gas to take into account that the molecules are indeed indistinguishable.

5.2 The density of states

Counting the microstates of a system is a central problem in statistical physics, as discussed in Chapter 3. When studying the classical ideal gas, we shall first consider the three-dimensional motion of a single particle. This motion is described by a path in a six-dimensional phase space (see Appendix 5.A). In this chapter we will disregard the spin states of a particle, which will be taken into account when studying the quantum gas in the next chapter. The state of motion of a system of N particles therefore requires a

6N-dimensional phase space for its specification. In classical mechanics any point in phase space represents a possible state of motion, which leads to complications when counting the states, as discussed in Appendix 5.A. This counting can be made easier in a quantum mechanical description of the motion, in which the states are no longer a continuum. Indeed, consider a particle that moves freely inside a box of volume

$$V = L_x L_y L_z. \tag{5.1}$$

Using the de Broglie relation $\mathbf{p} = \hbar \mathbf{k}$, the kinetic energy of the particle can be written as a function of its wavevector:

$$\varepsilon = \frac{p^2}{2m} = \frac{\hbar^2 k^2}{2m} = \frac{\hbar^2}{2m} \left(k_x^2 + k_y^2 + k_z^2 \right). \tag{5.2}$$

For plane waves obeying periodic boundary conditions, it can be shown (see Appendix 5.A) that, in each dimension, the wavevector can only take discrete values according to

$$k_x = \frac{2\pi}{l_x}, \quad k_y = \frac{2\pi}{l_y}, \quad k_z = \frac{2\pi}{l_z}, \quad l_{x,y,z} = 0, \pm 1, \pm 2, \ldots \tag{5.3}$$

The number Δl of possible l_x for which the wavevector takes values in $(k_x, k_x + dk_x)$, when the particle is bounded inside a length L_x, is given by

$$\Delta l_x = \frac{L_x}{2\pi} dk_x.$$

An analogous result follows for the y and z components of the wavevector. Consequently, the number of translation states for which the Cartesian components of the wavevector $\mathbf{k}$ take values in $(k_x, k_x + dk_x)$, $(k_y, k_y + dk_y)$ and $(k_z, k_z + dk_z)$ is

$$\Delta l_x \Delta l_y \Delta l_z = \frac{L_x L_y L_z}{(2\pi)^3} dk_x dk_y dk_z.$$

Using Eq. (5.1) we can write this number of states as

$$\frac{V}{(2\pi)^3} d^3 k = \rho_k d^3 k, \tag{5.4}$$

where

$$d^3 k = dk_x dk_y dk_z \tag{5.5}$$

is the volume element in $\mathbf{k}$-space and

$$\rho_k = \frac{V}{(2\pi)^3} \tag{5.6}$$

is the *density of states*, which is independent of $\mathbf{k}$. It is useful to write the number of translational states as a function of the linear momentum. Using the de Broglie relation, the right-hand side of Eq. (5.4) becomes

$$\rho_p d^3 p = \frac{V}{h^3} d^3 p, \tag{5.7}$$

where

$$\rho_p = \frac{V}{h^3} \tag{5.8}$$

is the corresponding density of states. Notice that this is simply the volume divided by h^3, i.e., the number of cubic 'cells' of edge length h contained in the system. In the classical limit, $h \to 0$ and the density of states goes to infinity. Thus quantum mechanics allows us to solve very naturally a problem that, in a purely classical treatment, would require us to introduce an *a priori* arbitrary factor with dimensions of energy×time. We shall discuss this matter further in Appendix 5.A.

It is sometimes useful to write the density of states in terms of the *magnitude* of the linear momentum. This is done by writing the element of volume in linear momentum space in spherical coordinates, when evaluating the density of states:

$$d^3 p = p^2 dp \sin\theta \, d\theta \, d\phi, \tag{5.9}$$

where θ and ϕ are, respectively, the polar and azimuthal angles that specify the orientation of **p** (see Figure 5.1).

Next we have to integrate the number of states, Eq. (5.7), over all possible directions of **p**:

$$\int_{\theta=0}^{\theta=\pi} \int_{\phi=0}^{\phi=2\pi} \rho_p d^3 p = \frac{V p^2 dp}{h^3} \int_0^\pi \sin\theta \, d\theta \int_0^{2\pi} d\phi = \frac{V}{h^3} 4\pi p^2 dp. \tag{5.10}$$

This is the number of individual states of a particle enclosed in volume V whose linear momentum has magnitude in $(p, p + dp)$. Equation (5.10) allows us to define the corresponding *density of states* $g(p)$ as

$$g(p) dp = \frac{V}{h^3} 4\pi p^2 dp. \tag{5.11}$$

In Eqs. (5.7) or (5.10) for the number of states, the volume appears explicitly, which requires some discussion. Note that the only constraint on the particle position is that it can be anywhere inside the volume V. Since the total number of states includes position states, this means that this number is proportional to V when the particle is free to move inside the volume. Hence the density of states given by Eqs. (5.8) or (5.11) applies to a

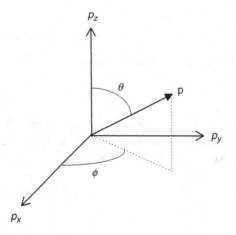

Figure 5.1   Coordinate frame in linear momentum space.

particle that can be found anywhere inside the volume V with equal probability. However, there are many practical problems in which the gas is subjected to a field (e.g., the gravitational field) and the particle positions will depend on their interaction with that field. Consequently, the corresponding probability distribution will not be spatially uniform. In this case, the density of states must include terms that depend on the position coordinates of the particles. Therefore, instead of Eq. (5.7), we have to start with the number of states $d\Gamma$ with position coordinates in d^3r and with linear momentum coordinates in d^3p (see Appendix 5.A):

$$d\Gamma = \frac{d^3r\, d^3p}{h^3}. \tag{5.12}$$

In this case the density of states is just

$$\rho = \frac{1}{h^3}. \tag{5.13}$$

If we now want the density of states in terms of the magnitude of the linear momentum and irrespective of the particle position inside the volume, we have to integrate $d\Gamma$ over the volume V and over all possible directions of $\mathbf{p}$. Using Eq. (5.9) we get

$$\int_{\theta=0}^{\theta=\pi}\int_{\phi=0}^{\phi=2\pi} d\Gamma = \frac{p^2 dp}{h^3}\int_0^\pi \sin\theta\, d\theta \int_0^{2\pi} d\phi \int_V d^3r = \frac{V}{h^3} 4\pi p^2 dp = g(p) dp. \tag{5.14}$$

whence Eq. (5.11) for the density of states again follows. We shall use this result later to find the single-particle partition function.

5.3 Maxwell–Boltzmann statistics

If the interactions between particles can be neglected, the energy of a gas is the sum of the energies of the individual particles. In the most general case, the energy of each particle comprises its kinetic energy of translation, its internal energy, and the energy of interaction of the particle with any applied field.

Assuming all particles to be identical, we denote the energy of a particle in quantum state j by ε_j, where

$$\varepsilon_1 \leq \varepsilon_2 \leq \cdots \leq \varepsilon_j \leq \cdots$$

For a gas in thermodynamic equilibrium at temperature T, enclosed in volume V and composed of N particles, the canonical ensemble is the appropriate (and traditional) statistical description. We shall specify the gas microstate by listing the number of particles in every possible state: n_1 particles in state 1, n_2 particles in state 2,..., n_j particles in state j,..., where n_j is the *occupation number* of state j. *The microstate of the gas is thus specified by the set of occupation numbers.* Consequently, the energy of the gas in a microstate specified by a given set of occupation numbers $(n_1, n_2, \ldots)$ is given by

$$E(n_1, n_2, \ldots) = \sum_j n_j \varepsilon_j, \tag{5.15}$$

where the sum is over all individual states. As the number of gas particles is fixed, the following constraint results for the occupation numbers:

$$N = \sum_j n_j, \tag{5.16}$$

which plays the role of a 'normalisation requirement' for the occupation numbers.

In a quantum description of a gas of identical particles, these behave as *fundamentally indistinguishable*, as we shall see in the next chapter. However, in a classical description, despite being identical and non-localised, the particles are considered to be distinguishable. In this case, for a given set of occupation numbers $\{n_1, n_2, \ldots\}$ the number of different ways to distribute N particles over the individual states is

$$\Gamma(n_1, n_2, \ldots) = \frac{N!}{n_1! n_2! \cdots}. \tag{5.17}$$

To find the partition function of the gas, we have to sum over all its microstates. Because (in a classical description) the particles are distinguishable, each set of occupation numbers (obeying Eq. (5.16)) corresponds to a distinct state of the gas. Hence, the entire collection of sets of occupation numbers describes all possible gas microstates. The partition function of an ideal gas of N particles, in thermal equilibrium at temperature T and enclosed in a volume V, is then written in terms of the occupation numbers as

$$Z_N \equiv Z(T, V, N) = \sum_{\{n_1, n_2, \ldots\}} \Gamma(n_1, n_2, \ldots) \exp[-\beta E(n_1, n_2, \ldots)], \tag{5.18}$$

where $\{n_1, n_2, \ldots\}$ means that the sum is over all sets of occupation numbers obeying constraint (5.16), i.e., over all possible values of n_1, of n_2, etc., such that the sum of all elements of each set equals N. It follows from Eqs. (5.15) and (5.17) that

$$Z_N = \sum_{\{n_1, n_2, \ldots\}} \frac{N!}{n_1! n_2! \cdots} \exp[-\beta(n_1 \varepsilon_1 + n_2 \varepsilon_2 + \cdots)]$$

$$= \sum_{\{n_1, n_2, \cdots\}} \frac{N!}{n_1! n_2! \cdots} [\exp(-\beta \varepsilon_1)]^{n_1} [\exp(-\beta \varepsilon_2)]^{n_2} \ldots \tag{5.19}$$

Taking into account Eq. (5.16), Eq. (5.19) is seen to be a polynomial expansion, leading to:

$$Z_N = \left(e^{-\beta \varepsilon_1} + e^{-\beta \varepsilon_2} + \ldots\right)^N = Z_1^N. \tag{5.20}$$

Since the sum is over individual states, Z_1 is the single-particle partition function:

$$Z_1 = \sum_j e^{-\beta \varepsilon_j}. \tag{5.21}$$

Result (5.20) is reminiscent of the partition function of N localised particles, such as spins in a crystal lattice in the quantum model of a paramagnetic solid of Section 4.3.1: because the particles are distinguishable, for statistical purposes they are 'countable' just as localised particles are (see Figure 5.2). But this gives an incorrect, description of physical reality, hence the need to develop a statistics of indistinguishable particles to describe the gas, as we shall see when working out the statistics of a quantum gas in Chapter 6.

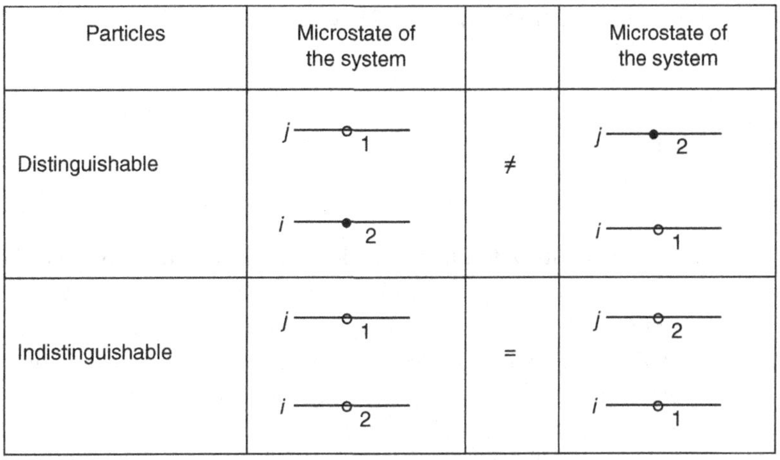

Particles	Microstate of the system		Microstate of the system
Distinguishable	$j \underset{1}{\circ}$ $i \underset{2}{\bullet}$	$\neq$	$j \underset{2}{\bullet}$ $i \underset{1}{\circ}$
Indistinguishable	$j \underset{1}{\circ}$ $i \underset{2}{\circ}$	$=$	$j \underset{2}{\circ}$ $i \underset{1}{\circ}$

Figure 5.2   Schematic representation of the microstates of a two-particle system that can be found in individual states i and j, with occupation numbers $n_i = n_j = 1$. Unlike in the case of distinguishable particles, the exchange of two indistinguishable particles leaves the system in the same state; hence in this case the particles do not need labels.

The probability to find the gas in a microstate specified by a given set of occupation numbers $n_1, n_2, \ldots$ is

$$p(n_1, n_2, \ldots) = \frac{\Gamma(n_1, n_2, \ldots) \exp[-\beta E(n_1, n_2, \ldots)]}{Z_N}, \tag{5.22}$$

with $\Gamma(n_1, n_2, \ldots)$ given by Eq. (5.17), $E(n_1, n_2, \ldots)$ given by Eq. (5.15), and Z_N given by Eq. (5.19):

$$p(n_1, n_2, \cdots) = \frac{\frac{N!}{n_1! n_2! \ldots} \exp[-\beta(n_1 \varepsilon_1 + n_2 \varepsilon_2 + \cdots)]}{\sum_{\{n_1, n_2, \cdots\}} \frac{N!}{n_1! n_2! \ldots} \exp[-\beta(n_1 \varepsilon_1 + n_2 \varepsilon_2 + \cdots)]}. \tag{5.23}$$

The mean occupation number of individual state i can be found from the partition function of the gas, Eq. (5.19):

$$\left(\frac{\partial \ln Z_N}{\partial \varepsilon_i}\right)_{T, \varepsilon_{j \neq i}} = \frac{1}{Z_N} \left\{ \sum_{\{n_1, n_2, \ldots\}} (-\beta n_i) \frac{N!}{n_1! n_2! \cdots} \exp[-\beta(n_1 \varepsilon_1 + n_2 \varepsilon_2 + \ldots)] \right\}$$

$$= -\beta \sum_{\{n_1, n_2, \ldots\}} n_i p(n_1, n_2, \cdots).$$

From Eqs. (5.19), (5.23) and the definition of mean value we get

$$\overline{n}_i = -\frac{1}{\beta} \left(\frac{\partial \ln Z_N}{\partial \varepsilon_i}\right)_{T, \varepsilon_{j \neq i}}. \tag{5.24}$$

We leave it as an exercise to check that the variance of n_i is given by

$$\sigma^2(n_i) = \frac{1}{\beta^2} \left(\frac{\partial^2 \ln Z_N}{\partial \varepsilon_i^2}\right)_{T, \varepsilon_{j \neq i}} = -\frac{1}{\beta} \left(\frac{\partial \overline{n}_i}{\partial \varepsilon_i}\right)_{T, \varepsilon_{j \neq i}}. \tag{5.25}$$

Equations (5.20), (5.24) and (5.21) yield

$$\overline{n}_i = N\frac{e^{-\beta\varepsilon_i}}{Z_1},\qquad(5.26)$$

where

$$\frac{e^{-\beta\varepsilon_i}}{Z_1} = p_i\qquad(5.27)$$

is the probability p_i to find a particle in state i. Equation (5.26) is the *Maxwell–Boltzmann statistic*, which gives *the mean occupation number of an individual state as the total number of particles multiplied by the probability of occupation of that same individual state.* Both $\overline{n}_i$ and p_i are independent of the state of the other particles in the gas because we are neglecting their interactions, i.e., the gas is ideal.

Finally, from Eq. (5.15) we get the mean energy of the gas in terms of the mean occupation numbers:

$$\overline{E} = \sum_j \overline{n}_j\varepsilon_j,\qquad(5.28)$$

and from Eq. (5.16) the constraint that the mean occupation numbers must obey:

$$\sum_j \overline{n}_j = N.\qquad(5.29)$$

In Eqs. (5.28) and (5.29), the allowed values of n_j are not independent of each other, since any set of occupation numbers must obey constraint (5.16), as the number of particles in the gas is fixed. In the grand canonical description of a gas this constraint no longer exists, since we are dealing with a system with a variable number of particles. In this case, the grand canonical partition function and the probability to find the gas with a given set of occupation numbers factorise into a product of terms each depending on one occupation number only. Consequently these terms are statistically independent, which translates into a much easier evaluation of the thermodynamic quantities, as we shall see in the chapter on the quantum ideal gas (Chapter 6).

5.4 Partition function of the ideal gas in the classical regime

From Eqs. (5.20) and (5.21), the partition function of the gas is

$$Z_N = \left(\sum_j e^{-\beta\varepsilon_j}\right)^N = \left(\sum_{j_1} e^{-\beta\varepsilon_{j_1}}\right)\left(\sum_{j_2} e^{-\beta\varepsilon_{j_2}}\right)\cdots\left(\sum_{j_N} e^{-\beta\varepsilon_{j_N}}\right)$$

$$\Rightarrow Z_N = \sum_{j_1,j_2,\ldots,j_N} e^{-\beta\left(\varepsilon_{j_1}+\varepsilon_{j_2}+\cdots+\varepsilon_{j_N}\right)},\qquad(5.30)$$

where the sum is now over all states of each particle. In this sum the particles are assumed distinguishable, as before. However, this counting of states leads to thermodynamic functions of the gas that are not properly intensive or extensive. For this reason, we shall need to correct the partition function in such a way as to make the particles indistinguishable. This will be the *first quantum correction to the classical partition function*. With this goal in mind, we first consider a classical system composed of two particles, for which Eq. (5.30) becomes

$$Z_2 = \left(\sum_{j_1} e^{-\beta \varepsilon_{j_1}} \right) \left(\sum_{j_2} e^{-\beta \varepsilon_{j_2}} \right) = \sum_{j} e^{-2\beta \varepsilon_j} + \sum_{j_1 \neq j_2} e^{-\beta \left(\varepsilon_{j_1} + \varepsilon_{j_2} \right)}.$$

The first term corresponds to having the two particles in the same state $j_1 = j_2 = j$, and the second term to having the two particles in different states. In the latter we are counting as different states of the system the state where particle 1 is in state j_1 and particle 2 is in state j_2, and the state where particle 2 is in state j_1 and particle 1 is in state j_2 (see Figure 5.2). Because the two particles are actually indistinguishable, this latter term will give the same state of the gas twice. It should therefore be divided by 2!, which is the number of terms that represent the same physical state of the system. Generalising this to a gas of N particles we get

$$(Z_N)_{\text{corrected}} = \sum_{j} e^{-N\beta \varepsilon_j} + \cdots + \frac{1}{N!} \sum_{j_1 \neq j_2 \neq \cdots \neq j_N} e^{-\beta \left(\varepsilon_{j_1} + \varepsilon_{j_2} + \cdots + \varepsilon_{j_N} \right)}, \qquad (5.31)$$

where the first term corresponds to having all the N particles in the same state j and the last term to having all the N particles in different states, while the remaining terms correspond to intermediate distributions of the particles over the states. Equation (5.31) is the partition function (5.30) corrected for indistinguishability, but has no practical interest since N is of the order of Avogadro's number. However, under given conditions of temperature and density (e.g., the standard conditions for temperature and pressure of common gases), the probability of occupation of the individual states is much smaller than 1 because, under these conditions, the number of individual states in a given energy range $(\varepsilon, \varepsilon + d\varepsilon)$ (i.e., the density of states) is much larger than the number of particles with energy in that range. With this fact in mind, we define a regime in which the corrected partition function is easily written: *A gas is in the classical regime when the probability that any individual state is occupied by more than one particle is very small.* We shall discuss later what physical requirements the classical regime must satisfy. For now, we note that under these conditions only the last term in Eq. (5.31) is important, and approximate the corrected partition function by

$$(Z_N)_{\text{corrected}} \simeq \frac{1}{N!} Z_1^N = \frac{Z_N}{N!}. \qquad (5.32)$$

In Eq. (5.32) the important terms (corresponding to all particles occupying different states) are correctly counted, whereas the unimportant terms in the classical regime are the most wrongly weighted. We therefore adopt as the partition function of the gas the Maxwell–Boltzmann partition function:

$$Z_{MB} = \frac{Z_N}{N!}, \quad Z_N = Z_1^N. \qquad (5.33)$$

The factor $1/N!$ does not introduce too many algebraic complications in what follows, while allowing the correct computation of the thermodynamic functions of the gas.

5.5 Single-particle partition function

To obtain the partition function of the gas (5.33) we need to find the single-particle partition function (5.21):

$$Z_1 = \sum_j e^{-\beta \varepsilon_j}, \tag{5.34}$$

where the sum is over all the states of a single particle. For particles without internal structure, the single-particle energy has the form

$$\varepsilon(\mathbf{r}, \mathbf{p}) = \frac{p^2}{2m} + u(\mathbf{r}). \tag{5.35}$$

where the first term is the translational kinetic energy and the second term is the potential energy in an external field; $\mathbf{r}$ and $\mathbf{p}$ are the position and linear momentum, respectively. Thus the energy of each particle is a function of its own coordinates (position and momentum) only.

For particles with internal structure, an internal energy term ε^{int} must be added to Eq. (5.35) (e.g., for diatomic molecules the rotational and vibrational kinetic energies – see Problems 5.7 and 5.8):

$$\varepsilon^{\text{total}} = \varepsilon(\mathbf{r}, \mathbf{p}) + \varepsilon^{\text{int}}, \tag{5.36}$$

$$\Rightarrow Z_1 = \sum_{m,n} \exp\left\{ -\beta \left[\varepsilon_m(\mathbf{r}, \mathbf{p}) + \varepsilon_n^{\text{int}} \right] \right\}$$

$$= \left\{ \sum_m \exp\left[-\beta \varepsilon_m(\mathbf{r}, \mathbf{p}) \right] \right\} \left[\sum_n \exp\left(-\beta \varepsilon_n^{\text{int}} \right) \right], \tag{5.37}$$

where indices m and n refer to internal energy levels and to field interaction energy levels, respectively. We define the partition function for translational motion and interaction with an external field as

$$z = \sum_m \exp\left[-\beta \varepsilon_m(\mathbf{r}, \mathbf{p}) \right], \tag{5.38}$$

and the partition function for the internal energy as

$$z^{\text{int}} = \sum_n \exp\left(-\beta \varepsilon_n^{\text{int}} \right). \tag{5.39}$$

The single-particle partition function (5.37) is then written:

$$Z_1 = z z^{\text{int}}. \tag{5.40}$$

Likewise, the partition function z, given by Eq. (5.38) with Eq. (5.35), also factorises into a product of terms for the translational kinetic energy and the potential energy in an external field. The single-particle partition function therefore factorises into a product of terms for the individual contributions to the single-particle energy:

$$Z_1 = z^{\text{tr}} z^{\text{pot}} z^{\text{int}}. \tag{5.41}$$

Once the partition function Z_1 is known, we shall be able to derive results that are independent of the internal particle structure, and so will be shared by all ideal gases when subjected to the same external field. The translational part will give universal results, as it describes the limiting behaviour of all gases if rarefied enough and in the absence of an external field.

We have defined the classical regime as that in which a range of particle energies $d\varepsilon$ contains a large number of individual states. Later we shall see that a sufficient condition for a gas to be in the classical regime is that its temperature should be high (for common gases, room temperature is usually high enough). Under these conditions, the energy difference $\Delta\varepsilon$ between two consecutive individual states is much smaller than the corresponding thermal energy $k_B T$. Consequently, as discussed in Section 4.2.1, the Boltzmann factor is in practice a continuous function, and the sum over states in Eq. (5.38) can be replaced by integrals (see also Problems 5.7 and 5.8):

$$\sum_m \ldots \to \int d^3 r \ldots \int d^3 p, \tag{5.42}$$

where

$$d^3 r = dx\,dy\,dz, \quad d^3 p = dp_x\,dp_y\,dp_z, \tag{5.43}$$

stand for the volume elements in position and in linear momentum space, respectively. In Eq. (5.42), the integral is over all possible values of the Cartesian components of the position and linear momentum. We know from quantum mechanics that, according to Eq. (5.8), a factor of $1/h^3$ must be inserted for a correct evaluation of the density of states when replacing the sum by an integral (see also Appendix 5.A). This is the *second quantum correction to classical statistics*, after the factor $1/N!$ inserted into the partition function of the gas, Eq. (5.33). We can thus write the partition function for a particle without internal structure in terms of the number of states $d\Gamma$ with position coordinates in $d^3 r$ and with linear momentum coordinates in $d^3 p$, as

$$z = \int d\Gamma\, e^{-\beta\varepsilon[\mathbf{r},\mathbf{p}]}, \tag{5.44}$$

with $d\Gamma$ given by Eq. (5.12) and $\varepsilon(\mathbf{r},\mathbf{p})$ given by Eq. (5.35), which yields

$$z = \frac{1}{h^3} \int d^3 p\, e^{-\beta p^2/2m} \int d^3 r\, e^{-\beta u(\mathbf{r})}. \tag{5.45}$$

For the translational part we get

$$z^{\mathrm{tr}} = \frac{1}{h^3} \int_{-\infty}^{+\infty} dp_x e^{-\beta p_x^2/2mk_B T} \int_{-\infty}^{+\infty} dp_y e^{-\beta p_y^2/2mk_B T} \int_{-\infty}^{+\infty} dp_z e^{-\beta p_z^2/2mk_B T}.$$

Each of the above integrals gives $(2\pi m k_B T)^{1/2}$ (see Appendix 1.A.2), whence

$$z^{\mathrm{tr}} = \left(\frac{2\pi m k_B T}{h^2} \right)^{3/2}. \tag{5.46}$$

The external field part depends on the form of the potential energy $u(\mathbf{r})$:

$$z^{\mathrm{pot}} = \int d^3 e^{-\beta u(\mathbf{r})}. \tag{5.47}$$

In the case of a particle moving freely inside a box of volume V, this equals

$$u(\mathbf{r}) = \begin{cases} 0, & \mathbf{r} \in V \\ \infty, & \mathbf{r} \notin V \end{cases} \Rightarrow e^{-\beta u(\mathbf{r})} = \begin{cases} 1, & \mathbf{r} \in V \\ 0, & \mathbf{r} \notin V \end{cases}. \tag{5.48}$$

The integration over position coordinates is restricted to V, whence the result:

$$z^{\text{pot}} = V. \tag{5.49}$$

Finally, it follows from Eqs. (5.45), (5.46) and (5.49) that the translational partition function of a free particle confined in a volume V is

$$z = V \left(\frac{2\pi m k_B T}{h^2} \right)^{3/2}. \tag{5.50}$$

Note that the partition function z, as given by Eq. (5.50), is (correctly) non-dimensional, because the factor $1/h^3$ was inserted when counting the states, thereby avoiding the need to include an *a priori* arbitrary factor with dimensions of (energy$\times$time)^3.

We remark that, when calculating the partition function (5.50), the integral was evaluated over $\mathbf{r}$ and $\mathbf{p}$ expressed in Cartesian coordinates, i.e., a *vector* integration was performed. Alternatively, a *scalar* integration can be performed that involves separating the integrals over the direction of and the magnitude of $\mathbf{p}$. To do so, in the calculation of z, Eq. (5.44), the volume element in linear momentum space in Cartesian coordinates has to be replaced by the volume element in spherical coordinates as in Eq. (5.9) (recall Figure 5.1).

Consider again the case of a free particle confined in a box of volume V, with the potential given by Eq. (5.48). Its partition function is

$$z = \int d\Gamma e^{-p^2/2mk_B T}. \tag{5.51}$$

Because the Boltzmann factor is a function of the linear momentum magnitude only, we can first integrate over the volume V and then over all directions of $\mathbf{p}$, which yields the density of states (5.14). We next integrate over the magnitude of $\mathbf{p}$ the Boltzmann factor (which pertains to translational motion only) multiplied by Eq. (5.14):

$$z = \int_0^{+\infty} g(p) dp \, e^{-p^2/2mk_B T} = V \left(\frac{2\pi m k_B T}{h^2} \right)^{3/2}, \tag{5.52}$$

where we have used the results of Appendix 1.A.2. This is again Eq. (5.50).

It is interesting to note that when changing from a vector integration (over the components of a vector) to a scalar integration (over the magnitude of the same vector), formally we are making the substitution

$$\int_{-\infty}^{+\infty} d^3 p \ldots = \int_0^{+\infty} 4\pi p^2 dp \ldots. \tag{5.53}$$

5.6 Thermodynamics of the classical ideal gas

In the canonical ensemble description of a gas, the thermodynamic functions are given in terms of the logarithm of the partition function. Hence from Eqs. (5.33) and (5.41), the different contributions to the thermodynamics of the gas can be studied separately. In this section we shall derive the thermodynamic functions pertaining to the translational motion of the particles of an ideal gas confined in a volume V, for which the partition function is

$$Z = \frac{1}{N!} z^N, \tag{5.54}$$

with z given by Eq. (5.50). The Helmholtz free energy can be found from Eq. (4.35) using the Stirling formula:

$$F = -k_B T \ln Z = k_B T[\ln(N!) - N\ln z] \simeq k_B T[N\ln(N/e) - N\ln z], \tag{5.55}$$

leading to

$$F = -Nk_B T \ln\left[\frac{eV}{N}\left(\frac{2\pi m k_B T}{h^2}\right)^{3/2}\right]. \tag{5.56}$$

Notice that the quantity in [ ] in Eq. (5.56) is proportional to V/N (this comes from the correction factor $1/N!$ in the gas partition function) and is therefore intensive, with the result that $F \propto N$ is correctly extensive.

The gas pressure is obtained from Eqs. (4.36) or (2.31) combined with Eq. (5.56):

$$P = -\left(\frac{\partial F}{\partial V}\right)_T = \frac{Nk_B T}{V}, \tag{5.57}$$

whence the *equation of state*

$$PV = Nk_B T, \tag{5.58}$$

as we know empirically from classical thermodynamics, is thus derived here from the postulates and formalism of statistical physics. This result establishes the equality of the (statistical) absolute temperature defined by Eq. (3.11) and the (thermodynamic) ideal gas temperature.

The chemical potential of the gas is obtained from the second equality in Eqs. (2.47) combined with Eq. (5.55):

$$\mu = \left(\frac{\partial F}{\partial N}\right)_T = -k_B T \ln\left(\frac{z}{N}\right), \tag{5.59}$$

where z is given by Eq. (5.50):

$$\mu = -k_B T \ln\left[\frac{V}{N}\left(\frac{2\pi m k_B T}{h^2}\right)^{3/2}\right]. \tag{5.60}$$

The quantity in [ ] is intensive, hence the chemical potential given by Eq. (5.60) is correctly intensive.

The mean translational kinetic energy of the gas is obtained from Eq. (4.16) combined with Eqs. (5.54) and (5.50):

$$\overline{E} = -\left(\frac{\partial \ln Z}{\partial \beta}\right)_V = \frac{3}{2}Nk_BT. \tag{5.61}$$

The mean translational kinetic energy per particle is then

$$\overline{\varepsilon} = \frac{3}{2}k_BT, \tag{5.62}$$

which means that each translational degree of freedom contributes $k_BT/2$ to the mean translational kinetic energy. This result is a special case of the *equipartition theorem* of classical statistical mechanics, which can be stated as follows: *In a thermodynamic system in equilibrium at temperature T, each generalised coordinate (position or momentum) that enters the energy of a microstate of the system (i.e., the Hamiltonian of the system) quadratically, contributes $k_BT/2$ to the mean energy of the system.* Notice that the theorem applies to *any* degree of freedom, translational, rotational, or other, provided that the generalised coordinate associated with that degree of freedom appears squared in the Hamiltonian of the system. In Appendix 5.C we give a proof of this theorem.

The translational kinetic energy contribution to the heat capacity at constant volume is, from Eq. (5.61),

$$C_V = \left(\frac{\partial \overline{E}}{\partial T}\right)_V = \frac{3}{2}Nk_B, \tag{5.63}$$

and the corresponding molar specific heat is given by

$$c_V = \frac{3}{2}R, \quad R = N_Ak_B, \tag{5.64}$$

which is the classical Dulong and Petit Law. The entropy follows from Eqs. (2.30) and (5.55):

$$S = -\left(\frac{\partial F}{\partial T}\right)_V = Nk_B\left[\ln\left(\frac{ez}{N}\right) + \frac{3}{2}\right], \tag{5.65}$$

with z given by Eq. (5.50), whence

$$S = Nk_B\left\{\ln\left[\frac{eV}{N}\left(\frac{2\pi mk_BT}{h^2}\right)^{3/2}\right] + \frac{3}{2}\right\}. \tag{5.66}$$

Again we notice that inclusion of the factor $1/N!$ in the partition function of the gas makes the entropy an extensive thermodynamic quantity, otherwise the statistical physics result would be thermodynamically inconsistent. This inconsistency of the classical result is known as the *Gibbs paradox* (Reif, 1985, chap. 7.3). On the other hand, the second correction to the classical statistics, i.e., inclusion of the factor $1/h^3$ in the number of states $d\Gamma$, allows the statistical formulae to be used for the computation of numerical values of the thermodynamic functions of the classical ideal gas: the Helmholtz free energy (5.56), the chemical potential (5.60) and the entropy (5.66).

However, the low-temperature limits of the heat capacity (5.63) and of the entropy (5.66) do not obey the third law, as they do not go to zero when $T \to 0$. The reason is that when

$T \to 0$ the particles tend to accumulate in the lower energy states, because in a gas of 'classical' particles there are no restrictions on the number of particles that can occupy each individual state, and this violates our assumption of validity of the classical regime. If there are no further restrictions on the occupation of the individual states (e.g., deriving from the Pauli exclusion principle), this particle accumulation can be understood on a microscopic level. We know from quantum mechanics that the transition energies between microstates are quantised. At very low temperatures, the thermal energy needed for certain transitions may not be available and, in this case, some degrees of freedom might stay 'inactive' or 'frozen' and thus not contribute to the total mean energy. This contradicts the equipartition theorem, which is a classical result.

5.7 Validity of the classical regime

We recall that, in the classical regime, the number of individual states in a given energy range$\{\varepsilon, \varepsilon + d\varepsilon\}$ is much larger than the number of particles with energy in that range, and therefore the probability that any individual state is occupied by more than one particle is very small. Consequently, most individual states will be empty, some will contain one particle, and a very small number will contain more than one particle. Hence a sufficient condition for the validity of the classical regime is that the mean occupation numbers should be much smaller than unity:

$$\overline{n}_i \ll 1, \quad \text{for all } i. \tag{5.67}$$

For free particles without internal structure in a box of volume V, from Eqs. (5.26), (5.50) and (5.67) we find

$$\frac{N}{V} \left(\frac{h^2}{2\pi m k_B T} \right)^{3/2} \exp\left(-\beta \varepsilon_i^{tr}\right) \ll 1 \, (\text{for all } i) \Rightarrow \frac{N}{V} \left(\frac{h^2}{2\pi m k_B T} \right)^{3/2} \ll 1. \tag{5.68}$$

This is the *criterion for the validity of the classical regime*. Requirement (5.68) is satisfied for low particle concentrations and/or high temperatures. Common gases, under standard conditions for temperature and pressure, usually satisfy Eq. (5.68), and therefore their thermodynamic properties can be studied using the Maxwell–Boltzmann statistics.

It is interesting to analyse the criterion for the validity of the classical regime from a quantum physics perspective. A particle with linear momentum p has a de Broglie wavelength given by

$$\lambda_{dB} = \frac{h}{p} = \frac{h}{\sqrt{2m\varepsilon^{tr}}}. \tag{5.69}$$

We can estimate this wavelength from the mean translational kinetic energy per particle (5.61), whence we obtain the *thermal wavelength*:

$$\lambda_T = \frac{h}{\sqrt{3mk_B T}} = \left(\frac{2\pi}{3} \right)^{1/2} \left(\frac{h^2}{2\pi m k_B T} \right)^{1/2}, \tag{5.70}$$

Then requirement (5.68) can be written in terms of this wavelength as

$$\lambda_T^3 \ll \frac{V}{N}. \tag{5.71}$$

The inverse particle density can be written in terms of the mean distance l between particles. If we assume that each particle sits at the centre of a cube of edge length l, we get

$$l^3 N = V \Rightarrow l^3 = \frac{V}{N}, \tag{5.72}$$

and it follows from Eq. (5.71) that

$$\lambda_T^3 \ll l^3. \tag{5.73}$$

Thus *in order for the classical regime to hold, the de Broglie wavelength of the particles must be much smaller than the mean interparticle distance.* Under these conditions, the distance between particles is large enough that quantum interference effects are negligible and the wave nature of the particles loses its importance. For a common gas under standard conditions for temperature and pressure, quantum effects are usually negligible, e.g., helium at room temperature and atmospheric pressure (see Problem 5.3(a)). However, if the thermal wavelength is of the same order of magnitude as the mean interparticle distance, then interference effects become important and a quantum statistical description is necessary. Think of the conduction electrons in metals as a gas of free particles confined to the volume of the sample – this is a model that we shall study in the next chapter. Noting that the mass of an electron is about seven thousand times smaller than the mass of a helium atom, and that in a solid the electron density is about one thousand times larger than the density of helium under standard conditions, it becomes obvious that quantum statistics must be used even at high temperatures (see Problem 5.3(b)).

5.8 The Maxwell–Boltzmann distribution

For an ideal gas in the classical regime, the probability to find a particle with position coordinates in $d^3 r$, linear momentum components in $d^3 p$, and in internal state s, is obtained by multiplying the canonical probability to find the particle in a given total energy state, Eq. (5.36), by the number of states $d\Gamma$ given by Eq. (5.12):

$$f_S(\mathbf{r}, \mathbf{p}) d^3 r d^3 p = \frac{e^{-\beta \varepsilon(\mathbf{r},\mathbf{p})} e^{-\beta \varepsilon_s^{\text{int}}}}{Z_1} \frac{d^3 r d^3 p}{h^3}, \tag{5.74}$$

with $\varepsilon(\mathbf{r}, \mathbf{p})$ given by Eq. (5.35) and Z_1 given by Eq. (5.41). The probability to find a particle with position coordinates in $d^3 r$ and linear momentum components in $d^3 p$ irrespective of its internal structure is obtained by summing Eq. (5.74) over all internal energy states. Consequently, from Eqs. (5.39) and (5.41) we get

$$f(\mathbf{r}, \mathbf{p}) d^3 r d^3 p = \frac{e^{-\beta \varepsilon(\mathbf{r},\mathbf{p})}}{z^{\text{tr}} z^{\text{pot}}} \frac{d^3 r d^3 p}{h^3}, \tag{5.75}$$

with z^{tr} and z^{pot} given by Eqs. (5.46) and (5.47), respectively, leading to the *Maxwell–Boltzmann (MB) probability distribution*:

$$f(\mathbf{r},\mathbf{p})d^3r d^3p = \left[\frac{e^{-p^2/2mk_BT}}{(2\pi mk_BT)^{3/2}}d^3p \right] \left[\frac{e^{-\beta u(\mathbf{r})}}{\int d^3r e^{-\beta u(\mathbf{r})}}d^3r \right]. \qquad (5.76)$$

This distribution can be thought of as the product of the probabilities of two statistically independent events, corresponding to the two terms in [ ], which is a consequence of the separability of the energy, given by Eq. (5.35). The first factor is the *Maxwell distribution* pertaining to translational motion, and gives the probability to find a particle with linear momentum components in d^3p. The second factor is the *Boltzmann distribution*, which depends on the form of the potential energy of interaction with an external field, and gives the probability to find a particle with position coordinates in d^3r. Both these probabilities are normalised, because each exponential factor is divided by the corresponding normalisation constant. If we integrate the MB distribution (5.76) over all possible values of the position coordinates we recover the Maxwell distribution, whereas if we integrate over all possible values of the linear momentum, we recover the Boltzmann distribution. In the language of probability theory, the MB distribution is a joint probability distribution, and the Maxwell and Boltzmann distributions are the corresponding marginal distributions (see Section 1.5.1).

In the absence of an external field, $u(\mathbf{r}) = 0$ and the particles are uniformly distributed in the volume V. Hence in Eq. (5.76) the second factor reduces to d^3r/V, whereupon the MB distribution becomes

$$f(\mathbf{r},\mathbf{p})d^3r d^3p = \frac{e^{-p^2/2mk_BT}}{V(2\pi mk_BT)^{3/2}}d^3r d^3p. \qquad (5.77)$$

As discussed earlier, upon integrating Eq. (5.76) or Eq. (5.77) over the volume we obtain the Maxwell distribution, which can be written

$$\frac{e^{-p_x^2/2mk_BT}}{(2\pi mk_BT)^{1/2}}dp_x\frac{e^{-p_y^2/2mk_BT}}{(2\pi mk_BT)^{1/2}}dp_y\frac{e^{-p_z^2/2mk_BT}}{(2\pi mk_BT)^{1/2}}dp_z = f(p_x)dp_x f(p_y)dp_y f(p_z)dp_z. \quad (5.78)$$

We are thus led to the conclusion that p_x, p_y and p_z are independent random variables, for which we can write

$$f(p_j)dp_j = \frac{e^{-p_j^2/2mk_BT}}{(2\pi mk_BT)^{1/2}}dp_j, \quad j = x,y,z. \qquad (5.79)$$

This is a Gaussian probability distribution (see Section 1.4.2), with mean value and variance given by

$$\begin{aligned} \overline{p_j} &= 0 \\ \sigma^2(p_j) &= mk_BT, \end{aligned} \qquad j = x,y,z. \qquad (5.80)$$

Upon integrating Eq. (5.76) or Eq. (5.77) over the volume and all possible orientations of $\mathbf{p}$, we obtain the Maxwell distribution as a function of the linear momentum magnitude.

Proceeding as in Section 5.4 for the alternative calculation of the single-particle partition function (which involves a scalar integration) and using Eq. (5.9), we find

$$f(p)dp = \frac{4\pi p^2 e^{-p^2/2mk_BT}}{(2\pi mk_BT)^{3/2}}dp, \tag{5.81}$$

which, unlike Eq. (5.79), is no longer a Gaussian distribution. Eq. (5.81) can be written in terms of the density of sates $g(p)$, Eq. (5.11), and of the single-particle partition function, z, Eq. (5.50):

$$f(p)dp = \frac{1}{z}g(p)e^{-p^2/2mk_BT}dp. \tag{5.82}$$

This equation, when written in terms of the translational kinetic energy $\varepsilon = p^2/2m$, is a practical realisation of Eq. (4.15). Substituting $p = mv$ in Eq. (5.81) we get the *Maxwell velocity distribution*:

$$f(v)dv = \left(\frac{m}{2\pi k_BT}\right)^{3/2} 4\pi v^2 e^{-mv^2/2k_BT}dv. \tag{5.83}$$

This is the probability that a gas particle has a velocity of magnitude in the interval $(v, v + dv)$. We leave it as an exercise (using the integrals in Appendix 1.A.2) to show that the Maxwell distribution has a maximum for

$$v_{max} = \sqrt{\frac{2k_BT}{m}}, \tag{5.84}$$

that the mean velocity of a gas particle is

$$\bar{v} = \sqrt{\frac{8k_BT}{\pi m}} = \frac{2}{\sqrt{\pi}}v_{max}, \tag{5.85}$$

and that the mean square velocity is

$$\overline{v^2} = \frac{3k_BT}{m} = \frac{3}{2}v_{max}^2, \tag{5.86}$$

whence we can define the root mean square velocity as

$$v_{rms} = \sqrt{\overline{v^2}} = \sqrt{\frac{3k_BT}{m}} = \sqrt{\frac{3}{2}}v_{max}. \tag{5.87}$$

As an example, consider nitrogen at room temperature (300 K). Using Eq. (5.87) and taking $m = 28/(6 \times 10^{23})$ g, we get $v_{rms} \simeq 500$ ms^{-1}.

5.9 Gas in a uniform external field

As an application of the Maxwell distribution we shall now study a gas in a uniform external field. Here we treat the case of a gravitational field; see Problem 5.9 for a gas of particles with permanent electric dipole moments in a uniform electric field.

Consider a gas in Earth's gravitational field. Taking the OZ axis along the vertical direction, the potential energy of a particle of mass m and altitude z is written

$$u(z) = mgz. \tag{5.88}$$

As the potential energy depends only on the position coordinate z, particles will be uniformly distributed over planes of constant z. Thus from the Boltzmann distribution in Eq. (5.76) the probability to find a particle between z and $z + dz$ is

$$f(z)dz = \frac{e^{-mgz/k_BT}dz}{\int e^{-mgz/k_BT}dz}, \tag{5.89}$$

where the integral is over all possible values of z. It is convenient to work with $n(z)$, the mean number of particles per unit volume at position coordinate z. This is obtained by multiplying the total number density by the probability (5.89):

$$n(z)dz = nf(z)dz = n\frac{e^{-mgz/k_BT}dz}{\int e^{-mgz/k_BT}dz}. \tag{5.90}$$

The quantity $n(z)dz$ represents the mean number of particles per unit area with z-coordinate between z and $z + dz$. Dividing $n(z)dz$ by $n_0 dz = n(z = 0)dz$, the mean number of particles per unit area in the $z = 0$ plane, we get

$$n(z) = n_0 e^{-mgz/k_BT}. \tag{5.91}$$

This formula, the *barometric law*, can also be derived direct from the Maxwell–Boltzmann statistics (5.26) by integrating over all translational states. It shows that the density of a gas in a gravitational field decreases according to an exponential law, with a characteristic length. For air at room temperature this length is of order 10^4 m. Notice that this result was obtained for a system in thermodynamic equilibrium and hence it must be applied with care to the atmosphere, which is not in equilibrium. Moreover, although air density decreases approximately exponentially with increasing altitude, Eqs. (5.89) and (5.90) were derived on the assumption that the gas temperature is constant, and as we know the temperature decreases strongly with increasing altitude.

5.10 The real gas[*]

To study phase transitions in a fluid we need to take into account the interactions between particles. Otherwise we will have an ideal gas, which, as we saw earlier, only exhibits a disordered gas phase. The Hamiltonian of a real gas is thus the sum of the kinetic and the interaction potential energies of the system:[1]

$$\mathcal{H}\left(\mathbf{r}^N, \mathbf{p}^N\right) = \frac{1}{2m}\sum_{i=1}^{N}\mathbf{p}_i^2 + U\left(\mathbf{r}^N\right), \tag{5.92}$$

---

[1] For simplicity we shall not consider any applied field, which is easily included, however, as we saw in the preceding section

where $\mathbf{r}^N$ represents the set of position coordinates $\mathbf{r}_1, \mathbf{r}_2, \dots, \mathbf{r}_N$, and $\mathbf{p}^N$ the set of linear momenta $\mathbf{p}_1, \mathbf{p}_2, \dots, \mathbf{p}_N$, of the N particles each of mass m. As described in Section 5.5, in the classical regime a very large number of individual states exist in every range of the system's energy. Consequently, the Boltzmann factor is effectively a continuous function of $\mathbf{r}^N$ and $\mathbf{p}^N$, and the canonical partition function is an integral over all these variables:

$$Z_N = \frac{1}{h^{3N}N!} \int \int e^{-\beta \mathcal{H}(\mathbf{r}^N, \mathbf{p}^N)} d\mathbf{r}^N d\mathbf{p}^N. \tag{5.93}$$

The integrals are over all possible values of the position coordinates and linear momentum components. Each Cartesian component of the linear momentum can take values between $-\infty$ and $+\infty$, whereas the Cartesian position coordinates can take values in the volume occupied by the system. Recall that we have introduced the factor $1/N!$ to account for particle indistinguishability. Moreover, the factor $1/h^{3N}$, where h is Planck's constant, comes from replacing the sums over the linear momentum components by integrals (see also Appendix 5.A).

Inserting Eq. (5.92) into Eq. (5.93) and integrating over the linear momentum components using the results of Section 5.5, we get

$$Z_N = \frac{1}{N!} \left(\frac{2\pi m k_B T}{h^2} \right)^{3N/2} Z_N^{\text{conf}} = \frac{Z_N^{\text{conf}}}{N! \Lambda^{3N}}, \tag{5.94}$$

where $\Lambda = (h^2/2\pi m k_B T)^{1/2}$ is the de Broglie thermal wavelength,[2] and

$$Z_N^{\text{conf}} = \int e^{-\beta U(\mathbf{r}^N)} d\mathbf{r}^N \tag{5.95}$$

is the *configurational integral* or *configurational partition function* of the system, i.e., that part of the partition function which depends only on the positions of the particles, but not on their velocities. If there are no interactions between the particles, then $U(\mathbf{r}^N) = 0$ and the configurational integral reduces to

$$Z_N^{\text{conf}} = \int d\mathbf{r}^N = \int d\mathbf{r}_1 \times \int d\mathbf{r}_2 \times \cdots \times \int d\mathbf{r}_N = V^N, \tag{5.96}$$

whence we recover the earlier results for the partition function and the equation of state of the classical ideal gas, Eqs. (5.54) and (5.58) respectively.

In what follows we shall consider only pairwise interactions between the particles, i.e., each particle interacts only with one other particle at a time (in other words, simultaneous interactions between three, four or more particles are neglected). We thus have

$$U(\mathbf{r}^N) = \frac{1}{2} \sum_{i,j=1}^{N} u(\mathbf{r}_i, \mathbf{r}_j) = \sum_{i<j}^{N} u(\mathbf{r}_i, \mathbf{r}_j). \tag{5.97}$$

Since each particle interacts with all other particles bar itself, there are $N(N-1)/2$ pair interactions. To ensure that each interaction between particles i and j is counted only once, the factor $1/2$ is introduced in the first sum in Eq. (5.97), where both indices run from 1 to

---

[2] This is a definition of the thermal wavelength alternative to Eq. (5.70), used in, e.g., Huang (1987).

N; and in the second sum we restrict $i < j$, the latter form being more convenient in what follows. Upon substitution of Eq. (5.97) into the configurational integral, Eq. (5.95), we get

$$Z_N^{\text{conf}} = \int \exp\left[-\beta \sum_{i<j}^N u\left(\mathbf{r}_i, \mathbf{r}_j\right)\right] d\mathbf{r}^N = \int \left[\prod_{i<j}^N e^{-\beta u(\mathbf{r}_i, \mathbf{r}_j)}\right] d\mathbf{r}^N$$

$$= \int \left\{\prod_{i<j}^N \left[1 + \Phi\left(\mathbf{r}_i, \mathbf{r}_j\right)\right]\right\} d\mathbf{r}^N, \tag{5.98}$$

where the *Mayer function* is defined as

$$\Phi\left(\mathbf{r}_i, \mathbf{r}_j\right) = e^{-\beta u(\mathbf{r}_i, \mathbf{r}_j)} - 1. \tag{5.99}$$

The number of Mayer functions in Eq. (5.98) equals the number of pairs of particles, $N(N-1)/2$, so by expanding the product we obtain

$$Z_N^{\text{conf}} = \int \left\{\prod_{i<j}^N \left[1 + \Phi\left(\mathbf{r}_i, \mathbf{r}_j\right)\right]\right\} d\mathbf{r}^N$$

$$= \int \left[1 + \frac{N(N-1)}{2} \Phi\left(\mathbf{r}_i, \mathbf{r}_j\right)\right] d\mathbf{r}^N + \cdots$$

$$= V^{N-2} \int \left[1 + \frac{N(N-1)}{2} \Phi\left(\mathbf{r}_i, \mathbf{r}_j\right)\right] d\mathbf{r}_i d\mathbf{r}_j + \cdots$$

$$= V^N \left[1 + \frac{N(N-1)}{2V^2} \int \Phi\left(\mathbf{r}_i, \mathbf{r}_j\right) d\mathbf{r}_i d\mathbf{r}_j\right] + \cdots \tag{5.100}$$

where $\cdots$ stands for integrals that contain more than one Mayer function, and the factor V^{N-2} results from integrating over the positions of all particles other than i and j. Given the definition of the Mayer function, Eq. (5.99), the integral of $\Phi\left(\mathbf{r}_i, \mathbf{r}_j\right)$ over $\mathbf{r}_i, \mathbf{r}_j$ measures the contribution of the interaction between particles i and j to the configurational partition function. Clearly, this integral is the same for any i and j. Likewise, integrals containing products of two or more Mayer functions, e.g., $\Phi\left(\mathbf{r}_i, \mathbf{r}_j\right) \Phi\left(\mathbf{r}_j, \mathbf{r}_k\right)$ or $\Phi\left(\mathbf{r}_i, \mathbf{r}_j\right) \Phi\left(\mathbf{r}_j, \mathbf{r}_k\right) \Phi(\mathbf{r}_k, \mathbf{r}_l)$, and so on, will represent the contributions to the configurational partition function of simultaneous pair interactions of three or four particles, respectively (recall that we are only considering pair interactions). It is reasonable to assume (and, in some cases, possible to prove) that the terms corresponding to interactions of three, four or more particles are only important at higher densities,[3] and therefore they will be neglected in a first approximation. Comparing Eqs. (5.100) and (5.96), we notice that Eq. (5.100) gives the first correction to the configurational partition function

---

[3] Of course, what a 'higher density' is will depend on the system under consideration.

of the classical ideal gas when there are pair interactions between particles. Because in general $N \gg 1$, Eq. (5.100) can be written as

$$Z_N^{\text{conf}} = V^N \left[1 + \frac{1}{2} \rho^2 \int \Phi \left(\mathbf{r}_i, \mathbf{r}_j \right) d\mathbf{r}_i d\mathbf{r}_j \right] + \cdots \tag{5.101}$$

where $\rho = N/V$ is the number density.

We can now find the Helmholtz free energy from Eqs. (4.35) and (5.94):

$$\begin{aligned} F &= -k_B T \ln Z_N \approx k_B T \ln N! + N k_B T \ln \Lambda^3 - k_B T \ln Z_N^{\text{conf}} \\ &= k_B T \ln N! + N k_B T \ln \Lambda^3 - N k_B T \ln V \\ &\quad - k_B T \ln \left[1 + \frac{1}{2} \rho^2 \int \Phi \left(\mathbf{r}_i, \mathbf{r}_j \right) d\mathbf{r}_i d\mathbf{r}_j \right]. \end{aligned} \tag{5.102}$$

Using the Stirling approximation $\ln N! \approx N \ln N - N$ (valid when $N \gg 1$) and the series expansion of the logarithm $\ln(1+x) \approx x$ (valid when $|x| \ll 1$), we get

$$\begin{aligned} \frac{F}{N k_B T} &\approx \ln N - 1 + \ln \Lambda^2 - \ln V - \frac{N}{2V^2} \int \Phi \left(\mathbf{r}_i, \mathbf{r}_j \right) d\mathbf{r}_i d\mathbf{r}_j \\ &= \ln \left(\Lambda^2 \rho \right) - 1 + \rho B_2, \end{aligned} \tag{5.103}$$

where the *second virial coefficient* B_2 is defined as

$$B_2 = -\frac{1}{2V} \int \Phi \left(\mathbf{r}_i, \mathbf{r}_j \right) d\mathbf{r}_i d\mathbf{r}_j. \tag{5.104}$$

It is straightforward to verify that the term containing the second virial coefficient gives the first correction (linear in the density) to the total free energy of the ideal gas (where there are no interparticle interactions, which is equivalent to zero density). Expanding the logarithm to derive Eq. (5.103) from Eq. (5.102) is then permissible when $\rho B_2 \ll 1$ (although it can still be useful even when this condition is not met). If we were to include the terms that $\cdots$ stands for in Eq. (5.101), we would obtain a formally exact expression for the Helmholtz free energy in the form of a *virial expansion*:

$$\frac{F}{N k_B T} = \ln \left(\Lambda^2 \rho \right) - 1 + \rho B_2 + \rho^2 B_3 + \cdots = \ln \left(\Lambda^2 \rho \right) - 1 + \sum_{n=2}^{\infty} B_n \rho^{n-1}. \tag{5.105}$$

However, calculation of the coefficients B_n becomes increasingly (indeed, prohibitively) complicated as n increases, and there are only a few systems for which virial coefficients of order higher than two can be found analytically.

We can now derive the equation of state for the real gas, in the low-density approximation, using the free energy given by Eq. (5.103) and assuming B_2 to be independent of the gas volume, which will be justified later:

$$P = -\left(\frac{\partial F}{\partial V} \right)_T = \frac{N k_B T}{V} \left(1 + \frac{N}{V} B_2 \right). \tag{5.106}$$

Free expansion of a gas

As an application of Eq. (5.106) we shall now study the free expansion of an isolated gas. It will be shown that the temperature of a real gas decreases in such a process, whereas for an ideal gas it does not. Even though this process is irreversible, it can be studied by thermodynamic methods because the initial and final states are equilibrium states. During the free expansion the internal energy of the gas is conserved:

$$dU(T, V) = 0 \Rightarrow \left(\frac{\partial U}{\partial T}\right)_V dT + \left(\frac{\partial U}{\partial V}\right)_T dV = 0. \tag{5.107}$$

The *Joule coefficient* is defined as

$$\alpha_J = \left(\frac{\partial T}{\partial V}\right)_U = -\left(\frac{\partial U}{\partial V}\right)_T / \left(\frac{\partial U}{\partial T}\right)_V$$

$$= -\frac{1}{C_V}\left[T\left(\frac{\partial P}{\partial T}\right)_V - P\right], \tag{5.108}$$

where Eq. (2.51) has been used to write the denominator in terms of C_V, and Eq. (2.55) to replace the numerator. From Eq. (5.108) it is straightforward to see that for an ideal gas the Joule coefficient vanishes. In the second virial approximation, it follows from Eq. (5.108) that

$$\alpha_J = -\frac{1}{C_V}\left(\frac{N}{V}\right)^2 k_B T^2 \frac{dB_2}{dT}. \tag{5.109}$$

We shall now find the second virial coefficient, Eq. (5.104), of the gas. We consider high enough temperature that $k_B T \gg u_0$, where $-u_0$ is the minimum of the intermolecular potential. As first noticed by Van der Waals, we can think of the gas molecules as composed of a hard core of radius r_0, below which they cannot be compressed, combined with a long-range attraction – i.e., such that $u(r) = \infty$ when $r < 2r_0$ and $u(r) \rightarrow 0^-$ when $r \rightarrow \infty$ (r is the distance between the centres of two molecules). From these considerations we arrive at a model real gas composed of weakly attracting hard molecules. Hence we can approximate the integrand in Eq. (5.104) with Eq. (5.99) by

$$\Phi(r) = \begin{cases} -1, & \text{if } 0 < r < 2r_0 \\ -\frac{u(r)}{k_B T}, & \text{if } r > 2r_0 \end{cases}, \quad r \equiv r_{ij} = ||\mathbf{r}_i - \mathbf{r}_j||. \tag{5.110}$$

We calculate B_2 from Eq. (5.104) with Eq. (5.110) and making the changes of variables $\mathbf{r} = \mathbf{r}_i - \mathbf{r}_j$ and $\mathbf{R} = (\mathbf{r}_i + \mathbf{r}_j)/2$. The integral over $\mathbf{R}$ gives a factor of V that cancels the $1/V$ in Eq. (5.104). Next, in the integral over $\mathbf{r}$, we first integrate over all possible orientations of $\mathbf{r}$ and then over its magnitude, with the result:

$$B_2 = 2\pi \int_0^\infty \left[1 - e^{-\beta u(r)}\right] r^2 dr = 2\pi \frac{(2r_0)^3}{3} + \frac{2\pi}{k_B T} \int_{2r_0}^\infty r^2 u(r) dr. \tag{5.111}$$

On defining

$$a = -2\pi \int_{2r_0}^{\infty} r^2 u(r) dr > 0, \qquad (5.112)$$

$$b = 4 \left(\frac{4\pi r_0^3}{3} \right) = 4v_0, \qquad (5.113)$$

where v_0 is the volume of one molecule, the second virial coefficient can be written:

$$B_2(T) = b - \frac{a}{k_B T}, \qquad (5.114)$$

where coefficient a comes from the weak, long-range intermolecular attractions (Van der Waals forces).

We can now find the Joule coefficient using Eqs. (5.109) and (5.114):

$$\alpha_J = \left(\frac{\partial T}{\partial V} \right)_U = -\frac{a}{C_V} \left(\frac{N}{V} \right)^2 < 0, \qquad (5.115)$$

whence it follows that the gas temperature drops as the gas expands freely. Assuming that coefficient a remains constant throughout the expansion, we can derive the corresponding temperature change in terms of the specific heat and specific volume:

$$\int_{T_1}^{T_2} dT = -\frac{a}{c_v} \int_{v_1}^{v_2} \frac{dv}{v^2} \Rightarrow T_2 - T_1 = \frac{a}{c_v} \left(\frac{1}{v_2} - \frac{1}{v_1} \right), \quad v = \frac{V}{N}. \qquad (5.116)$$

In practice, the internal energy of the container enclosing the gas also decreases by $C_r(T_2 - T_1)$, where C_r is the heat capacity of the container. To take this into account, we must replace C_V by $C_V + C_r$, which yields a much smaller change in gas temperature, as the heat capacity of the container is usually very large, $C_r \gg C_V$. This difficulty can be overcome by replacing the free expansion by a continuous expansion of a fluid under flow, such that the container walls remain at constant temperature after the steady state has been reached. This cooling process, known as the *Joule–Thomson effect*, is commonly used in refrigerators and air conditioners and is discussed in most books on applied thermodynamics (see e.g., Reif (1985, chap. 5.10)).

5.A Phase space. Density of states

In classical mechanics the state of motion of a particle can, in principle, be determined by specifying its position $\mathbf{r}$ and linear momentum $\mathbf{p}$ at each instant. One-dimensional motion of a particle, e.g., along the OX axis, is described by a trajectory in the plane having x and p_x as orthogonal axes, which is called the *phase space* of the particle. In this space, each point represents one state of the particle motion, as sketched in Figures 5.3, 5.4, and 5.5.

Three-dimensional particle motion is described as a trajectory in a six-dimensional phase space. In classical mechanics, any point in phase space represents a possible state of

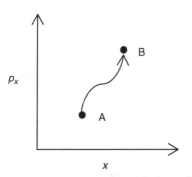

Figure 5.3   The motion of a particle in one dimension along the *OX* axis is described in phase space by a trajectory in the plane defined by *x* and its conjugate momentum p_x as orthogonal axes. The trajectory sketched corresponds to the classical description of a particle evolving from state of motion *A* to state of motion *B*.

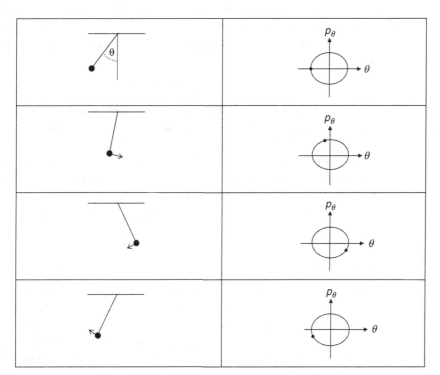

Figure 5.4   Sketches of the phase space trajectories (right panels) representing the motion of a free pendulum (left panels). The angular momentum p_θ is the (generalised) momentum conjugate to the angular coordinate θ (see, e.g. Goldstein (1980, chap. 8.1), or Mandl (1988, chap. 7.9)). The location of a point in phase space specifies the dynamical state of the pendulum in each instant.

motion; thus the number of states in any given region of phase space is proportional to the six-dimensional-volume of that region. The state of motion of a system of *N* particles requires a 6*N*-dimensional phase space; hence the number of states in a given region of the phase space of a system is proportional to the 6*N*-dimensional volume of that region.

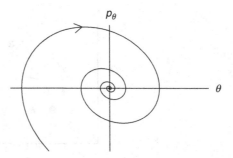

Figure 5.5   A damped pendulum: the phase space trajectories are spirals that converge to the origin, which is a stable equilibrium point (attractor).

For a single particle, the number of states $d\Gamma$ with position coordinates in $(x, x + dx)$, $(y, y + dy)$, $(z, z + dz)$, i.e., in the volume element $d^3r = dxdydz$, and with linear momentum components in $(p_x, p_x + dp_x), (p_y, p_y + dp_y), (p_z, p_z + dp_z)$, i.e., in the volume element $d^3p = dp_x dp_y dp_z$, is proportional to the corresponding volume element in the phase space of the particle:

$$d\Gamma \propto d^3r d^3p. \tag{5.117}$$

In classical mechanics, the proportionality constant is undetermined because there is a continuum of possible states. Any constant with dimensions of (angular momentum)$^{-3}$ can be chosen, in order to make $d\Gamma$ non-dimensional.

In quantum mechanics, it is not possible to specify simultaneously and exactly the position and the conjugate momentum components of a particle. If motion is along the OX axis, the position and linear momentum of the particle can only be known with accuracies Δx and Δp, respectively, that satisfy the Heisenberg uncertainty relation:

$$\Delta x \Delta p \geq \frac{\hbar}{2}. \tag{5.118}$$

A quantum state of the particle is associated with possible values of its position and linear momentum according to this relation; therefore a state of motion of the particle occupies a volume in two-dimensional space, i.e., an area not smaller than $\hbar/2$ in phase space, since a smaller area would correspond to a better specification than allowed by quantum mechanics. It can be shown that this lower bound for the area is in fact h, whence we can imagine that phase space is divided into cells of area h each (see Figure 5.6). In the phase space of a particle moving in three dimensions, the volume of each cell is h^3, whence the density of states (5.13) is seen to follow. For a system of N particles, the volume of each cell in the corresponding $6N$-dimensional phase space is h^{3N}. Consequently, for each particle, we can go from the classical continuous phase space to the quantum cellular phase space by dividing the classical result, Eq. (5.117), by h^3 (this was the method used by Planck when deriving his theory of blackbody radiation):

$$d\Gamma = \frac{d^3r d^3p}{h^3}, \tag{5.119}$$

and the corresponding density of states is therefore $\rho = 1/h^3$.

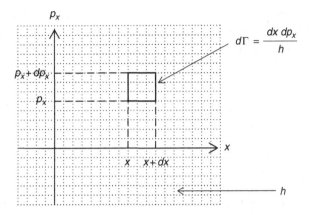

Figure 5.6 Schematic representation of the quantum phase space for the one-dimensional motion of a particle along *OX*. With each cell of area *h* is associated one quantum state, hence the density of states is 1/*h*.

Alternatively, we can arrive at Eq. (5.119) by calculating the number of allowed states with energy less than or equal to some ε. First, we divide the volume in phase space bounded by the surface $\varepsilon = $ constant by the volume available to each state. Recall that, in the case of a free particle in a volume V studied at the beginning of this chapter, by Eq. (5.2) the surface $\varepsilon = $ constant defines a sphere of radius k in k-space. Now from Schrödinger's equation the motion of a particle confined in a volume $V = L_x L_y L_z$ is described by a stationary plane wave with wavevector k. For each direction (e.g., along OX), the stationary wave equation is written

$$\frac{\partial^2 \phi}{\partial x^2} + k_x^2 \phi = 0,$$

the general solution of which is

$$\phi = A \sin(k_x x) + B \cos(k_x x),$$

as can be verified by substitution. A and B are integration constants. If we impose periodic boundary conditions over a segment of length L_x, we get

$$\phi(0) = \phi(L_x) \Rightarrow B = A \sin(k_x L_x) + B \cos(k_x L_x)$$

$$\Rightarrow \begin{cases} A = 0 \\ \cos(k_x L_x) = 1 \end{cases} \Rightarrow k_x = \frac{2\pi}{L_x} l_x, \quad l_x = 0, \pm 1, \pm 2, \ldots$$

Generalising this to three dimensions, we obtain Eq. (5.3). From this result, the available volume (in k space) per state is $(2\pi)^3/L_x L_y L_z = (2\pi)^3/V$.

Dividing the volume $4\pi k^3/3$ of the sphere of radius k by the available volume per state gives the number of allowed states with wavevector less than or equal to k:

$$\Omega(k) = \frac{V k^3}{6\pi^2}. \tag{5.120}$$

Of course, dividing a spherical volume by a parallelepipedic volume is an approximation that works better the larger the density of points, i.e., the larger L_x, L_y, L_z are. The number of states with wavevector magnitude in $(k, k + dk)$ follows by differentiating $\Omega(k)$:

$$f(k)dk = \frac{d\Omega(k)}{dk}dk = \frac{Vk^2}{2\pi^2}dk. \qquad (5.121)$$

This can be expressed in terms of the magnitude of the linear momentum using the de Broglie relation, with the result

$$g(p)dp = \frac{V}{h^3}4\pi p^2 dp, \qquad (5.122)$$

which is identical to Eq. (5.11). Finally, to arrive at Eq. (5.119) we have to calculate, starting from Eq. (5.117), the volume in phase space corresponding to a particle confined in volume V and whose linear momentum has magnitude in $(p, p + dp)$:

$$\int_V \int_{\theta=0}^{\pi} \int_{\phi=0}^{2\pi} d^3 r d^3 p = V 4\pi p^2 dp, \quad d^3 p = p^2 dp \sin\theta d\theta d\phi. \qquad (5.123)$$

Comparing Eqs. (5.122) and (5.123) leads to Eq. (5.119).

5.B Liouville's theorem in classical mechanics

Consider an isolated classical system specified by f generalised coordinates and momenta $q_1, \ldots, q_f, p_1, \ldots, p_f$. If we assume that the system is described statistically by an ensemble, let ρ denote the density of ensemble systems in phase space. The number of systems in the ensemble with coordinates and conjugate momenta in phase space volume element $(dq_1 \cdots dq_f dp_1 \cdots dp_f)$, i.e., with coordinates in $(q_1, q_1 + dq_1)$, $\ldots$, $(q_f, q_f + dq_f)$, and momenta in $(p_1, p_1 + dp_1)$, $\ldots$, $(p_f, p_f + dp_f)$, at time t, is given by

$$\rho(q_1, \ldots, q_f, p_1, \ldots, p_f) (dq_1 \cdots dq_f dp_1 \cdots dp_f).$$

Any system in the ensemble will evolve in time according to the classical equations of motion that, in Hamiltonian mechanics, are written

$$\dot{q}_i = \frac{\partial \mathcal{H}}{\partial p_i}, \quad \dot{p}_i = -\frac{\partial \mathcal{H}}{\partial q_i}, \qquad (5.124)$$

where $\mathcal{H}(q_1, \ldots, q_f, p_1, \ldots, p_f)$ is the Hamiltonian of the system. As a result of this evolution, the density of systems in phase space changes with time. Our aim is to find $\partial \rho / \partial t$ at a given point in phase space.

Consider a volume element $dq_1 \cdots dp_f$ in phase space, fixed and bounded by q_1 and $q_1 + dq_1$, $\ldots$, q_f and $q_f + dq_f$, p_1 and $p_1 + dp_1$, $\ldots$, p_f and $p_f + dp_f$ (see Figure 5.7). The number of systems in this volume changes as their coordinates and momenta evolve according to Eqs. (5.124). In a time interval dt the change in the number of systems contained in this volume of phase space is given by

$$\frac{\partial \rho}{\partial t} dt dq_1 \cdots dp_f.$$

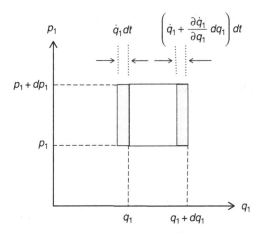

Figure 5.7　A fixed volume element $dq_1 dp_1$ in a two-dimensional phase space. The shaded areas show the number of systems that enter (left) or exit (right) this volume in the time dt.

This change is a consequence of systems entering or exiting the volume in the time dt. The number of systems that enter through side $q_1 = $ constant equals the number of systems contained in volume $v = \dot{q}_1 dt dq_2 \cdots dp_f$, i.e., it equals $\rho(q_1,\dots,p_f)v$. The number of systems that exit through side $q_1 + dq_1 = $ constant is given by the same expression, but where now ρ and $\dot{q}_1$ must be evaluated at $q_1 + dq_1$. The balance between the number of systems that enter volume $(dq_1 \cdots dp_f)$ through side $q_1 = $ constant and the number of systems that exit through side $q_1 + dq_1 = $ constant is then

$$(\rho \dot{q}_1)\, dt dq_2 \cdots dp_f - \left[(\rho \dot{q}_1) + \frac{\partial}{\partial q_1}(\rho \dot{q}_1)\, dq_1\right] dt dq_2 \cdots dp_f$$

$$= -\frac{\partial}{\partial q_1}(\rho \dot{q}_1)\, dt dq_1 dq_2 \cdots dp_f. \tag{5.125}$$

We obtain the total change in the number of systems in this volume during time dt by summing the previous result over all 'sides', labelled $q_1,\dots,q_f,p_1,\dots,p_f$, which gives

$$\frac{\partial \rho}{\partial t} dt dq_1 \cdots dp_f = -\left[\sum_{i=1}^{f} \frac{\partial}{\partial q_i}(\rho \dot{q}_i) + \sum_{i=1}^{f} \frac{\partial}{\partial p_i}(\rho \dot{p}_i)\right] dt dq_1 \cdots dp_f$$

$$\Rightarrow \frac{\partial \rho}{\partial t} = -\sum_{i=1}^{f}\left[\frac{\partial}{\partial q_i}(\rho \dot{q}_i) + \frac{\partial}{\partial p_i}(\rho \dot{p}_i)\right]. \tag{5.126}$$

This result can be rewritten as

$$\frac{\partial \rho}{\partial t} = -\sum_{i=1}^{f}\left[\left(\frac{\partial \rho}{\partial q_i}\dot{q}_i + \frac{\partial \rho}{\partial p_i}\dot{p}_i\right) + \rho\left(\frac{\partial \dot{q}_i}{\partial q_i} + \frac{\partial \dot{p}_i}{\partial p_i}\right)\right]. \tag{5.127}$$

Noting that from the equations of motion (5.124) we have

$$\frac{\partial \dot{q}_i}{\partial q_i} + \frac{\partial \dot{p}_i}{\partial p_i} = \frac{\partial^2 \mathcal{H}}{\partial q_i \partial p_i} - \frac{\partial^2 \mathcal{H}}{\partial p_i \partial q_i} = 0, \tag{5.128}$$

Eq. (5.127) becomes

$$\frac{\partial \rho}{\partial t} = - \sum_{i=1}^{f} \left(\frac{\partial \rho}{\partial q_i} \dot{q}_i + \frac{\partial \rho}{\partial p_i} \dot{p}_i \right). \tag{5.129}$$

We identify the right-hand side of this equation with the implicit time variation of the density when coordinates (q_i, p_i) change with time, and consequently the point representing the system moves in phase space. On the other hand, the left-hand side is the explicit time dependence. The total variation of the density is given by the total derivative, which comprises these two terms:

$$\frac{d\rho}{dt} = \frac{\partial \rho}{\partial t} + \sum_{i=1}^{f} \left(\frac{\partial \rho}{\partial q_i} \dot{q}_i + \frac{\partial \rho}{\partial p_i} \dot{p}_i \right) = 0. \tag{5.130}$$

This result is known as *Liouville's theorem*, and it asserts that *the density of system points in the vicinity of a given system point in phase space is constant while the system evolves over time*. Equation (5.129) is an alternative statement of Liouville's theorem.[4] Our derivation follows essentially that of Reif (1985, App. A.13).

The ensemble of points evolving through phase space is analogous to a fluid flowing in a multidimensional space. In this analogy, the total derivative can be identified as the *material derivative* that gives the density variation when we follow the motion of a fluid element over time. On the other hand, the partial derivative is evaluated at a fixed point, i.e., when we sit at a given location in phase space and measure the density as the ensemble of points flows past us.

Equation (5.129) can be written in terms of the Hamiltonian using Eqs. (5.124):

$$\frac{\partial \rho}{\partial t} = - \sum_{i=1}^{f} \left(\frac{\partial \rho}{\partial q_i} \frac{\partial \mathcal{H}}{\partial p_i} - \frac{\partial \rho}{\partial p_i} \frac{\partial \mathcal{H}}{\partial q_i} \right). \tag{5.131}$$

If we further define the *Poisson brackets* as (Goldstein, 1980, chap. 9)

$$[\rho, \mathcal{H}] = \sum_{i=1}^{f} \left(\frac{\partial \rho}{\partial q_i} \frac{\partial \mathcal{H}}{\partial p_i} - \frac{\partial \rho}{\partial p_i} \frac{\partial \mathcal{H}}{\partial q_i} \right), \tag{5.132}$$

the alternative form of the Liouville theorem, Eq. (5.129), becomes

$$\frac{\partial \rho}{\partial t} = -[\rho, \mathcal{H}]. \tag{5.133}$$

---

[4] We note, as a curiosity, that the French mathematician Joseph Liouville proved the theorem that bears his name in 1838, in the context of the solution of a given class of differential equations, without any reference to mechanics or to phase space, a concept that did not even exist at that time. The relevance of the Liouville theorem to statistical physics was only established later with the works of Jacobi and Boltzmann (Nolte, 2010).

At equilibrium, the distribution over states of the systems in the ensemble does not change in time, i.e., the density of system points in a given region of phase space is constant. The time variation of the density is given by its partial derivative, hence at equilibrium this derivative must vanish. Consequently, from Eq. (5.133) the equilibrium condition is

$$[\rho, \mathcal{H}] = 0. \tag{5.134}$$

In conservative systems the energy is a constant of the motion. In cases where the density only depends on the energy E, we have

$$\frac{\partial \rho}{\partial q_i} = \frac{\partial \rho}{\partial E}\frac{\partial E}{\partial q_i} = 0; \quad \frac{\partial \rho}{\partial p_i} = \frac{\partial \rho}{\partial E}\frac{\partial E}{\partial p_i} = 0, \tag{5.135}$$

and the Poisson brackets of ρ with $\mathcal{H}$ vanish, which ensures equilibrium. In particular, this is consistent with a microcanonical ensemble description, where $\rho = $ constant (i.e., the systems are uniformly distributed over all phase space) when $E_0 < E < E_0 + \delta E$, and $\rho = 0$ otherwise. Note that, if we want to ensure equilibrium in any other case, we can choose a constant of the motion other than the energy.

We have thus demonstrated the usefulness of Liouville's theorem and of the Poisson brackets in classical statistical mechanics. This can be translated into the quantum realm using the density matrix formalism, which is outside the scope of this book (Huang, 1987).

5.C The equipartition theorem

The equipartition theorem is a general result of classical statistical mechanics with important applications in statistical physics and thermodynamics.

Consider, again, a system in equilibrium with a heat reservoir and specified classically by f generalised coordinates and f generalised momenta $q_1, \ldots, q_f, p_1, \ldots, p_f$, and with Hamiltonian $\mathcal{H}(q_1, \ldots, q_f, p_1, \ldots, p_f)$. Such a system is said to have f *degrees of freedom*. The Hamiltonian represents the energy of the system in the microstate in question. The probability to find the system in this particular microstate is given by the canonical expression

$$p(q_1, \ldots, q_f, p_1, \ldots, p_f) = \frac{e^{-\beta \mathcal{H}(q_1, \ldots, q_f, p_1, \ldots, p_f)}}{Z}, \tag{5.136}$$

where the partition function is

$$Z = \int \cdots \int e^{-\beta \mathcal{H}(q_1, \ldots, q_f, p_1, \ldots, p_f)} dq_1 \cdots dq_f dp_1 \cdots dp_f. \tag{5.137}$$

Suppose that the Hamiltonian contains a term that is quadratic in one of the f coordinates or momenta $(q_1, \ldots, q_f, p_1, \ldots, p_f)$. We denote this particular coordinate or momentum by ξ and write the Hamiltonian as:

$$\mathcal{H}(q_1, \ldots, q_f, p_1, \ldots, p_f) = A\xi^2 + \mathcal{H}', \tag{5.138}$$

where A and $\mathcal{H}'$ are independent of ξ, but may depend on any other coordinates or momenta. Substituting Eq. (5.138) into Eqs. (5.136) and (5.137), the probability that the chosen coordinate or momentum will take the value ξ, for arbitrary fixed values of the remaining coordinates or momenta, is

$$p(\xi) = \frac{e^{-\beta A \xi^2}}{\int e^{-\beta A \xi^2} d\xi}. \tag{5.139}$$

The contribution of the ξ-term to the mean energy, while keeping all the other coordinates and momenta constant, is

$$\overline{A\xi^2} = \frac{\int A \xi^2 e^{-\beta A \xi^2} d\xi}{\int e^{-\beta A \xi^2} d\xi} = -\frac{\partial}{\partial \beta} \left[\ln \left(\int e^{-\beta A \xi^2} d\xi \right) \right]. \tag{5.140}$$

The limits of integration depend on system shape and dimensions. However, if we assume that βA is large enough that the Gaussian factor in the integrand decays fast enough when ξ differs substantially from zero, we can take the limits of integration as $-\infty$ and $+\infty$. Using the results of Appendix 1.A.2, Chapter 1, we get

$$\overline{A\xi^2} = \frac{1}{2\beta} = \frac{k_B T}{2}. \tag{5.141}$$

We conclude that each quadratic term in the Hamiltonian contributes $k_B T/2$ to the mean energy of the system, which is a statement of the *equipartition theorem*. Notice that this result does not depend on A, which can be a function of the remaining generalised coordinates and momenta. The derivation of Eq. (5.141) follows essentially Mandl (1988, chap. 7.9).

Now, in many physical systems the Hamiltonian can be cast in diagonal form:

$$\mathcal{H}(q_1,\dots,q_f,p_1,\dots,p_f) = \mathcal{H}(Q_1,\dots,Q_f,P_1,\dots,P_f) = \sum_{i=1}^{f} A_i P_i^2 + B_i Q_i^2, \tag{5.142}$$

where A_i, B_i are constants, and $(Q_1,\dots,Q_f,P_1,\dots,P_f)$ are related to $(q_1,\dots,q_f,p_1,\dots,p_f)$ by an appropriate coordinate transformation. It then follows from the equipartition theorem that

$$\overline{\mathcal{H}(q_1,\dots,q_f,p_1,\dots,p_f)} = f k_B T, \tag{5.143}$$

i.e., that each degree of freedom contributes $k_B T$ to the mean energy of the system. This is exemplified in Problems 5.7 and 5.8

Problems

5.1 Derive Eq. (5.25).

5.2 As an application of Eq. (5.91), show that:

(a) The weight W_h of a column of gas of identical molecules of mass m each in a container of height h and cross-sectional area A, held at temperature T, is given by

$$W_h = Ak_B T n_0 \left(1 - e^{-mgh/k_B T}\right) = Ak_B T (n_0 - n_h),$$

where n_0 and n_h are the densities at the bottom and at the top of the container respectively.

(b) The pressures at the bottom and at the top of the container are, respectively, $P_0 = n_0 k_B T$ and $P_h = P_0 e^{-mgh/k_B T}$. *Hint:* note that $P_0 - P_h = W_h/A$.

5.3 Consider Eq. (5.73).

(a) Show that this equation is satisfied for helium at room temperature ($T = 300$ K) and standard atmospheric pressure (10^5 Pa). The molecular mass of helium is $m = 6.7 \times 10^{-24}$ g. *Hint:* use the ideal gas equation of state to find the number density.

(b) Show that for a free electron gas in a metal at room temperature the same equation is not satisfied. The electron rest mass is $m = 9.1 \times 10^{-28}$ g. Assume that there is one conduction electron per atom and the interatomic distance is 2 Å. Show also that only at a temperature $T \gg 3 \times 10^5$ K would the electron gas be in the classical regime.

5.4 Show that for an ideal gas obeying a Maxwell–Boltzmann distribution, the most probable velocity is given by Eq. (5.84), the mean velocity is given by Eq. (5.85), and the mean square velocity is given by Eq. (5.86). Find the mean translational kinetic energy per molecule and discuss your result in connection with the equipartition theorem.

5.5 Assume that the vibrational motion of a diatomic molecule can be modelled as a one-dimensional harmonic oscillator, whose energy levels are quantised according to

$$\varepsilon_r = \hbar\omega \left(r + \frac{1}{2}\right), \quad r = 0,1,2,\ldots$$

Assume hydrogen to be an ideal gas and compute the relative populations of the first three excited vibrational states at 1000 K and 100 K. Take for the angular frequency of the H_2 molecule $\omega = 1.32 \times 10^{14}$ rad s^{-1}.

5.6 According to quantum mechanics, the rotational energy levels of a rigid diatomic molecule are

$$\varepsilon_r = \frac{\hbar^2}{2I} r(r+1), \quad r = 0,1,2,\ldots$$

where I is the molecular moment of inertia. Level r has degeneracy $g_r = 2r+1$. Assuming ideal gas behaviour, find the most populated rotational level of CO at 20°C. The molecular moment of inertia of CO is $I = 1.46 \times 10^{-48}$ kg m^2.

5.7 Using the model of the preceding problem, find the internal partition function corresponding to the rotational motion of a diatomic ideal gas. Next, calculate the mean rotational kinetic energy per particle, and the molar specific heat at constant volume in the high-temperature limit. Discuss your results in connection with the equipartition theorem. *Hint:* at high enough temperatures the sum can be replaced by an integral.

5.8 Using the model of Problem 5.5, find the internal partition function corresponding to the vibrational motion of a diatomic ideal gas. Next, calculate the mean vibrational kinetic energy per particle, and the molar specific heat at constant volume in the high-temperature limit. Discuss your results, and that for the total c_v (translation + rotation + vibration), in connection with the equipartition theorem.

5.9 Consider an ideal gas composed of N identical molecules with permanent electric dipole moment **p** in an external electric field **E**. The gas is at temperature T and its volume is V.

(a) Find the mean magnitude of the total electric dipole moment of the gas, $\mathcal{P} = \sum_{i=1}^{N} p_i$, using the equation of state

$$\mathcal{P} = \left(\frac{\partial F}{\partial E} \right)_{T,V},$$

where F is the Helmholtz free energy, and $\mathcal{P}$ and E are the magnitudes of the total dipole moment and of the electric field, respectively.

(b) The electric susceptibility χ_E is defined, for linear media, as $\mathscr{P} = \chi_E E$, where $\mathscr{P} = \mathcal{P}/V$ is the polarisation. Because $p \sim 10^{-30}$ C m, we can replace the sum by an integral when calculating the partition function. Show that for reasonable field strengths and temperatures not too low, the electric susceptibility of the gas is given by the classical Langevin formula:

$$\chi_E = \frac{Np^2}{3Vk_BT}.$$

References

Goldstein, H. 1980. *Classical Mechanics*, 2nd edition. Addison-Wesley.
Huang, K. 1987. *Statistical Mechanics*, 2nd edition. Wiley.
Mandl, F. 1988. *Statistical Physics*, 2nd edition. Wiley.
Nolte, D. D. 2010. The tangled tale of phase space. *Physics Today* **63**, 33–38.
Reif, F. 1985. *Fundamentals of Statistical and Thermal Physics*. McGraw-Hill.

6 The quantum ideal gas

6.1 Introduction

As an introduction to the study of the quantum ideal gas we give a brief qualitative description of quantum systems of identical particles and show that such particles are genuinely indistinguishable according to quantum mechanics. We then derive the quantum statistics of ideal gases in the grand canonical ensemble, since dropping the constraints of a fixed number of particles (as in the canonical ensemble) or a fixed energy (as in the microcanonical ensemble) substantially simplifies the treatment. Next, we show that in the classical limit of quantum statistics we recover some of the results derived for the classical ideal gas in the preceding chapter using the canonical ensemble description. As applications of quantum statistics, we study the free electron gas in metals (Fermi gas), Bose–Einstein condensation, thermal vibrations of the crystal lattice (Debye theory), and blackbody radiation (Planck's law).

6.2 Systems of identical particles*

Consider a gas of identical particles without internal structure that move freely in a volume V. Let $\Psi_{n_1,n_2,\ldots,n_j,\ldots,n_N}(q_1,q_2,\ldots,q_j,\ldots,q_N)$ be the wavefunction describing a microstate of the gas with N particles (i.e., an eigenfunction of the Hamiltonian of the gas). Generalised coordinate q_j stands for the three position coordinates and spin coordinate of particle j, and n_j denotes the quantum numbers that characterise the individual state of particle j (i.e., that label a state of given linear momentum and spin). From quantum field theory we know that *a system of identical particles with integer spin is described by a symmetric wavefunction* – such particles are called *bosons*; and that *a system of identical particles with half-integer spin is described by an anti-symmetric wavefunction* – such particles are called *fermions*. The different symmetry properties of the wavefunction give rise to different statistics; hence there is a *connection between spin and statistics*.

A *symmetric* wavefunction is invariant under exchange of the coordinates of any two particles:

$$\Psi_S(\ldots,q_i,\ldots,q_j,\ldots) = \Psi_S(\ldots,q_j,\ldots,q_i,\ldots), \tag{6.1}$$

whereas an *anti-symmetric* wavefunction changes sign under the same operation:

$$\Psi_A(\ldots,q_i,\ldots,q_j,\ldots) = -\Psi_A(\ldots,q_j,\ldots,q_i,\ldots). \tag{6.2}$$

On the other hand, when particles i and j are in the same state, the wave function should be invariant under coordinate exchange, as the two states are identical: $\Psi_A(\ldots,q_i,\ldots,q_j,\ldots) = \Psi_A(\ldots,q_j,\ldots,q_i,\ldots)$. However, this relation, together with Eq. (6.2), implies that $\Psi_A = 0$: *a microstate of the gas of identical particles with half-integer spin in which two or more particles are in the same individual state is not allowed*. This is the *Pauli exclusion principle*.

Example: consider a system of two identical, non-interacting and non-localised particles. Let $\hat{H}_{gas}$ be the Hamiltonian of the system. Let $\phi_a(1)$ be the eigenfunction of the Hamiltonian $\hat{H}_1$ of particle 1 associated with the individual energy eigenvalue ε_1, and let $\phi_b(2)$ be the eigenfunction of the Hamiltonian $\hat{H}_2$ of particle 2 associated with the eigenvalue ε_2. For non-interacting particles we have

$$\hat{H}_{gas} = \hat{H}_1 + \hat{H}_2.$$

A wavefunction of the two-particle system can be obtained by multiplying the wavefunctions of the two individual particles:

$$\Psi_{ab}(1,2) = \phi_a(1)\phi_b(2).$$

Using these two relations and noting that each individual Hamiltonian $\hat{H}_i$ acts only on wavefunction $\phi(i)$, we can write the eigenvalue equation for the Hamiltonian of the system in terms of the eigenvalues of the energy of the individual particles:

$$\hat{H}_{gas}\Psi(1,2) = \left(\hat{H}_1 + \hat{H}_2\right)\phi_a(1)\phi_b(2) = (\varepsilon_1 + \varepsilon_2)\,\phi_a(1)\phi_b(2) = E_{gas}\Psi_{ab}(1,2),$$

whence it follows that $E_{gas} = \varepsilon_1 + \varepsilon_2$ is the eigenvalue of the energy of the system associated with the eigenfunction $\Psi_{ab}(1,2) = \phi_a(1)\phi_b(2)$ – as expected, since the two particles do not interact.

We now repeat the above procedure for a symmetric wavefunction:

$$\Psi_S(1,2) = \Psi_{ab}(1,2) + \Psi_{ab}(2,1) = \phi_a(1)\phi_b(2) + \phi_a(2)\phi_b(1),$$

$$\hat{H}_{gas}\Psi_S(1,2) = \hat{H}_{gas}\Psi_{ab}(1,2) + \hat{H}_{gas}\Psi_{ab}(2,1)$$

$$= E_{gas}\Psi_{ab}(1,2) + E_{gas}\Psi_{ab}(2,1) = E_{gas}\Psi_S(1,2).$$

On the other hand,

$$\Psi_S(1,2) = \Psi_S(2,1) \Rightarrow \hat{H}_{gas}\Psi_S(2,1) = E_{gas}\Psi_S(2,1),$$

whence particle exchange does not lead to a new state of the system: *the particles are genuinely indistinguishable in the quantum description*.

Finally, for an anti-symmetric wavefunction we have

$$\Psi_A(1,2) = \Psi_{ab}(1,2) - \Psi_{ab}(2,1) = \phi_a(1)\phi_b(2) - \phi_a(2)\phi_b(1),$$

$$\hat{H}_{gas}\Psi_A(1,2) = \hat{H}_{gas}\Psi_{ab}(1,2) - \hat{H}_{gas}\Psi_{ab}(2,1)$$

$$= E_{gas}\Psi_{ab}(1,2) - E_{gas}\Psi_{ab}(2,1) = E_{gas}\Psi_A(1,2).$$

On the other hand,

$$\Psi_A(1,2) = -\Psi_A(2,1) \Rightarrow \hat{H}_{gas}\Psi_A(2,1) = E_{gas}\Psi_A(2,1),$$

whence also in this case particle exchange does not lead to a new state of the system, and again *the particles are indistinguishable*. However, if both particles are in the same individual state, say state a, then:

$$\Psi_A(1,2) = \Psi_{aa}(1,2) - \Psi_{aa}(2,1) = \phi_a(1)\phi_a(2) - \phi_a(2)\phi_a(1) = 0.$$

We conclude that there is no anti-symmetric microstate of the system in which both particles occupy the same individual state. This is a very simple example of the Pauli exclusion principle.

6.3 Quantum statistics

In the grand canonical ensemble description of the quantum ideal gas, a macrostate of the gas is specified by its temperature, its chemical potential and its volume, while the mean energy and the mean number of particles are to be calculated. However, each microstate of the gas is specified by its energy and its number of particles (as well as by its volume, which is a fixed parameter). In order to simplify our study, we shall only consider the case of a gas of particles without internal structure that move freely inside a volume V, in the absence of any external field. Therefore only the translational kinetic energy needs to be considered.

As in the preceding chapter, we shall describe the microstates of the gas in terms of the occupation numbers of the individual states (see Section 5.3). A microstate of the gas is then specified by a given set of occupation numbers $n_1, n_2, \ldots, n_i, \ldots$ of the individual states, with energies

$$\varepsilon_1, \varepsilon_2, \ldots, \varepsilon_i, \ldots.$$

The energy of the gas microstate specified by a given set of occupation numbers $\{n_1, n_2, \ldots\}$ is thus

$$E(n_1, n_2, \ldots) = \sum_j n_j \varepsilon_j, \tag{6.3}$$

where the sum is over all the individual states of each particle, which comprise both translational states and spin states. For each microstate, the total number of particles is given by

$$N(n_1, n_2, \ldots) = \sum_j n_j. \tag{6.4}$$

This relation means that the number of particles in a given microstate should be understood as a (random) variable, as is the energy of the microstate, in contradistinction to the fixed value it has in the canonical ensemble description of the classical ideal gas, given in the preceding chapter. Consequently, Eq. (6.4) no longer plays the role of a constraint in the calculations that follow.

Using Eqs. (6.3) and (6.4), the grand partition function of the gas, Eq. (4.107),

$$Y = \sum_{N=0}^{\infty} \left[\sum_{r=1}^{\infty} e^{\beta(\mu N - E_{Nr})} \right],$$

where r is label of a microstate, becomes

$$Y = \sum_{n_1, n_2, \ldots} \exp \left\{ \beta \left[\mu \underbrace{(n_1 + n_2 + \cdots)}_{N} - \underbrace{(n_1 \varepsilon_1 + n_2 \varepsilon_2 + \cdots)}_{E(n_1, n_2, \ldots)} \right] \right\}, \tag{6.5}$$

where the sum is over all the possible values of n_1, of n_2, etc. From Eq. (4.108), the probability to find n_1 particles in individual state 1, n_2 particles in individual state 2, etc., is given by

$$p_{Nr} = \frac{e^{\beta(\mu N - E_{Nr})}}{Y},$$

with Eqs. (6.3)–(6.5):

$$p(n_1, n_2, \ldots) = \frac{\exp\{\beta [\mu (n_1 + n_2 + \cdots) - (n_1 \varepsilon_1 + n_2 \varepsilon_2 + \cdots)]\}}{\sum_{n_1, n_2, \ldots} \exp\{\beta [\mu (n_1 + n_2 + \cdots) - (n_1 \varepsilon_1 + n_2 \varepsilon_2 + \cdots)]\}}. \tag{6.6}$$

Regrouping terms in the exponents, this probability writes as

$$p(n_1, n_2, \ldots) = \frac{\exp\{\beta [(\mu - \varepsilon_1) n_1 + (\mu - \varepsilon_2) n_2 + \cdots]\}}{\sum_{n_1, n_2, \ldots} \exp\{\beta [(\mu - \varepsilon_1) n_1 + (\mu - \varepsilon_2) n_2 + \cdots]\}}.$$

Factoring the exponentials and noting that

$$\left(\sum_r A_r \right) \left(\sum_s B_s \right) = (A_1 + A_2 + \cdots)(B_1 + B_2 + \cdots) = \sum_{r,s} A_r B_s,$$

we can rewrite Eq. (6.6) as

$$p(n_1, n_2, \ldots) = \frac{e^{\beta(\mu - \varepsilon_1)n_1} e^{\beta(\mu - \varepsilon_2)n_2} \cdots}{\sum_{n_1} e^{\beta(\mu - \varepsilon_1)n_1} \sum_{n_2} e^{\beta(\mu - \varepsilon_2)n_2} \cdots}. \tag{6.7}$$

Further defining the *grand partition function of the single-particle state j* as

$$Y_j = \sum_{n_j} e^{\beta(\mu - \varepsilon_j)n_j}, \tag{6.8}$$

and using Eqs. (6.5) and (6.7), the grand partition function of the gas can be written as a product of the grand partition functions of each single-particle state:

$$Y = \prod_j Y_j. \tag{6.9}$$

We likewise obtain, for the probability (6.7):

$$p(n_1, n_2, \ldots) = \prod_j \frac{e^{\beta(\mu - \varepsilon_j)n_j}}{Y_j} = \prod_j p_j(n_j). \tag{6.10}$$

This means that the probability that the gas is in a given microstate, i.e., has a given set of occupation numbers, equals a product of factors, each of which depends only on a single individual state, because there are no inter-particle interaction terms in the expression of the energy microstates, i.e., as a consequence of the fact that the gas is ideal. *The probability $p_j(n_j)$ of finding n_j particles in state j is thus independent of the occupation numbers $n_{i \neq j}$ of all the other states.*

From the grand canonical general form of equation of state (4.122), we find the mean occupation number of state j:

$$\overline{n_j} = \frac{1}{\beta} \left(\frac{\partial \ln Y_j}{\partial \mu} \right)_{T,V}. \tag{6.11}$$

In order to find $\overline{n_j}$, we need to calculate Y_j from Eq. (6.8). Depending on the allowed values of the occupation numbers n_j, two cases can be distinguished that will result in different statistics. In the case of a gas of particles with integer spin, i.e., a gas of bosons or Bose gas, there are no restrictions on the number of particles that can occupy each individual state; hence the occupation numbers can take any integer value:

$$n_j = 0, 1, 2, 3, \ldots \quad \text{for all } j \qquad \textit{Bosons.} \tag{6.12}$$

In this case the series in Eq. (6.8) converges if

$$e^{\beta(\mu - \varepsilon_j)} < 1 \Rightarrow \mu < \varepsilon_j, \text{ for all } j.$$

Choosing the zero of the individual energy such that

$$0 \leq \varepsilon_1 \leq \varepsilon_2 \leq \cdots \leq \varepsilon_j \leq \cdots, \tag{6.13}$$

a sufficient condition for convergence is that $\mu < 0$. Under these conditions the series sums to

$$Y_j = \frac{1}{1 - e^{\beta(\mu - \varepsilon_j)}} \qquad \textit{Bosons.} \tag{6.14}$$

And from Eq. (6.11) with Eq. (6.14) we obtain the *Bose–Einstein (BE) statistics*:

$$\overline{n_j} = \frac{1}{e^{\beta(\varepsilon_j - \mu)} - 1} \qquad \textit{Bose–Einstein.} \tag{6.15}$$

In the case of a gas of half-integer spin particles, i.e., a gas of fermions, it follows from the Pauli exclusion principle that there can be at most one particle in each individual state; hence the occupation numbers can take only two values:

$$n_j = 0, 1 \quad \text{for all } j \qquad \textit{Fermions.} \tag{6.16}$$

And from Eq. (6.8) with Eq. (6.16) we find

$$Y_j = 1 + e^{\beta(\mu - \varepsilon_j)} \qquad \textit{Fermions.} \tag{6.17}$$

Substituting this expression into Eq. (6.11) we obtain the *Fermi–Dirac (FD) statistics*:

$$\overline{n_j} = \frac{1}{e^{\beta(\varepsilon_j - \mu)} + 1} \qquad \textit{Fermi–Dirac.} \tag{6.18}$$

The two quantum statistics can be given as a single expression:

$$\bar{n}_j = \frac{1}{e^{\beta(\varepsilon_j - \mu)} \pm 1} \quad \begin{cases} + : & \text{FD} \\ - : & \text{BE} \end{cases} . \tag{6.19}$$

Summing $\bar{n}_j$ over all the individual states, we obtain the mean total number of particles of the gas in the macrostate specified by T, V and μ:

$$\bar{N}(T, V, \mu) = \sum_j \bar{n}_j = \sum_j \frac{1}{e^{\beta(\varepsilon_j - \mu)} \pm 1} \quad \begin{cases} + : & \text{FD} \\ - : & \text{BE} \end{cases} . \tag{6.20}$$

Alternatively, this expression can be obtained directly from the grand canonical equation of state (4.122), combined with Eqs. (6.9), (6.11) and (6.19):

$$\bar{N} = \frac{1}{\beta} \left(\frac{\partial \ln Y}{\partial \mu} \right)_{T,V} = \sum_j \frac{1}{\beta} \left(\frac{\partial \ln Y_j}{\partial \mu} \right)_{T,V} = \sum_j \bar{n}_j.$$

Note that, as discussed in Section 4.4, for macroscopic systems the mean number of particles of the gas can be taken, in practice, as the effective number of particles. This will be very important for understanding the results of quantum statistics applied to particular systems later in this chapter.

The mean energy of the quantum ideal gas can be calculated from Eq. (6.3) with Eq. (6.19):

$$\bar{E} = \sum_j \bar{n}_j \varepsilon_j = \sum_j \frac{\varepsilon_j}{e^{\beta(\varepsilon_j - \mu)} \pm 1} \quad \begin{cases} + : & \text{FD} \\ - : & \text{BE} \end{cases} . \tag{6.21}$$

As mentioned at the beginning of this section, in the above expressions the sums over j include both the translational states, of wavevector $\mathbf{k}$, and the spin states σ:

$$\sum_j \cdots = \sum_{\sigma = -S}^{S} \sum_{\mathbf{k}} \cdots , \tag{6.22}$$

where S is the magnitude of the spin. Since the energy of a free particle does not depend on its spin, the sum over σ gives a factor of $2S + 1$, because $\sigma = -S, -S + 1, \ldots, S - 1, S$. This factor is important, as it means that each translational state may be occupied by $2S + 1$ particles in different spin states.

We shall conclude our study of quantum statistics by considering the case of gases for which the number of particles is not conserved. It is easily shown that the chemical potential of such a gas must vanish. Indeed, consider how the chemical potential is related to the Helmholtz and Gibbs free energies, Eq. (2.47), and the thermodynamic equilibrium conditions in terms of the same potentials, Eqs. (3.58) and (3.64); we get

$$\mu = \left(\frac{\partial F}{\partial N} \right)_{T,V} = \left(\frac{\partial G}{\partial N} \right)_{T,P} = 0. \tag{6.23}$$

This means that the Helmholtz potential (at constant temperature and volume) and the Gibbs potential (at constant temperature and pressure) of the gas are independent of the number of particles. We shall study later two examples of this: the phonon gas and the photon gas, which are both boson gases . We shall therefore use Bose–Einstein statistics (6.15) with $\mu = 0$:

$$\overline{n}_j = \frac{1}{e^{\beta \varepsilon_j} - 1} \qquad Planck. \tag{6.24}$$

This is called the *Planck statistics*.

Finally, note that if the number of particles is not conserved, then the canonical and grand canonical partition functions of the gas are identical:

$$Z(T,V) = \sum_r e^{-\beta E_r(V)} = Y(T,V,\mu = 0). \tag{6.25}$$

6.4 The classical limit

In Chapter 5 we discussed the criterion for the validity of the classical regime. When this criterion is satisfied, both quantum statistics have the same limiting behaviour, known as the *classical limit*. We recall that a sufficient condition for the validity of the classical regime is

$$\overline{n}_j \ll 1, \text{for all } j. \tag{6.26}$$

From the mean occupation number given by Eq. (6.19), clearly this condition is satisfied if

$$e^{-\beta \mu} \gg 1. \tag{6.27}$$

We shall discuss the physical meaning of this condition in the next section. Choosing the zero of the single-particle energy according to Eq. (6.13), then $e^{\beta \varepsilon_j} \geq 1$ and $e^{\beta(\varepsilon_j - \mu)} \geq e^{-\beta \mu} \gg 1$. Under these conditions, the quantum statistics, Eq. (6.19), reduces to

$$\overline{n}_j = e^{\beta(\mu - \varepsilon_j)} \qquad Maxwell\text{–}Boltzmann\ statistics \tag{6.28}$$

This is the classical limit of the quantum statistics and is known as the *Maxwell–Boltzmann statistics*. We now show this is consistent with the corresponding expression derived in Chapter 5 in the canonical ensemble formalism:

$$\overline{n}_j = N\frac{e^{-\beta \varepsilon_j}}{Z_1} = Np_j, \quad Z_1 = \sum_j e^{-\beta \varepsilon_j}, \tag{6.29}$$

where Z_1 is the single-particle partition function and p_j is the canonical probability to find one particle in individual state j. From Eq. (6.28) we can write the chemical potential as a function of the mean total number of particles:

$$\overline{N} = \sum_j \overline{n}_j = e^{\beta \mu} \sum_j e^{-\beta \varepsilon_j} = e^{\beta \mu} Z_1 \Rightarrow e^{\beta \mu} = \frac{\overline{N}}{Z_1}. \tag{6.30}$$

We eliminate the chemical potential by substituting this expression into Eq. (6.28):

$$\overline{n}_j = \overline{N}\frac{e^{-\beta \varepsilon_j}}{Z_1}, \tag{6.31}$$

whence Eq. (6.29) follows upon identifying the mean total number of particles $\overline{N}$ with the thermodynamic number of particles N.

6.5 Continuum-states approximation

In most practical applications, the spacing between translational energy levels is much smaller than the thermal energy and, consequently, the translational energy spectrum can be regarded, to a good approximation, as continuous. In this case, as discussed in Section 5.2, the sum over translational states with linear momentum $\mathbf{p}$ can be replaced by an integral, as follows:

$$\sum_{\mathbf{p}} \cdots \rightarrow \frac{V}{h^3} \int_{-\infty}^{+\infty} d^3 p \cdots = \frac{V}{h^3} \int_0^{+\infty} 4\pi p^2 dp \cdots,$$

whence

$$\sum_{\mathbf{p}} \overline{n}_{\mathbf{p}} \rightarrow \frac{4\pi V}{h^3} \int_0^{+\infty} \overline{n}(p) p^2 dp.$$

In this expression we identify the density of translational states as a function of the linear momentum magnitude: $g(p)dp$ given by Eq. (5.11). We can therefore write, for spinless particles:

$$\overline{N} = \int_0^{+\infty} \overline{n}(p) g(p) dp, \quad g(p)dp = \frac{4\pi V}{h^3} p^2 dp, \tag{6.32}$$

leading to the following conclusion: *in the continuum-states approximation, we replace the mean number of particles in individual state j, $\overline{n}_j$, given by the quantum statistical distribution (6.19) or by the classical limit (6.28), with the mean number of particles with energies in the interval $(\varepsilon, \varepsilon + d\varepsilon)$:*

$$dN(\varepsilon) = \overline{n}(\varepsilon) g(\varepsilon) d\varepsilon, \tag{6.33}$$

where $\overline{n}(\varepsilon)$ follows from Eq. (6.19) or (6.28), taking ε as a continuous variable. In Eq. (6.33), $g(\varepsilon)d\varepsilon$ is the number of states with energies in $(\varepsilon, \varepsilon + d\varepsilon)$, with $g(\varepsilon)$ the appropriate density of states. To calculate the mean total number of particles in the gas, $\overline{N} = \sum_j \overline{n}_j$ is replaced by

$$\overline{N} = (2S+1) \int_0^{+\infty} dN(\varepsilon), \tag{6.34}$$

where the factor $(2S+1)$ counts the number of spin states. For a boson gas, a non-negligible fraction of the particles may occupy the state of zero linear momentum, which is not included in Eq. (6.34) because the density of translational states vanishes when $p = 0$. In this case, the missing term, i.e., the mean number of particles in the ground state, has to be added to the right-hand side of Eq. (6.34). We shall follow this procedure when studying Bose–Einstein condensation later in this chapter.

Likewise, the mean energy of the gas is obtained by replacing the sum in Eq. (6.21) with an integral. Defining the mean energy in the interval $(\varepsilon, \varepsilon + d\varepsilon)$ as

$$dE(\varepsilon) = \varepsilon \bar{n}(\varepsilon)g(\varepsilon)d\varepsilon, \tag{6.35}$$

we get

$$\bar{E} = (2S+1)\int_0^{+\infty} dE(\varepsilon). \tag{6.36}$$

6.5.1 Classical limit of the quantum ideal gas

We recall that we are considering the case of a gas of particles without internal structure and in the absence of any external fields. The particles are free to move inside a volume V; hence the individual states are only translational motion states.

We wish to find the mean total number of particles. This is

$$\bar{N} = (2S+1)V\underbrace{\left(\frac{2\pi m k_B T}{h^2}\right)^{3/2}}_{z} e^{\beta\mu}, \tag{6.37}$$

where z is the translational partition function (5.50) of a particle that moves freely in a box of volume V, which we derived in the preceding chapter. To prove Eq. (6.37), start with Eqs. (6.34) and (6.33). then use the Maxwell–Boltzmann statistics (6.28) written in terms of the continuous variable ε:

$$\bar{n}_{MB}(\varepsilon) = e^{\beta(\mu-\varepsilon)}, \tag{6.38}$$

and the density of states (5.11) with $p = \sqrt{2m\varepsilon}$ (non-relativistic particles):

$$g(\varepsilon)d\varepsilon = \frac{2\pi V}{h^3}(2m)^{3/2}\varepsilon^{1/2}d\varepsilon. \tag{6.39}$$

We further leave it as an exercise to show that, from Eqs. (6.35), (6.36), (6.38), (6.39) and (6.37), the mean total energy of the gas is

$$\bar{E} = \frac{3}{2}\bar{N}k_B T. \tag{6.40}$$

As discussed before, for macroscopic systems we can equate the mean total number of gas particles with its thermodynamic value, and therefore set $\bar{N} = N$. This allows us to identify Eq. (6.40) as the mean energy of the classical ideal gas (5.61), derived in Chapter 5.

Proceeding as in the discrete case, from Eq. (6.37) we can write the chemical potential of the gas in terms of the mean total number of particles:

$$e^{-\beta\mu} = (2S+1)\frac{z}{\bar{N}}, \quad z = V\left(\frac{2\pi m k_B T}{h^2}\right)^{3/2}. \tag{6.41}$$

Setting $\overline{N} = N$ in this equation, in the classical limit of quantum statistics, Eq. (6.27), we recover the criterion for the validity of the classical regime, Eq. (5.68):

$$e^{-\beta\mu} = (2S+1)\frac{V}{N}\left(\frac{2\pi m k_B T}{h^2}\right)^{3/2} \gg 1 \Rightarrow \frac{V}{N}\left(\frac{2\pi m k_B T}{h^2}\right)^{3/2} \gg 1. \tag{6.42}$$

As discussed in the preceding chapter, this criterion is satisfied at high temperatures and low densities N/V.

Equation (6.41) can be solved for the chemical potential of the gas with $\overline{N} = N$, yielding a generalisation of Eq. (5.60) that takes into account the number of spin states:

$$\mu = -k_B T \ln\left[(2S+1)\frac{V}{N}\left(\frac{2\pi m k_B T}{h^2}\right)^{3/2}\right]. \tag{6.43}$$

Finally, we write the mean number of particles with energies in $(\varepsilon, \varepsilon + d\varepsilon)$ by combining Eq. (6.33) with Eqs. (6.38), (6.39) and (6.41), and writing $\varepsilon = p^2/2m$:

$$dN(p) = \frac{\overline{N} 4\pi p^2 e^{-p^2/2mk_B T}}{(2\pi m k_B T)^{3/2}} dp. \tag{6.44}$$

Dividing this equation by $\overline{N}$, we obtain the Maxwell distribution as a function of the magnitude of the linear momentum, Eq. (5.81):

$$f(p)dp = \frac{4\pi p^2 e^{-p^2/2mk_B T}}{(2\pi m k_B T)^{3/2}} dp. \tag{6.45}$$

We have thus reproduced the results derived in Chapter 5 for the classical ideal gas with a fixed number of particles.

The thermodynamic quantities can also be easily calculated in the grand canonical description. We start with the grand canonical partition function of the gas written in the form (4.109):

$$Y(T,V,\mu) = \sum_N Z(T,V,N)e^{\beta\mu N}, \tag{6.46}$$

where $Z(T,V,N)$ is the partition function for each microstate of the gas with N particles. In Chapter 5 we showed that a gas in the classical regime is well described by the Maxwell–Boltzmann partition function (5.33):

$$Z(T,V,N) = \frac{z^N}{N!}, \tag{6.47}$$

where $z \equiv z(T,V)$ is the single-particle partition function. Combining Eqs. (6.46) and (6.47), we get

$$Y(T,V,\mu) = \sum_N \frac{1}{N!}\left[z(T,V)e^{\beta\mu}\right]^N = \exp\left[z(T,V)e^{\beta\mu}\right]. \tag{6.48}$$

From Eq. (4.112) together with Eqs. (6.48) and (4.115), the grand canonical potential is written:

$$J(T,V,\mu) = -PV = -k_B T \ln Y(T,V,\mu) = -k_B T\left[z(T,V)e^{\beta\mu}\right]. \tag{6.49}$$

On the other hand, from equation of state (4.121) with Eq. (6.49) we find

$$N = -\left(\frac{\partial J}{\partial \mu}\right)_{T,V} = z(T,V)e^{\beta\mu}, \qquad (6.50)$$

which, upon substitution in Eq. (6.49), leads to the ideal gas equation of state (5.58), without the need to calculate the single-particle partition function:

$$PV = Nk_B T. \qquad (6.51)$$

As discussed in the preceding chapter, the derivation of equation of state (6.51) allows us to identify the statistical temperature with the ideal gas temperature, i.e., the thermodynamic temperature.

6.6 The ideal Fermi gas

We shall call a system of non-interacting particles obeying the Fermi–Dirac statistics an *ideal Fermi gas*. We wish to study the properties of the FD function (6.18) that gives the mean occupation number of microstate j of energy ε_j at temperature T:

$$\overline{n}_j = \frac{1}{e^{\beta(\varepsilon_j - \mu)} + 1}.$$

Clearly, from Eq. (6.16) we must have

$$0 \le \overline{n}_j \le 1. \qquad (6.52)$$

We define the *Fermi energy* ε_F of the gas as the value of its chemical potential at $T = 0$, i.e., $\beta = \infty$. This will be shown to be positive when studying the free electron gas later in this chapter. We get

$$\overline{n}_j = \begin{cases} 1, & \text{if } \varepsilon_j < \varepsilon_F \\ 0, & \text{if } \varepsilon_j > \varepsilon_F \end{cases}. \qquad (6.53)$$

Hence, in the limit $T = 0$ all states with energy less than ε_F are occupied and all states with energy greater than ε_F are empty. In this limit, the Fermi–Dirac function becomes a step function. The Fermi energy is then the *highest energy level that is occupied at the absolute zero of temperature*.

In Figure 6.1b we show how the Fermi–Dirac function changes when the temperature departs from zero: as the temperature increases, only those fermions that are near the top of the occupied states at $T = 0$ K are excited by the thermal energy $k_B T$ and 'jump' to the next unoccupied states of energies higher than ε_F, with a consequent rounding of $\overline{n}_j(\varepsilon)$ around the Fermi energy.

A Fermi gas at $T = 0$ K is said to be *completely degenerate*. The *Fermi temperature* T_F is defined as

$$\varepsilon_F = k_B T_F. \qquad (6.54)$$

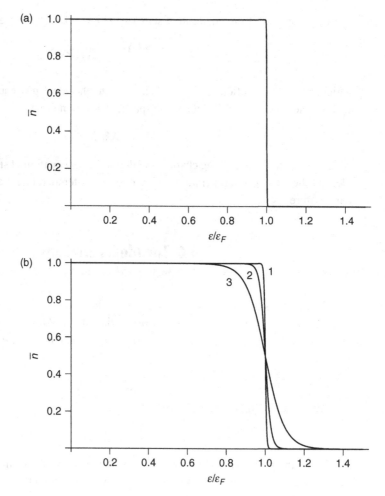

Figure 6.1 Plot of the Fermi–Dirac function (6.18) vs $\varepsilon/\varepsilon_F$, at: (a) $T=0$; (b) $T = 3.7 \times 10^{-3}T_F$ (curve 1), $T = 1.47 \times 10^{-2}T_F$ (curve 2), and $T = 5 \times 10^{-2}T_F$ (curve 3)(see Section 6.6.1 for a discussion of the temperatures used in the case of copper).

For a gas at a temperature much lower than T_F, the mean occupation number $\bar{n}(\varepsilon)$ behaves very much as it does at $T = 0$ K, as shown in Figure 6.1b. A Fermi gas under these conditions is said to be *extremely degenerate*. This has important consequences for the description of the conduction properties of metals at room temperatures using the free-electron model, for which the Fermi temperatures are of order 10^4-10^5 K, as we shall see later.

For macroscopic systems the energy can often be regarded as a continuous variable, as discussed before. Thus in the continuum-states approximation, from Eqs. (6.33), (6.39) and

$$\bar{n}(\varepsilon) = \frac{1}{e^{\beta(\varepsilon-\mu)} + 1},$$

(6.55)

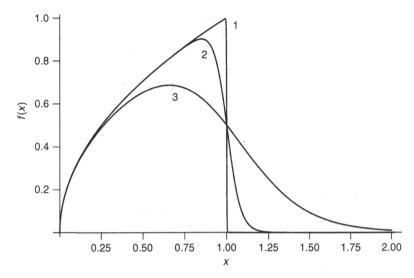

Figure 6.2 Plots of the dimensionless function $f(x)$, Eqs. (6.57) and (6.58), vs x for (1) $k_B T = 0$, (2) $k_B T = 0.04\mu$ and (3) $k_B T = 0.2\mu$, See the text for details.

the mean number of fermions with (non-relativistic) kinetic energy in the interval $(\varepsilon, \varepsilon + d\varepsilon)$ is

$$dN(\varepsilon) = \frac{2\pi V}{h^3}(2m)^{3/2}\frac{\varepsilon^{1/2}}{e^{\beta(\varepsilon-\mu)} + 1}d\varepsilon. \tag{6.56}$$

This is called the *Fermi–Dirac distribution*. We illustrate it in Figure 6.2, where we plot the non-dimensional function $f(x)$, defined for $T = 0$ as

$$f(x) \equiv \frac{h^3}{2\pi V(2m\varepsilon_F)^{3/2}}\frac{dN(x)}{dx} = \begin{cases} x^{1/2}, & 0 \leq x < 1 \\ 0. & x > 1 \end{cases}, \quad x = \varepsilon/\varepsilon_F; \tag{6.57}$$

and for $T \neq 0$ as

$$f(x) \equiv \frac{h^3}{2\pi V(2m\mu)^{3/2}}\frac{dN(x)}{dx} = \frac{x^{1/2}}{e^{\beta\mu(x-1)} + 1}, \quad x = \varepsilon/\mu, \tag{6.58}$$

At $T = 0$ K, states with energy greater than the Fermi energy are unoccupied and the mean occupation number of states with $\varepsilon \leq \varepsilon_F$ is 1; hence the mean total number of particles in the gas, equation (6.34), is found from

$$\overline{N} = (2S + 1)\frac{2\pi V}{h^3}(2m)^{3/2}\int_0^{\varepsilon_F}\varepsilon^{1/2}d\varepsilon. \tag{6.59}$$

Setting $\overline{N} = N$ and solving for ε_F after performing the integral, we get for the Fermi energy of the gas:

$$\varepsilon_F = \frac{h^2}{2m}\left[\frac{6}{8\pi(2S+1)}\frac{N}{V}\right]^{2/3} = \frac{\hbar^2}{2m}\left(\frac{6\pi^2}{2S+1}\frac{N}{V}\right)^{2/3}. \tag{6.60}$$

This result shows that the Fermi energy depends on both the density and the mass of the gas particles.

Calculation of the mean kinetic energy at zero temperature follows along the same lines as that of the mean total number of particles. From Eq. (6.36) with Eqs. (6.35) and (6.39), and recalling that $\bar{n}(\varepsilon) = 1$ for $\varepsilon \leq \varepsilon_F$ and $\bar{n}(\varepsilon) = 0$ for $\varepsilon > \varepsilon_F$, we find

$$\overline{E_0} = (2S+1) \int_0^{\varepsilon_F} \varepsilon g(\varepsilon) d\varepsilon. \tag{6.61}$$

We leave it as an exercise for readers to show that, setting $\overline{N} = N$ and using Eq. (6.60), Eq. (6.61) yields

$$\overline{E_0} = \frac{3}{5} N \varepsilon_F. \tag{6.62}$$

Note that the mean energy of a Fermi gas, unlike that of the classical ideal gas, is non-vanishing at the absolute zero of temperature. Whereas in the classical gas all the particles can reside simultaneously in the ground state, which is usually taken as the zero of energy, this is not possible for fermions, as it would violate the Pauli exclusion principle.

We also leave it as an exercise to show that the pressure of a Fermi gas at $T = 0$ K is

$$P_0 = \frac{2}{5} \frac{N}{V} \varepsilon_F. \tag{6.63}$$

Thus the pressure of the Fermi gas, like its energy, does not vanish at $T = 0$ K, again unlike that of the classical ideal gas. This *Fermi pressure* explains why ordinary matter is rather incompressible, even though the internal structure of solids and liquids is very different from an ideal Fermi gas at the absolute zero of temperature. In rocky planets like Earth, the incompressibility of ordinary matter prevents it from collapsing under gravitation. In neutron stars the balance between the Fermi pressure of the neutrons and the gravitational pressure plays a similar role. On the other hand, in stars like the Sun, plasma at $T \sim 5 \times 10^7$ K generates a classical kinetic pressure that balances the gravitational pressure.

The *Fermi wavevector* k_F is defined by

$$\varepsilon_F = \frac{\hbar^2 k_F^2}{2m}. \tag{6.64}$$

From Eq. (6.60) for electrons ($S = 1/2$) we get

$$k_F = \left(3\pi^2 \frac{N}{V}\right)^{2/3}. \tag{6.65}$$

The Fermi wavevector is thus a function of the particle density only, and defines the *Fermi sphere* of radius k_F in k-space. In the ground state of a free electron gas (see the next section), the occupied states fill this sphere.

6.6.1 The free electron gas

Some important physical properties of metals can be understood on the basis of the free electron gas model. We know from quantum physics that the valence electrons in metals tend to be delocalised, i.e., not bound to any particular atom. Consequently, when a potential difference is applied to a metal sample, the valence electrons acquire a collective

Table 6.1 Calculated Fermi energies and Fermi temperatures of some metals (Kittel, 1995, chap. 6).

Element	N/V (10^{22} cm^{-3})	ε_F (eV)	T_F (10^4 K)
Lithium	4.70	4.72	5.48
Copper	8.45	7.00	8.12
Silver	5.85	5.48	6.36
Zinc	13.10	9.39	10.90

motion in a given direction and an electric current is thus established. These electrons are therefore called *conduction electrons*. In a first approximation the conduction electrons in metals can be considered to move freely inside the sample, because their motion is nearly unperturbed by the ions of the crystal lattice and they suffer few collisions with other conduction electrons. In a quantum description of the motion of electrons in metals, the electron wavefunctions propagate almost freely in the periodic potential of the ions of a perfect crystal lattice at $T = 0$ K. Lattice imperfections such as defects, impurities or thermal vibrations perturb the propagation of the electronic wavefunctions. In a very pure sample and at low temperature, the mean free path of an electron can become much larger than the interatomic separation. On the other hand, when two electrons collide they are scattered into two new individual energy states, which is only possible if there are unoccupied states, according to the Pauli exclusion principle. However, as a consequence of the Fermi–Dirac distribution, most of the accessible states are already occupied; hence the probability that two conduction electrons will collide is very low. For a more thorough discussion, see, e.g., Kittel (1995, chap. 6).

It follows from the above that, to a first approximation, the conduction electrons in a metal can be regarded as a quantum ideal gas of particles that move freely inside the volume of the sample and obey Fermi–Dirac statistics. Because electrons have spin $S = 1/2$, the free electron gas is studied by setting $2S + 1 = 2$ in the expressions derived earlier for the general case of an ideal Fermi gas. In this book we are mainly concerned with the basic features of the free electron model; further applications to the (electric and thermal) conductivity of metals are left to a course in solid state physics.

We can estimate the order of magnitude of ε_F and T_F in metals from Eqs. (6.54) and (6.60): taking $N/V \simeq 5 \times 10^{22}$ cm^{-3} (corresponding to an atomic spacing $(V/N)^{1/3} = 2.7$ Å and one conduction electron per atom), we get $\varepsilon_F = 4.5$ eV and $T_F = 5.48 \times 10^4$ K. Table 6.1 lists the Fermi energies and Fermi temperatures of a few metals calculated using the free electron gas model.

For copper ($T_F = 8.12 \times 10^4$ K), the three curves shown in Figure 6.1b correspond to (1) $T \simeq 300$ K, (2) $T \simeq 1200$ K and (3) $T \simeq 4 \times 10^3$ K. As the melting point of copper is 1356 K, curve (3) is not physical for this material.

In view of the order of magnitude of the Fermi temperatures of ordinary metals, at room temperatures the mean occupation number of the individual states behaves almost like at $T = 0$ K (recall Figure 6.1b). Only at high temperatures do we expect a significant departure from zero-temperature behaviour. This has important consequences for the description of the conduction properties of metals at room temperature using the free electron gas model.

Indeed, because the Fermi temperatures of metals are of order 10^4–10^5 K, we expect that a good approximation for the physical properties of metals at room temperature can be obtained by retaining just the lower-order terms in their series expansions. For this purpose, consider the integral

$$I = \int_0^{+\infty} \frac{\phi(\varepsilon)d\varepsilon}{e^{\beta(\varepsilon-\mu)}+1},$$ (6.66)

where $\phi(\varepsilon)$ is a continuously differentiable function of ε. At low enough temperatures this integral can be approximated by a series expansion about $T=0$ truncated at second order, known as the *Sommerfeld expansion* (see e.g., Christman (1988, app. D)):

$$I \simeq \int_0^\mu \phi(\varepsilon)d\varepsilon + \frac{\pi^2}{6}(k_BT)^2 \frac{d\phi(\varepsilon)}{d\varepsilon}\bigg|_{\varepsilon=\mu} + \mathcal{O}\left(T^4\right).$$ (6.67)

We now use this expansion to calculate the mean total number of electrons in the gas for $T \ll T_F$, from Eqs. (6.34) and (6.56) with $S = 1/2$. Since at low enough temperatures $\mu \simeq \varepsilon_F$, in a first approximation we can replace the derivative at $\varepsilon = \mu$ in the second-order term by the derivative at $\varepsilon = \varepsilon_F$, whence we get

$$\overline{N} \simeq \frac{4\pi V}{h^3}(2m)^{3/2}\left[\frac{2}{3}\mu^{3/2} + \frac{\pi^2}{6}(k_BT)^2\frac{1}{2}\varepsilon_F^{-1/2}\right].$$ (6.68)

Setting $T = 0$ in the above equation we conclude that $\mu(0) > 0$, and consequently that the Fermi energy is a positive quantity, which fact was used when deriving Eq. (6.53).

As discussed earlier, in Eq. (6.68) we can set $\overline{N} = N$ with negligible error. Since N is temperature-independent, we can equate Eq. (6.68) with the result of integrating Eq. (6.59) with $S = 1/2$ at $T = 0$ K, and solve the resulting equation to find the chemical potential as a function of the temperature. Using the definition of Fermi temperature, Eq. (6.54), we obtain

$$\mu = \varepsilon_F\left[1 - \frac{3}{2}\frac{\pi^2}{12}\left(\frac{T}{T_F}\right)^2\right]^{2/3} \simeq \varepsilon_F\left[1 - \frac{\pi^2}{12}\left(\frac{T}{T_F}\right)^2\right], T \ll T_F$$ (6.69)

where the second, approximate, equality follows from expanding the power $2/3$ to first order in $(T/T_F)^2$.

Next, we again use expansion (6.67) to calculate the mean energy from Eqs. (6.35), (6.36), (6.39) and (6.55) with $S = 1/2$. Further replacing the derivative at $\varepsilon = \mu$ by the derivative at $\varepsilon = \varepsilon_F$ in the second-order term, we obtain

$$\overline{E} = \frac{4\pi V}{h^3}(2m)^{3/2}\left[\int_0^\mu \varepsilon^{3/2}d\varepsilon + \frac{\pi^2}{6}(k_BT)^2\frac{3}{2}\varepsilon_F^{1/2}\right].$$ (6.70)

Using the first equality in Eq. (6.69) for the chemical potential, the integral in Eq. (6.70) yields

$$\int_0^\mu \varepsilon^{3/2}d\varepsilon = \frac{2}{5}\mu^{5/2} = \frac{2}{5}\varepsilon_F^{5/2}\left[1 - \frac{3}{2}\frac{\pi^2}{12}\left(\frac{T}{T_F}\right)^2\right]^{5/3}$$

$$\simeq \frac{2}{5}\varepsilon_F^{5/2}\left[1 - \frac{5}{2}\frac{\pi^2}{12}\left(\frac{T}{T_F}\right)^2\right], \quad T \ll T_F.$$

Inserting this into Eq. (6.70), we find

$$\overline{E} = \frac{4\pi V}{h^3}(2m)^{3/2}\left[\frac{2}{5}\varepsilon_F^{5/2} + \frac{\pi^2}{6}(k_BT)^2\varepsilon_F^{1/2}\right]. \qquad (6.71)$$

If we now write the Fermi energy as a function of $\overline{N}$ using Eq. (6.60), and combine this with the definition of Fermi temperature and Eq. (6.62) for the mean energy at absolute zero $\overline{E_0}$ in terms of ε_F, then Eq. (6.71) can be rewritten as

$$\overline{E} = \overline{E_0} + \frac{\pi^2}{4}\left(\frac{T}{T_F}\right)^2 N\varepsilon_F, \quad T \ll T_F. \qquad (6.72)$$

The heat capacity at constant volume of the free electron gas in the same approximation is easily derived from the mean energy:

$$C_V = \left(\frac{\partial \overline{E}}{\partial T}\right)_V = \frac{\pi^2}{2}Nk_B\frac{T}{T_F} = \frac{\pi^2}{2}Nk_B\frac{k_BT}{\varepsilon_F}, \quad T \ll T_F, \qquad (6.73)$$

where the definition of T_F, Eq. (6.54) was used for the final result. Because $k_BT \ll \varepsilon_F$, this means that only a small fraction of the electrons are thermally excited, i.e., only those electrons near the Fermi surface. This is reflected in the shape of the Fermi–Dirac distribution for $T \ll T_F$, as can be seen in curve 2 of Figure 6.2, and is a consequence of the Pauli exclusion principle, as discussed in the preceding section. Note that $C_V \propto Nk_BT/T_F \ll Nk_B$, thus is much smaller than the classical prediction.

From Eq. (6.73) we see that the contribution of the conduction electrons to the heat capacity of the metal is proportional to the temperature at low temperatures. In addition to this contribution there is a term arising from the thermal vibrations of the crystal lattice. Debye theory predicts this latter term to be proportional to T^3 at low temperatures (see Section 6.7.2). Fitting a function of the form

$$C_V = \alpha T + \gamma T^3, \qquad (6.74)$$

to experimental data for the heat capacity vs temperature, the Fermi energy can be found from α using Eq. (6.73). Figure 6.3 shows one such plot.

6.7 The ideal Bose gas

In 1924, the Indian physicist Satyendra Nath Bose sent Einstein an article in which Bose derived the Planck law of radiation by treating photons as a gas of identical particles, a method that we will follow here and that we shall also apply to the study of thermal vibrations of the crystal lattice in solids (phonon gas). Einstein generalised Bose's theory to describe an ideal gas of atoms or molecules and predicted that, at low enough temperatures, the particles would be confined to the lowest quantum state of the system (Pais, 1982, chap. 23). This phenomenon occurs for particles of integer spin only, i.e., for bosons, and is known as *Bose–Einstein condensation*, which will be discussed later.

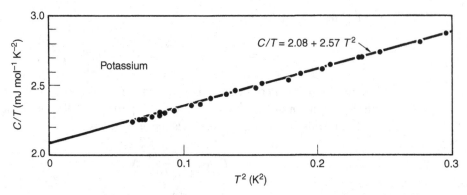

Figure 6.3 Experimentally measured heat capacity of potassium, plotted as C_V/T vs T^2. The straight line is the fit to Eq. (6.74) (From Kittel (1995, chap. 6). Figure adapted from Lien, W. H., and Phillips, N. E. 1964. Low-temperature heat capacities of potassium, rubidium, and cesium. *Phys. Rev.* **133**, A1370–A1377. © APS, https://journals.aps.org/pr/abstract/10.1103/PhysRev.133.A1370.)

An *ideal Bose gas* is a system of indistinguishable non-interacting bosons. To such a system the Bose–Einstein statistics (6.15) applies:

$$\overline{n}_j = \frac{1}{e^{\beta(\varepsilon_j - \mu)} - 1}.$$

The properties of this function are such that

$$\overline{n}_j \geq 0 \Rightarrow e^{\beta(\varepsilon_j - \mu)} \geq 1 \Rightarrow \varepsilon_j \geq \mu. \tag{6.75}$$

If this condition is satisfied in the ground state, then it will be satisfied in all other states. For a gas obeying Bose–Einstein statistics, if we set the energy of the ground state to zero, then from Eq. (6.13) it follows that $\mu < 0$; hence condition (6.75) is satisfied.

In the continuum-states approximation, from Eqs. (6.33), (6.39) and

$$\overline{n}(\varepsilon) = \frac{1}{e^{\beta(\varepsilon - \mu)} - 1}, \tag{6.76}$$

the mean number of bosons with (non-relativistic) kinetic energy in the interval $(\varepsilon, \varepsilon + d\varepsilon)$ is

$$dN(\varepsilon) = \frac{2\pi V}{h^3} (2m)^{3/2} \frac{\varepsilon^{1/2}}{e^{\beta(\varepsilon - \mu)} - 1} d\varepsilon. \tag{6.77}$$

As discussed in Section 6.5, in the ideal Bose gas a finite fraction of the particles can occupy the state of zero kinetic energy, which is not included in Eq. (6.77) because the density of translational states (6.39) vanishes when $\varepsilon = 0$. Therefore, for a correct calculation of the mean total number of particles in the gas, the mean number of particles in the ground state, $\overline{n}_0$, must be added. This quantity can be evaluated from Eq. (6.11) using the grand canonical partition function of a single boson in the ground state, Y_0 from Eq. (6.14) (i.e., setting $\varepsilon_0 = 0$). We find

$$\overline{n}_0 = \frac{1}{\beta} \left(\frac{\partial \ln Y_0}{\partial \mu} \right) = \frac{e^{\beta \mu}}{1 - e^{\beta \mu}}. \tag{6.78}$$

The mean total number of particles in the gas is obtained from Eq. (6.34) with (6.77), to which $\overline{n_0}$ given by (6.78) must be added:

$$\overline{N} = (2S+1) \left[\frac{2\pi V}{h^3} (2m)^{3/2} \int_0^{+\infty} \frac{\varepsilon^{1/2}}{e^{\beta(\varepsilon-\mu)} - 1} d\varepsilon + \frac{e^{\beta\mu}}{1 - e^{\beta\mu}} \right]. \tag{6.79}$$

The mean kinetic energy of the Bose gas is likewise derived from Eq. (6.36), with Eqs. (6.35), (6.39) and (6.76):

$$\overline{E} = (2S+1) \frac{2\pi V}{h^3} (2m)^{3/2} \int_0^{+\infty} \frac{\varepsilon^{3/2}}{e^{\beta(\varepsilon-\mu)} - 1} d\varepsilon. \tag{6.80}$$

6.7.1 Bose–Einstein condensation

Bose–Einstein condensation is the most remarkable property of a Bose gas. It is a phase transition related to the fact that a finite fraction of the particles can occupy the state with zero linear momentum. Bose–Einstein condensation is underpinned by the temperature dependence of the chemical potential: when the chemical potential tends to zero (from below), it follows from Eq. (6.78) that the mean number of particles in the ground state diverges, and therefore the corresponding term in the mean total number of particles (6.79) will dominate. We shall perform a quantitative analysis of this transition by setting $\overline{N} = N$; for simplicity, we consider zero-spin particles (which are defined to be bosons). We thus set $S = 0$ in all formulas derived in Section 6.7. Furthermore, we define:

$$\phi = e^{\beta\mu} \quad \text{(fugacity)}, \tag{6.81}$$

$$v = \frac{V}{N} \quad \text{(specific volume)}, \tag{6.82}$$

$$\lambda = \sqrt{\frac{2\pi \hbar^2}{mk_B T}} \quad \text{(thermal wavelength)}, \tag{6.83}$$

$$x = \frac{p}{\sqrt{2mk_B T}} \quad \text{(reduced linear momentum)}, \tag{6.84}$$

$$\mathrm{Li}_{3/2}(\phi) = \frac{4}{\sqrt{\pi}} \int_0^{+\infty} \frac{x^2}{\phi^{-1} e^{x^2} - 1} dx$$

$$= \sum_{k=1}^{\infty} \frac{\phi^k}{k^{3/2}} \quad \text{(polylogarithm function)}. \tag{6.85}$$

With these definitions, Eq. (6.79) is written:

$$\frac{1}{v} = \frac{1}{\lambda^3} \mathrm{Li}_{3/2}(\phi) + \frac{1}{V} \frac{\phi}{1 - \phi}. \tag{6.86}$$

Using definition (6.81) and Eq. (6.78), this becomes

$$\lambda^3 \frac{\overline{n_0}}{V} = \frac{\lambda^3}{v} - \mathrm{Li}_{3/2}(\phi). \tag{6.87}$$

From Eq. (6.81), $\mu \to 0 \Rightarrow \phi \to 1$, hence the fugacity is such that $\mu < 0 \Rightarrow 0 \le \phi \le 1$. In the upper limit, $\phi = 1$ and the polylogarithm function reduces to

$$\text{Li}_{3/2}(1) = \sum_{k=1}^{\infty} \frac{1}{k^{3/2}} = \zeta\left(\frac{3}{2}\right) \simeq 2.612, \tag{6.88}$$

where ζ is the Riemann zeta function. When the temperature and the specific volume are such that

$$\frac{\lambda^3}{v} > \zeta\left(\frac{3}{2}\right), \tag{6.89}$$

we have $\overline{n_0} > 0$. This means that *a finite fraction of the particles occupy the $p = 0$ state.* Condition (6.89) defines a sub-space in the thermodynamic space $P - v - T$ of the boson gas that we shall refer to as the *Bose–Einstein condensation region*, bounded by the surface

$$\frac{\lambda^3}{v} = \zeta\left(\frac{3}{2}\right). \tag{6.90}$$

For a given specific volume v, Eq. (6.90) defines a critical wavelength λ_c and, from Eq. (6.83), a critical temperature T_c:

$$\lambda_c^3 = v\zeta\left(\frac{3}{2}\right), \tag{6.91}$$

$$k_B T_c = \frac{2\pi \hbar^2 / m}{[v\zeta(3/2)]^{2/3}}. \tag{6.92}$$

Thus at the critical temperature, the thermal wavelength is of the order of the mean inter-particle distance ($\sim v^{1/3}$). On the other hand, at some temperature T, Eq. (6.91) defines a critical specific volume v_c:

$$v_c = \frac{\lambda^3}{\zeta(3/2)}. \tag{6.93}$$

Condensation therefore occurs when $T < T_c$ or $v < v_c$.

Equation (6.87) can be solved graphically for ϕ as a function of T and v. We shall do this in the *thermodynamic limit* by taking $V \to \infty$ at constant v, i.e., at constant density, which implies $N \to \infty$, This approximation introduces an error of order $1/V$ (Huang, 1987, chap. 12.3).

As can be seen from Figure 6.4, the solutions of Eq. (6.87) with Eq. (6.88) are

$$\phi = \begin{cases} 1, & \text{if } \frac{\lambda^3}{v} \ge \zeta\left(\frac{3}{2}\right) \\[2ex] \text{root of } \frac{\lambda^3}{v} = \zeta\left(\frac{3}{2}\right), & \text{if } \frac{\lambda^3}{v} < \zeta\left(\frac{3}{2}\right) \end{cases}. \tag{6.94}$$

In the first of the above cases, $\text{Li}_{3/2}(\phi)$ diverges when $\phi \to 1$; hence the solution is

$$\phi = 1 \Rightarrow \text{Li}_{3/2}(\phi) = \zeta\left(\frac{3}{2}\right).$$

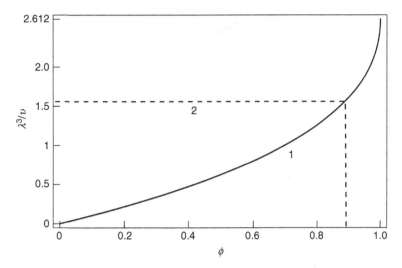

Figure 6.4 Graphical solution of Eq. (6.87) in the limit $V \to \infty$ while keeping v constant. Curve 1 is $\mathrm{Li}_{3/2}(\phi)$; the horizontal dashed line 2 represents a given value of λ^3/v for which we wish to find ϕ.

In the second case the solution is found numerically; see, e.g., Huang (1987). Here we are only interested in the fraction of particles occupying the $p = 0$ state, as a function of T or v. To accomplish this, we go back to Eq. (6.87), which, using Eq. (6.82), can be written as

$$\frac{\overline{n_0}}{N} = 1 - \frac{v}{\lambda^3}\mathrm{Li}_{3/2}(\phi). \tag{6.95}$$

From Eqs. (6.94) and (6.95) we get

$$\frac{\overline{n_0}}{N} = 1 - \frac{v}{\lambda^3}\zeta\left(\frac{3}{2}\right) \quad \text{if} \quad \frac{\lambda^3}{v} \geq \zeta\left(\frac{3}{2}\right). \tag{6.96}$$

Using Eqs. (6.91), (6.83) (6.92) and (6.93) in succession, we obtain

$$\frac{\overline{n_0}}{N} = 1 - \left(\frac{\lambda_c}{\lambda}\right)^3 \Rightarrow \begin{cases} \frac{\overline{n_0}}{N} = 1 - \left(\frac{T}{T_c}\right)^{3/2} & (\text{fixed } v) \\ \frac{\overline{n_0}}{N} = 1 - \frac{v}{v_c} & (\text{fixed } T) \end{cases}. \tag{6.97}$$

Figure 6.5 shows that, if v is kept fixed, when $T < T_c$ there is a finite fraction of gas particles in the state of zero linear momentum, and at $T = 0$ all the particles occupy this state. In this temperature range the system can be considered as a mixture of two thermodynamic phases: one consisting of particles with zero linear momentum and another consisting of particles with non-zero linear momentum. Bose–Einstein condensation can therefore be described as a 'condensation in linear momentum space' of a system of non-interacting particles, to stress its distinctiveness from, e.g., ordinary gas–liquid condensation, which is driven by interactions between the particles. Both transitions are first-order, on thermodynamic grounds (Huang, 1987, chap. 12.3).

For several decades, the only physical evidence of Bose–Einstein condensation came from studies of superfluid liquid helium (see, e.g., Pitaevsky and Stringari (2003, chap. 7)). However, the strong interactions between particles in a liquid change the nature of the

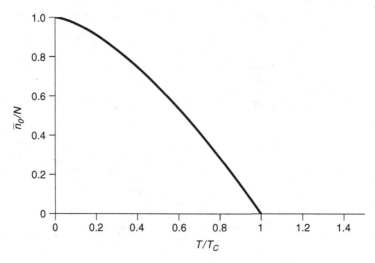

Figure 6.5 Fraction of particles in ground state $\overline{n_0}/N$ given by Eq. (6.97) vs reduced temperature T/T_c, at fixed v.

transition. For this reason, experimental physicists had long sought to create a Bose–Einstein condensate (BEC) in a dilute gas. The main difficulty lay in cooling the gases to temperatures of the order of a microkelvin or even lower, which are necessary conditions for a BEC, while at the same time preventing the atoms from condensing into a liquid or freezing into a solid. Finally, in 1995, two groups, one of the National Institute of Standards and Technology and the University of Colorado in Boulder, USA, and another of the Massachusetts Institute of Technology (MIT), USA, experimentally achieved Bose–Einstein condensation in dilute gases (Figure 6.6). For this work, Eric Cornell and Carl Wieman (Boulder) and Wolfgang Ketterle (MIT) received the 2001 Nobel Prize in Physics. Experimental research on BECs has made significant advances since. A particularly interesting finding is that the atoms in a BEC are 'laser-like', i.e., the condensate wavefunctions are coherent. The MIT group has pioneered direct observation of this coherence, constructing an 'atom laser' that generates a beam of coherent atoms, by analogy with the coherent emission of photons in an optical laser.

For further reading, see the above reference and the following web sites:

- 'Bose–Einstein condensation'. Christopher Townsend, Wolfgang Ketterle and Sandro Stringari, in *Physics World*, March 1997
  http://physicsworld.com/cws/article/print/1997/mar/01/bose-einstein
  -condensation
- 'A NEW FORM OF MATTER: Bose–Einstein Condensation and the Atom Laser'. Wolfgang Ketterle
  http://online.itp.ucsb.edu/online/plecture/ketterle/
- The atomic lab: BEC – University of Colorado at Boulder
  http://www.colorado.edu/physics/2000/bec/

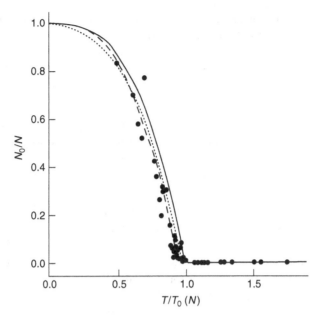

Figure 6.6 BEC Fraction N_0/N vs reduced temperature T/T_0, as measured by the Boulder group (symbols). Here, T_0 is the critical temperature predicted for the ideal boson gas in the thermodynamic limit, denoted T_c in the text. The solid line is the predicted dependence in the thermodynamic limit from Figure 6.5. The dotted line includes a correction for the finite number of atoms (4000) in the condensate. The dashed line is a fit to the experimental data points. The measured results are consistent with the theory of the ideal boson gas, within experimental uncertainty. (From *Physics World*, March 1997 (see text for URL).)

6.7.2 The phonon gas

Here we revisit the subject of thermal vibrations of the crystal lattice in monatomic solids of which we gave a simplified description in Section 4.3.3. We shall derive the Debye model that allows quantitative evaluation of the heat capacity of monatomic solids as a function of the temperature.

Consider a linear chain of length L, composed of N identical atoms of mass m. At equilibrium, the interatomic distance is the lattice constant a. A harmonic interaction potential between nearest-neighbour atoms is assumed:

$$V = \sum_{n=1}^{N} \frac{1}{2} C (u_n - u_{n+1})^2, \qquad (6.98)$$

where C is the force constant and u_n is the (small) displacement of the nth atom relative to its equilibrium position. This is a classic problem of mechanics: the study of the motion of N masses connected by springs. By writing the equations for vibrational motion of the lattice using a force that is proportional to the displacement, we are assuming harmonic motion, which is valid only for small displacements around the equilibrium positions, hence for not too large vibrational energies, i.e., for not too high temperatures. The equation of motion for the nth atom is thus:

$$m\frac{\partial^2 u_n}{\partial t^2} = F_n = -\frac{\partial V}{\partial u_n} = -C(u_n - u_{n+1}) + C(u_{n-1} - u_n)$$

$$= C(u_{n+1} - 2u_n + u_{n-1}). \tag{6.99}$$

This is identifiable as a wave equation: the right-hand side is a discretised second derivative with respect to the position coordinate. We assume a periodic solution appropriate to the periodicity a of the crystal lattice:

$$u_n = Ae^{i(\omega t - kna)}, \tag{6.100}$$

where $\omega = 2\pi f$ is the angular frequency (f is the frequency) $k = 2\pi/\lambda$ is the magnitude of the wavevector (λ is the wavelength) and $A = B^{i\phi}$, with B and ϕ real numbers (where ϕ is a phase factor).[1] Though only the real part of u_n is of interest, the complex form is convenient for the calculation that follows. Upon substitution into the equation of motion (6.99), the following relation is obtained:

$$-m\omega^2 = C\left(e^{ika} + e^{-ika} - 2\right) = 2C(\cos ka - 1). \tag{6.101}$$

Solving for the angular frequency, we find the dispersion relation (see Figure 6.7):

$$\omega = \sqrt{\frac{2C}{m}}\sqrt{1 - \cos ka} = \omega_0\left|\sin\frac{ka}{2}\right|, \quad \omega_0 = \sqrt{\frac{4C}{m}}. \tag{6.102}$$

The speed at which energy is transmitted in the material is the *group velocity*,

$$v_g = \frac{d\omega}{dk}, \tag{6.103}$$

which for dispersion relation (6.102) equals

$$v_g = \sqrt{\frac{Ca^2}{m}}\cos\frac{ka}{2}, \tag{6.104}$$

which shows that the group velocity vanishes at the boundary of the first Brillouin zone (see Figure 6.7). This corresponds to a standing wave. To study travelling waves in the crystal we set periodic boundary conditions (as if the chain were closed on itself, i.e., into a loop):

$$u_1 = u_{N+1} \Rightarrow e^{ikaN} = 1 \Rightarrow \cos kaN = 1$$

$$\Rightarrow k \equiv k_l = \frac{2\pi}{aN}l = \frac{2\pi}{L}l, \tag{6.105}$$

where l is a (positive or negative) integer. Since $-\pi/a < k \leq \pi/a$, it follows that

$$l = 0, \pm 1, \pm 2, \ldots, +\frac{N}{2}, \tag{6.106}$$

---

[1] Substituting $x_n = na$ into Eq. (6.100), we get $u_n = Ae^{i(\omega t - kx_n)}$: this is a monochromatic plane wave propagating along the OX axis, which describes the oscillation of the atom at spatial coordinate x_n.

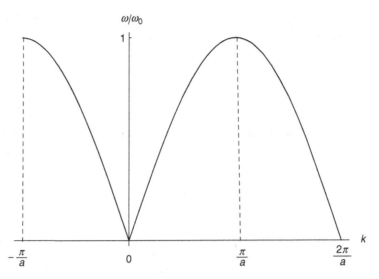

Reduced angular frequency ω/ω_0 vs wavevector k for a monatomic linear lattice. The physically significant values of k are in the range $-\pi/a < k \leq \pi/a$, called the *first Brillouin zone*. Values of k outside the first Brillouin zone replicate lattice motions described by the k values inside it. See, e.g., Kittel (1995, chap. 4) for details.

i.e., there are N discrete values for k, the vibrational normal modes, with separation $\Delta k = 2\pi/L$. From Eqs. (6.102) and (6.105), the corresponding characteristic angular frequencies are

$$\omega_l = \omega_0 \left| \sin\left(\frac{\pi l}{N} \right) \right|. \tag{6.107}$$

The energy of each lattice vibrational normal mode, or elastic mode, is quantised. The quantum of vibration is called the *phonon*, by analogy with the photon, which is the quantum of electromagnetic energy. Thermal vibrations in crystals can be described as thermally excited phonons, just as electromagnetic blackbody radiation can be described as thermally excited photons in a cavity, as we shall see in the next subsection. In the harmonic regime, the elastic modes are independent; hence phonon, like photons, can be described as non-interacting quantum harmonic oscillators.

According to quantum mechanics, each mode has the energy

$$\varepsilon_n = \left(n + \frac{1}{2} \right) \hbar\omega, \quad n = 0, 1, 2, \ldots, \tag{6.108}$$

where ω is the angular frequency of the mode. In the phonon picture, ε_n is the energy of the mode occupied by n phonons. We can therefore take as a model of the thermal vibrations of the crystal lattice of monatomic solids a set of $3N$ identical and localised harmonic oscillators. The factor 3 takes explicitly into account the number of oscillators in three dimensions. There are thus $3N$ types of phonon, with characteristic frequencies $\omega_1, \omega_2, \ldots, \omega_{3N}$.

Consider a vibrational state of the solid with n_1 phonons of angular frequency ω_1, n_2 of angular frequency ω_2, etc. The energy of a state characterised by this set of occupation numbers is

$$E(n_1, n_2, \ldots) = \sum_{k=1}^{3N} \left(n_n + \frac{1}{2} \right) \hbar \omega_k. \tag{6.109}$$

The partition function of the system is obtained by summing the Boltzmann factor over all the sets of occupation numbers, the energy of which is given by Eq. (6.109):

$$Z = \sum_{\{n_1, n_2, \ldots\}} e^{-\beta E(n_1, n_2, \ldots)} = \prod_{k=1}^{3N} \left[\sum_{n_k=0}^{\infty} e^{-\beta \hbar \omega_k \left(n_k + \frac{1}{2} \right)} \right]$$

$$= \prod_{k=1}^{3N} e^{-\beta \hbar \omega_k/2} \left(\sum_{n_k=0}^{\infty} e^{-\beta \hbar \omega_k n_k} \right) \Rightarrow Z = \prod_{k=1}^{3N} \frac{e^{-\beta \hbar \omega_k/2}}{1 - e^{-\beta \hbar \omega_k}}. \tag{6.110}$$

Since the number of phonons is not fixed, there are no restrictions on the occupation numbers $n_1, n_2, \ldots$, i.e., $\sum_k n_k \neq$ const.. The mean occupation number of each individual state i is found using Eq. (5.24) with Eq. (6.110):

$$\overline{n}_i = -\frac{1}{\beta} \left(\frac{\partial \ln Z}{\partial (\hbar \omega_i)} \right)_{T, \omega_{k \neq i}} = \frac{1}{2} + \frac{1}{e^{\beta \hbar \omega_i} - 1}. \tag{6.111}$$

This result shows that the phonon gas obeys the Bose–Einstein statistics with zero chemical potential, i.e., the Planck statistics, Eq. (6.24). This is a consequence of the number of phonons not being constant, as discussed at the end of Section 6.3. The factor $1/2$ comes from the zero-point energy of the oscillators.

From Eqs. (6.109) and (6.111) the mean total energy of the system is

$$\overline{E} = \sum_{i=1}^{3N} \overline{n}_i \hbar \omega_i = \sum_{i=1}^{3N} \hbar \omega_i \left(\frac{1}{2} + \frac{1}{e^{\beta \hbar \omega_i} - 1} \right) = \sum_{i=1}^{3N} \varepsilon_i, \tag{6.112}$$

where ε_i is the mean energy of each oscillator, Eq. (4.96). This result is the starting point for the derivation of the Debye model of thermal vibrations of the crystal lattice of monatomic solids.

Debye theory

To simplify evaluation of the right-hand side of Eq. (6.112), Debye suggested taking the frequency spectrum as a continuum, which allows us to replace the sum with an integral. The physical basis for this approximation is that we treat the lattice vibrations as long-wavelength ($\lambda \gg a$) waves propagating in the material. Under these conditions, the discrete nature of the crystal lattice can be neglected and the solid thought of as a continuous elastic medium. Consequently, in the Debye approximation lattice vibrations are elastic waves in a solid, i.e., sound waves. In the continuum limit $\lambda \gg a \Rightarrow ka \ll 1$; hence from Eq. (6.102) we find

$$\omega = \sqrt{\frac{Ca^2}{m}}k. \tag{6.113}$$

From the definition of group velocity (6.103), it follows that the velocity of sound does not depend on the frequency, and therefore these waves propagate in a non-dispersive medium:

$$v = \frac{\omega}{k} = \sqrt{\frac{Ca^2}{m}}. \tag{6.114}$$

Elastic waves in crystalline solids have three independent polarisation modes for each wavevector k: one longitudinal mode and two transverse modes. For longitudinal modes, the displacement of the lattice planes is in the direction of propagation, corresponding to compression waves. In this case, the wave given by Eq. (6.100) describes, e.g., the displacement of planes parallel to YOZ in the x direction. Now m and C are the mass and the elastic constant of a single plane, with the ratio C/m kept fixed. For tranverse modes, the displacement of the planes occurs in two mutually orthogonal directions and orthogonal to the direction of wave propagation (shear waves); see Figures 6.8 and 6.9.

To calculate the mean energy using Eqs. (6.35) and (6.111) and taking the frequency to be a continuous variable, we need to know the number of states $g(\omega)d\omega$ with frequencies in the interval $(\omega, \omega + d\omega)$, where $g(\omega)$ is the density of states:

$$\overline{E} = \int_0^\infty \hbar\omega \left(\frac{1}{2} + \frac{1}{e^{\beta\hbar\omega} - 1} \right) g(\omega)d\omega. \tag{6.115}$$

The density of states as a function of the angular frequency is obtained from the general form of the density of states as a function of the wavevector given by Eq. (5.121) in

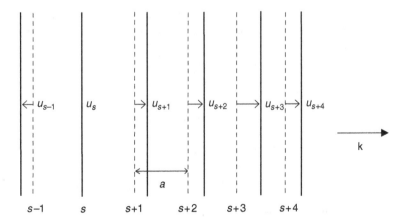

Figure 6.8 Broken lines: lattice planes at equilibrium. Solid lines: lattice planes displaced by a longitudinal wave. Coordinate u measures the displacement of the planes. Adapted from Kittel (1995, chap. 4, fig. 2).

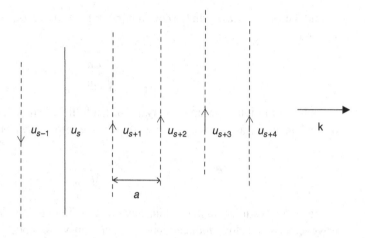

Figure 6.9 Lattice planes displaced by a transverse wave, in the mode corresponding to an oscillation in the plane of the figure (up and down). There is a second transverse mode, corresponding to an oscillation perpendicular to the plane of the figure (in and out), not shown. Adapted from Kittel (1995, chap. 4, fig. 3).

Appendix 5.A of Chapter 5:

$$g(k)dk = \frac{Vk^2}{2\pi^2}dk,$$ (6.116)

which, combined with Eqs. (6.113) and (6.114), and the definition of group velocity, Eq. (6.103), yields

$$g(\omega)d\omega = g(k)\frac{dk}{d\omega}d\omega = \frac{V\omega^2}{2\pi^2 v^2 v_g}d\omega.$$ (6.117)

In a non-dispersive medium, $v_g = v$, with the result

$$g(\omega)d\omega = \frac{V\omega^2}{2\pi^2 v^3}d\omega,$$ (6.118)

where $g(\omega)$ is the density of states for each polarisation mode of the elastic wave.

In what follows we shall assume that, for each mode, the velocity of sound is independent of the direction of propagation of the elastic wave. However, longitudinal waves and transverse waves have different velocities, v_L and v_T respectively. Taking into account the three independent modes we get

$$g(\omega)d\omega = \frac{V\omega^2}{2\pi^2}\left(\frac{1}{v_L^3} + \frac{2}{v_T^3}\right)d\omega.$$ (6.119)

If we now define a mean velocity $\overline{v}$ as

$$\frac{3}{\overline{v}^3} = \frac{1}{v_L^3} + \frac{2}{v_T^3},$$ (6.120)

the density of states can be written as

$$g(\omega)d\omega = \frac{3V\omega^2}{2\pi^2\overline{v}^3}d\omega.$$ (6.121)

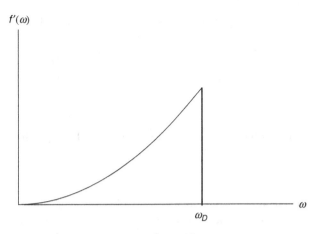

$f'(\omega)$

ω_D

ω

Figure 6.10   Reduced Debye density of states $g'(\omega) = g(\omega)/(9N/\omega_D^3) = \omega^2$.

This equation is valid for long wavelengths, i.e., low frequencies.[2] In the Debye theory this expression is applied in a frequency range with an upper bound, called the *Debye frequency* ω_D, which is a consequence of the restriction that there are only $3N$ independent elastic modes:

$$\int_0^{\omega_D} g(\omega)d\omega = 3N. \tag{6.122}$$

Using Eq. (6.121) and performing the integral, the Debye frequency is found to be a function of the atomic density N/V and the mean velocity of sound:

$$\omega_D^3 = 6\pi^2 \frac{N}{V}\bar{v}^3. \tag{6.123}$$

The density of states (6.121) can thus be written in terms of ω_D, which defines the *Debye frequency spectrum*:

$$g(\omega)d\omega = \begin{cases} \frac{9N\omega^2}{\omega_D^3}d\omega, & \omega \le \omega_D \\ 0, & \omega > \omega_D \end{cases}, \tag{6.124}$$

(see Figure 6.10). Inserting Eq. (6.124) into Eq. (6.115), the total mean energy of the phonon gas follows:

$$\bar{E} = \frac{9}{8}N\hbar\omega_D + \frac{9N\hbar}{\omega_D^3}\int_0^{\omega_D} \frac{\omega^3}{e^{\beta\hbar\omega}-1}d\omega. \tag{6.125}$$

From the mean energy we obtain the heat capacity at constant volume:

$$C_V = \left(\frac{\partial\bar{E}}{\partial T}\right)_V.$$

[2] See, e.g., Kittel (1995, chap. 5) for the general derivation of the density of states.

For the calculation that follows, we choose a new integration variable:

$$x \equiv \beta \hbar \omega = \frac{\hbar \omega}{k_B T},$$ (6.126)

such that

$$x_D = \frac{\hbar \omega_D}{k_B T} = \frac{\Theta_D}{T},$$ (6.127)

which defines the *Debye temperature*, Θ_D. The heat capacity is then found to be

$$C_V = 3Nk_B \left[\frac{3}{x_D^3} \int_0^{x_D} \frac{x^4 e^x dx}{(e^x - 1)^2} \right] = 9Nk_B \left(\frac{T}{\Theta_D} \right)^3 \int_0^{x_D} \frac{x^4 e^x dx}{(e^x - 1)^2}.$$ (6.128)

The function in square brackets is called the *Debye function*: tabulations of it can be found in the literature. Here we are interested in the limiting behaviour of the integral. In the low-temperature limit $T \ll \Theta_D$ and the upper limit of integration in Eq. (6.128) can be taken to be infinity, whereupon the integral evaluates to $4\pi^4/15$ (see Reif (1985, Appendix A.11) or Mandl (1988, Appendix A.2)), whence the heat capacity is

$$C_V = \frac{12\pi^4}{5} Nk_B \left(\frac{T}{\Theta_D} \right)^3, \quad T \ll \Theta_D.$$ (6.129)

The Debye approximation works fairly well at low enough temperatures, because at such temperatures only long-wavelength acoustic modes are excited. We leave it as an exercise for the reader to prove that, in the high-temperature limit, the Debye specific heat at constant volume is $c_V = 3R$, i.e., the same as in Einstein's theory. Figure 6.11 shows the good agreement between Debye theory and experiment. The Debye temperature is the fitting

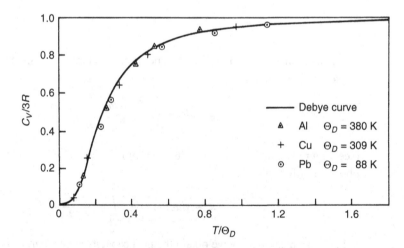

Figure 6.11 Debye reduced molar heat capacity, Eq. (6.129) (solid line). The experimental data points for aluminium, copper and lead (symbols) fall on the universal curve obtained by reducing the absolute temperature by the Debye temperature. (From Mandl (1988, chap. 6). Data from *A Compendium of the Properties of Materials at Low Temperatures, Part II: Properties of Solids*, Wadd Technical Report 60–56, Part II, 1960.)

parameter to the low-temperature experimental results. Alternatively, it can be calculated from knowledge of the elastic constants of the solid and Eqs. (6.127) and (6.123). The Debye temperature is basically a measure of the stiffness of the crystal lattice. See Kittel (1995, chap. 5) for more details.

6.7.3 The photon gas

Consider a body in thermal equilibrium with electromagnetic radiation (e.g., a stone left in the sun), and suppose that the body does not reflect any of the incident radiation. Therefore the energy absorbed by the body is emitted in the form of thermal energy only. Such a system is called a *black body* (e.g., a black stone in the sunshine). The total energy of the radiation emitted by a black body is a function of its temperature only, and as such is independent of any other property of the black body. For this reason, the radiation emitted by a black body is called *thermal radiation*. To study the physical properties of this radiation, it is convenient to consider an equivalent system. If we take electromagnetic radiation inside an opaque cavity with a non-reflecting external surface (e.g., a bread oven) and internal reflecting walls, and kept at a uniform temperature T, the radiation emitted by the walls will attain thermal equilibrium with the walls, and the radiation will then acquire well-defined properties. A small hole is now drilled in the cavity wall. If the hole is small enough, it will not disturb the thermodynamic equilibrium inside the cavity, and the radiation incident on the walls inside the cavity will exit through the hole with the same properties as inside the cavity. This radiation has the same properties as the radiation emitted by a black body at the same temperature as the cavity. Therefore this cavity radiation is equivalent to the blackbody radiation. The window of a house as seen from a distance is an example of a 'small hole into a cavity': the inside of the house seems to us very dark, since most of the light going into it is not coming back out.

We shall derive the Planck law of blackbody radiation using a method originally proposed by Bose (1924), which consists in treating the cavity radiation as an ideal gas of photons in thermal equilibrium (see Pais (1982, chap. 23) for a discussion of the original derivation). A photon is a quantum of electromagnetic energy. According to the quantum theory of radiation, photons behave like massless particles that travel at the speed of light. Both the Compton effect and the photoelectric effect are explained on the basis of photons as particles that carry linear momentum. Photons have spin 1 (in units of $\hbar$) and are therefore bosons. Since photons do not interact with each other, a photon gas is an ideal boson gas, obeying Bose–Einstein statistics. As photons are emitted and absorbed by the cavity walls, their number is not conserved. Under these conditions, the Planck statistics, Eq. (6.24), holds.

In his theory of blackbody radiation (1900), Planck posited that radiation was absorbed and emitted in multiples of a fundamental unit, the *electromagnetic quantum*, with energy proportional to the frequency of the radiation:

$$\varepsilon = h\nu = \hbar\omega, \tag{6.130}$$

where the proportionality constant is called the *Planck constant*, $h = 6.626 \times 10^{-34}$ J s, $\hbar = h/2\pi$ and $\omega = 2\pi\nu$. Using the Planck relation and Eq. (6.24), the photon statistics is written:

$$\bar{n}_j = \frac{1}{e^{\beta\varepsilon_j} - 1}, \quad \varepsilon_j = h\nu_j = \hbar\omega_j, \tag{6.131}$$

where ν_j is the frequency.

In non-dispersive media, for photons with wavelength λ and frequency ν, the relation $\lambda\nu = c \Rightarrow \omega = ck$ holds, where c is the speed of light in vacuum and k is the magnitude of the wavevector. Using the de Broglie relation $p = h/\lambda = \hbar k$, we get

$$p = \frac{h\nu}{c} = \frac{\hbar\omega}{c}. \tag{6.132}$$

Using Planck's postulate (6.130), it is clear that Eq. (6.132) is the relativistic relation between linear momentum and energy for particles of zero rest mass travelling at the speed of light:

$$\varepsilon = pc. \tag{6.133}$$

From Eqs. (5.11) and (6.132), and noting that there are two independent polarisation states of an electromagnetic plane wave, the density of translational states is in terms of the angular frequency

$$g(\omega)d\omega = \frac{V\omega^2}{\pi^2 c^3}d\omega. \tag{6.134}$$

Using Eqs. (6.33) and (6.130), and replacing ω_j with the continuous variable ω, the mean number of photons with angular frequency in the interval $(\omega, \omega + d\omega)$ is given by

$$dN(\omega) = \bar{n}(\omega)g(\omega)d\omega = \frac{V}{\pi^2 c^3} \frac{\omega^2}{e^{\beta\hbar\omega} - 1}d\omega. \tag{6.135}$$

The corresponding expression for the radiation energy is

$$dE(\omega) = \hbar\omega dN(\omega) = \frac{V\hbar}{\pi^2 c^3} \frac{\omega^3}{e^{\beta\hbar\omega} - 1}d\omega. \tag{6.136}$$

Dividing this by the volume, we finally get

$$u(\omega, T)d\omega = \frac{\hbar}{\pi^2 c^3} \frac{\omega^3}{e^{\beta\hbar\omega} - 1}d\omega. \tag{6.137}$$

This equation is *Planck's law of blackbody radiation*, which gives the spectral energy density in a cavity in thermal equilibrium at temperature T. We plot the Planck law in non-dimensional form,

$$f(x,t)dx \equiv \left(\pi^2 c^3 \hbar^3 \beta^4\right)u(\omega, T)d\omega = \frac{x^3}{e^x - 1}dx, \quad x = \frac{\hbar\omega}{k_B T}, \tag{6.138}$$

in Figure 6.12, and as a function of the frequency ν,

$$u(\nu, T)d\nu = \frac{8\pi h}{c^3} \frac{\nu^3}{e^{h\nu/k_B T} - 1}d\nu, \tag{6.139}$$

at several temperatures, in Figure 6.13.

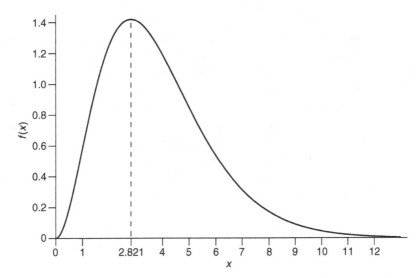

Figure 6.12   Plot of the Planck law in non-dimensional form, Eq. (6.138). The curve has a maximum for $x = 2.821$: this gives the temperature dependence of the frequency for which the density of radiation energy is maximal, known as the *Wien displacement law* (see the text and Figure 6.13).

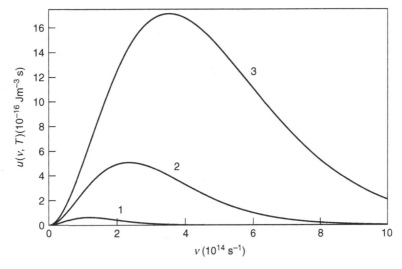

Figure 6.13   The Planck law plotted vs frequency ν, Eq. (6.139), for (1) $T = 2000$ K; (2) $T = 4000$ K; (3) $T = 6000$ K.

The Planck law provides a complete description of blackbody radiation in excellent agreement with experiment. The total energy density in the cavity is obtained by integrating the spectral density over all frequencies, with the result

$$u(T) = \int_0^\infty u(\omega, T)d\omega = aT^4, \tag{6.140}$$

$$a = \frac{\hbar}{\pi^2 c^3} \left(\frac{k_B}{\hbar} \right)^4 \int_0^\infty \frac{x^3 dx}{e^x - 1} = \frac{\pi^2 k_B^4}{15\hbar^3 c^3}, \quad x = \frac{\hbar\omega}{k_B T}.$$

The fourth-power law in the absolute temperature was first discovered experimentally by Stefan (1879) and later derived theoretically by Boltzmann (1884) using the methods of classical thermodynamics (see Problem 6.10), but only Planck's statistical physics method (1900) made it possible to calculate the prefactor, which contains Planck's constant. To derive the *Stefan–Boltzmann law* from Eq. (6.140), we calculate the power of the radiation emitted through a small hole in the cavity wall. This power equals the energy density current per unit time that is incident upon the cavity wall, multiplied by the area A of the hole. In the calculation of this current density $u(T)v$, only the component of the velocity v of the incident photons perpendicular to the wall at the hole is relevant, i.e., $\mathbf{v} \cdot \mathbf{n} = c\cos\theta$, where $\mathbf{n}$ is the wall normal and θ is the angle between $\mathbf{v}$ and $\mathbf{n}$. Since the directions of the photon velocities are distributed statistically, we average the contribution from directions between $\theta = 0$ and $\theta = \pi/2$:

$$P(T) = u(T)cA \times \frac{1}{4\pi} \int_0^{2\pi} d\phi \int_0^1 \cos\theta\, d(\cos\theta) = \frac{1}{4}cAu(T).$$

The *Stefan–Boltzmann law* gives the emitted power per unit area:

$$p(T) \equiv \frac{P(T)}{A} = \sigma T^4, \tag{6.141}$$

$$\sigma = \frac{c}{4}a = \frac{\pi^2 k_B^4}{60\hbar^3 c^2} = 5.67 \times 10^{-8}\,\mathrm{Wm^{-2}K^{-4}}\ \textit{Stefan's constant.}$$

For real objects, the emissivity ϵ of the surface of the object is defined as taking values between 0 and 1, where $\epsilon = 1$ corresponds to a black body. The power radiated per unit area then equals $p(T) = \sigma\epsilon T^4$.

It can be seen in Figure 6.13 that the maximum of $u(v, T)$ is displaced to higher frequencies as the temperature increases. The frequency that maximises $u(v, T)$ satisfies the following transcendental equation:

$$(3 - x_{max})e^{x_{max}} = 3, \quad x_{max} = \frac{\hbar\omega_{max}}{k_B T}. \tag{6.142}$$

This can be solved numerically to yield the result known as the *Wien displacement law*:

$$x_{max} = 2.821, \tag{6.143}$$

which agrees with experiment. In astronomy, this result allows the temperatures of stars to be estimated from observation of their colours.

In Figure 6.13 note that 6000 K is the temperature of the Sun's surface (the photosphere), whose radiative properties are approximately those of a black body at the same temperature. The visible spectrum spans a frequency range roughly between 4×10^{14} and 8×10^{14} $\mathrm{s^{-1}}$, hence is centred near the maximum of the energy distribution.

It is possible to derive analytically the low- and high-frequency limits of the Planck law (6.139) shown in Figure 6.13. These are given respectively by

$$u(v, T)dv = \frac{8\pi k_B T}{c^3}v^2 dv, \quad hv \ll k_B T \tag{6.144}$$

$$u(v, T)dv = \frac{8\pi h}{c^3}v^3 e^{-hv/k_B T}dv, \quad hv \gg k_B T. \tag{6.145}$$

These limits were already known in classical physics. The low-frequency limit is the classical statistical physics result, known as the *Rayleigh–Jeans law*: comparing Eq. (6.144) with the density of states (6.134), Eq. (6.144) is seen to be the spectral density of states when the mean energy of a photon is $k_B T$. The high-frequency limit, known as *Wien's law*, was proposed empirically to apply to the whole frequency range (Wien, 1898), although it is only correct in the high-frequency limit. As can be inferred from the behaviour of the curves in Figure 6.13, the classical law (6.144) diverges at high frequencies, in contradiction with experiment. This incorrect prediction was called at the end of the nineteenth century 'the ultraviolet catastrophe'. The correct treatment of blackbody radiation by Planck, on the assumption that absorption and emission of electromagnetic energy are quantised, and which underpins the statistics (6.131), was the beginning of the quantum revolution in physics. An informative review of this fascinating period in the history of Physics can be found in Pais (1982, chap. 19).

Advanced topic 6.1

Equation (6.131) can also be derived by a method analogous to that applied to the phonon gas: the total energy of the state of the electromagnetic field in which there are n_1 photons with angular frequency ω_1 and polarisation m, n_2 photons with angular frequency ω_2 and polarisation m, etc., is given by

$$E(\{n_{k,m}\}) = \sum_{\mathbf{k},m} n_{k,m} \hbar \omega, \quad n_{k,m} = 0, 1, 2, \ldots \quad \omega = ck. \qquad (6.146)$$

Since the number of photons is not fixed, there are no restrictions on the $\{n_{k,m}\}$. The partition function of the gas is then

$$Z = \sum_{\{n_{k,m}\}} e^{\beta E(\{n_{k,m}\})} = \prod_{\mathbf{k},m} \left(\sum_{n=0}^{\infty} e^{-\beta \hbar \omega n} \right) = \prod_{\mathbf{k},m} \frac{1}{1 - e^{-\beta \hbar \omega}}, \qquad (6.147)$$

and the mean occupation number for photons with wavevector $\mathbf{k}$ is

$$\overline{n_{\mathbf{k}}} = -\frac{1}{\beta} \left(\frac{\partial \ln Z}{\partial (\hbar \omega)} \right) = \frac{2}{e^{\beta \hbar \omega} - 1}, \qquad (6.148)$$

where the factor 2 comes from the fact that there are two possible polarisation states. This result shows that the photon gas obeys Bose–Einstein statistics with zero chemical potential, i.e., Planck statistics.

The cosmic background radiation

According to the Big Bang cosmological model, the universe is approximately 13 800 million years old.[3] The universe has been expanding and cooling down all this time, leaving the *cosmic background radiation* as the afterglow left over from the hot Big Bang. This

---

[3] As of 29 October 2013, the Wilkinson Microwave Anisotropy Probe data give the age of the Universe as 13 770 million years, with an uncertainty of just 0.4%: see http://map.gsfc.nasa.gov/universe/uni_age.html.

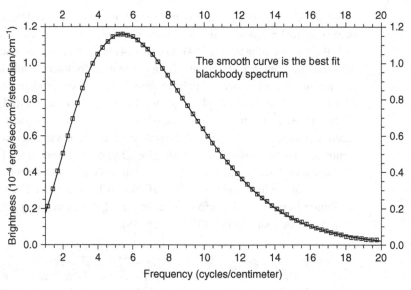

Figure 6.14   The FIRAS (Far InfraRed Absolute Spectrophotometer of the COBE satellite) measured the cosmic microwave background radiation. This figure shows the first FIRAS results accumulated over 9 minutes. The small squares are data points with an error estimate of 1%. The solid line is a fit to the blackbody radiation formula. (From J. C. Mather *et al.* 1990. A preliminary measurement of the cosmic microwave background spectrum by the Cosmic Background Explorer (COBE) satellite. *Astrophys. J.* **354**, L37–L40. ©AAS. Reproduced with permission. NASA and the COBE Science Working Group are gratefully acknowledged.)

radiation ranges from microwave to infrared in the electromagnetic spectrum and pervades the entire universe. The COBE (COsmic Background Explorer) satellite measurements fit perfectly to a blackbody spectrum with a temperature of 2.735 ± 0.060 K, in agreement with Big Bang theory predictions (see Figure 6.14). The COBE project and its results won the 2006 Nobel Prize in Physics for John C. Mather, NASA Goddard Space Flight Center, Greenbelt, MD, USA, and George F. Smoot, University of California, Berkeley, CA, USA, 'for their discovery of the blackbody form and anisotropy of the cosmic microwave background radiation'.

Problems

6.1 The pressure of a classical gas can be calculated using equation of state (6.51), with N given by Eq. (6.37). Show that it can be written as

$$P(T,\mu) = (2S+1)e^{\beta\mu} (k_B T)^{5/2} \left(\frac{2\pi m}{h^2} \right)^{3/2}.$$

6.2 Show that the mean energy of a classical gas is given by Eq. (6.40):

$$\overline{E} = \frac{3}{2}\overline{N}k_B T.$$

6.3 Consider a gas of N fermions of spin 1/2 in a volume V at the absolute zero of temperature. Show that the following quantities can be written in terms of the Fermi energy:

   a. Mean energy: $\overline{E} = \frac{3}{5}N\varepsilon_F$
   b. Pressure: $P = \frac{2}{5}n\varepsilon_F$, where $n = N/V$
   c. Isothermal compressibility: $\kappa_T = \frac{3}{2n\varepsilon_F}$.

6.4 From Eqs. (6.62) and (6.63) we find the following relation between the mean total kinetic energy and the pressure of a Fermi gas at $T = 0$ K:

$$P_0 = \frac{2}{3}\frac{\overline{E_0}}{V}.$$

The above result again follows.by comparing the expressions for the pressure of the classical gas, as obtained from equation of state (6.51), and for its mean total energy, Eq. (6.40),

   a. Show that this is a general (non-relativistic) relation, by calculating the mean pressure of a gas from the quantum expression of the kinetic energy states of a non-relativistic free particle in a box of volume $V = L^3$. This relation can also be obtained from the kinetic theory of gases (Reif, 1985, chap. 7.13).
   b. Repeat the above calculation in the extreme relativistic limit (see Problem 6.5) and show that in this case:

$$P = \frac{1}{3}\frac{\overline{E}}{V}.$$

   These relations involving P and $\overline{E}$ are a consequence of the form of the energy dependence of the linear momentum of individual particles, and are therefore independent of statistics. Hence they are also valid for the ideal boson gas.

6.5 Consider a fully degenerate free electron gas. As the gas is compressed, the kinetic energy of the electrons increases (why?). When this energy becomes of order mc^2, relativistic effects became important. The relativistic energy of an electron, ε, is given in terms of its linear momentum p by

$$\varepsilon^2 = c^2 p^2 + m^2 c^4,$$

where m is the electron mass and c is the speed of light in vacuum. When the electron energy is large compared to mc^2, in the *extreme relativistic limit*, the kinetic energy is much larger than the rest energy and $\varepsilon \approx cp$. In this case, calculate:

   a. The Fermi energy of the gas
   b. The total energy at the absolute zero of temperature, in terms of ε_F
   c. The pressure of the gas at the absolute zero of temperature.

6.6 Consider a two-dimensional ideal gas of zero-spin bosons, confined within a surface S.

    a. Calculate the mean number of particles in the gas. The following result may be useful:

$$\int_0^{+\infty} \frac{1}{e^{x-b}-1}\,dx = -\ln\left(1-e^b\right).$$

    b. Show that Bose–Einstein condensation is not possible in such a gas (consider the thermodynamic limit).

6.7 Show that, in the high-temperature limit, the Debye specific heat is $3R$.

6.8 Consider electromagnetic radiation confined in a cavity of volume V, in equilibrium at temperature T. Using relations of the type

$$\bar{n}(\varepsilon) = -\frac{\partial}{\partial(\beta\varepsilon)}\ln Y(\varepsilon),$$

where $Y(\varepsilon)$ is the grand canonical partition function of photons whose one-particle state has energy ε:

    a. Calculate the mean number of photons with angular frequency ω (this is the *Planck statistics*).

    b. Calculate the relative fluctuation of the number of photons with angular frequency ω. Discuss the behaviour of this quantity as a function of the ratio $\hbar\omega/k_B T$.

6.9 Consider a spherical cavity of radius $R = 10$ cm at $T = 1000$ K. The cavitiy has reflective internal walls and a small enough hole that the radiation coming out is approximately that of a black body at the same temperature.

    a. Calculate the total number of photons in the cavity.

    b. Calculate the total internal energy of this photon gas.

    c. Find the frequency that maximises the distribution of energy emitted by the cavity. What region of the electromagnetic spectrum does this correspond to?

    You may find the following results useful:

$$\int_0^{+\infty} \frac{x^2}{e^x-1}\,dx \simeq 2.4, \quad \int_0^{+\infty} \frac{x^3}{e^x-1}\,dx = \frac{\pi^4}{15}.$$

6.10 Consider electromagnetic radiation in equilibrium at temperature T inside a cavity of volume V. We know from Problem 6.4 that the radiation pressure is given by $P = (1/3)u(T)$, where $u(T)$ is the energy density. Using thermodynamic methods, show that:

    a. $u(T) = aT^4$ (the Stefan–Boltzmann law).

    b. The entropy density is $s = \frac{4}{3}aT^3$.

References

Christman, J. R. 1988. *Fundamentals of Solid State Physics*. Wiley.

Huang, K. 1987. *Statistical Mechanics, 2nd edition*. Wiley.

Kittel, C. 1995. *Introduction to Solid State Physics*, 7th edition. Wiley.

Mandl, F. 1988. *Statistical Physics*, 2nd edition. Wiley.

Pais, A. 1982. *Subtle is the Lord*. Oxford University Press.

Pitaevsky, L., and Stringari, S. 2003. *Bose–Einstein Condensation*. Oxford University Press.

Reif, F. 1985. *Fundamentals of Statistical and Thermal Physics*. McGraw-Hill.

Magnetism

7.1 Introduction

In this chapter we extend the study of the magnetic properties of solids beyond the simple, ideal-gas-type paramagnetic models of previous chapters. Our purpose is to describe more complex magnetic phenomena, such as ferromagnetism, antiferromagnetism, and elementary excitations in solids. In addition to its intrinsic interest, magnetism here plays an important role in illustrating how the methods of statistical physics can be applied to systems of interacting particles and to phase transitions, which are of great importance in condensed matter physics.

Electrons in solids generate magnetic fields, since they are moving charges with intrinsic magnetic dipole moments due to their spin. The magnetic fields generated by the nuclear spins are much smaller than those of the electrons, and their contribution to the magnetisation of a solid can be neglected. If the atoms in a solid have permanent magnetic dipole moments, the solid will be paramagnetic or ferromagnetic. More complex behaviours such as antiferromagnetism and ferrimagnetism are also possible (see Figure 7.1). The main contributions to the magnetisation of a solid can come either from the interaction between the electrons and an applied magnetic field, or from interactions between the electrons themselves. If the latter are weak enough to be negligible, the system of magnetic moments can, to a first approximation, be regarded as ideal. This was the subject of our study of paramagnetism in Chapter 3, where we treated a paramagnetic solid as an isolated system, and in Chapter 4, where we treated it as a system in equilibrium with a heat reservoir. Otherwise, when the interactions between electrons are to be taken into account, the collective behaviour of a large number of magnetic moments has to be studied in the more general framework of non-ideal systems. Such is the case with ferromagnetism and antiferromagnetism, for which we shall develop models in this chapter.

A simplified model of a paramagnetic solid was introduced in Chapter 4 that consisted of N identical, ideal spins located at the sites of a lattice and in equilibrium with a heat reservoir, which is the sample itself. In the case of a paramagnet, the local field is approximately the same as the applied field, and it follows from Eqs. (4.52) and (4.63) that the magnetisation is a function of the ratio of the applied field to the sample temperature, which vanishes for zero applied field. In real systems, as opposed to ideal systems, the interaction between magnetic moments can lead to a zero-field sample magnetisation, in which case the material is ferromagnetic or ferrimagnetic. In this context the relevant interactions between the spin are: (i) the exchange interaction, a short-range interaction that prevails at small separations; and (ii) the direct magnetic dipolar interaction, a long-range interaction much

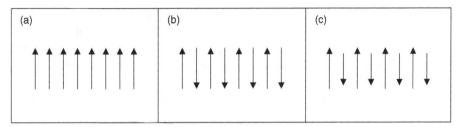

Figure 7.1 Schematic ordered arrangements of electron spins: (a) ferromagnetic; (b) antiferromagnetic; (c) ferrimagnetic.

Table 7.1 Curie temperature T_c and saturation magnetisation $\mathcal{M}_0 = \mathcal{M}(T = 0)$ for some ferromagnetic materials (Christman, 1988, chap. 11).

Material	T_c (K)	$\mathcal{M}_0$ (10^6 A m^{-1})
Iron	1043	1.75
Cobalt	1404	1.45
Nickel	631	0.51
Gadolinium	289	2.00
Dysprosium	85	2.01

weaker than the exchange interaction, but dominant at larger separations. Our treatment of ferromagnetism will focus on the exchange interaction between localised electrons. In the next section we show that this is an indirect interaction between spins, originating in the electrostatic repulsion between a pair of electrons and in the Pauli exclusion principle. Hence the magnetism of matter is essentially of a quantum nature.

At high temperatures, a ferromagnetic material will become paramagnetic. This phase transition occurs at a critical temperature T_c, the *Curie temperature*. The simplified models of magnetism to be introduced in this chapter provide approximate descriptions of this. Table 7.1 collects the Curie temperatures and saturation magnetisations (to be defined later) of a few ferromagnetic materials.

7.2 The Heisenberg model*

In this introductory section we treat the magnetism of solids using a simplified model that considers only the interactions between the spins of two unpaired electrons, localised at neighbouring atoms or ions in the crystal lattice. From this analysis follows the Heisenberg model, which is our starting point for the study of the magnetic properties of matter. This approach is adequate for materials such as the rare earth metals gadolinium and dysprosium, in which magnetism arises from the interaction between localised electrons. However, it is not always adequate, in particular for transition metals such as nickel, copper or iron, in which ferromagnetism derives instead from the interaction between partially itinerant electrons. In this case, a band theory such as the Stoner model is preferable. Band theories require more advanced knowledge of the quantum theory of solids and are usually taught in courses on solid state physics (see, e.g., Ibach and Lüth (2003, chap. 8)). However, both the Heisenberg model of localised electrons and the Stoner model of itinerant electrons lead, under appropriate approximations, to the Weiss model, which provides a phenomenological description of the magnetism of matter at different temperatures and which will be derived in the next section.

7.2.1 Exchange interaction

In a real crystal, the spins of unpaired electrons localised at the atoms or ions of the crystal lattice will interact with one another. In what follows we shall assume there is one unpaired electron per atom. As stated above, our study of ferromagnetism will be based on the exchange interaction between localised electrons. We shall now show that this interaction comes from the electrostatic repulsion between electrons and that it is a consequence of the Pauli exclusion principle, which is reflected in the symmetry of the wavefunction of the electrons. Consider a two-electron system, for which we write the wavefunction as

$$\psi(1,2) = \phi(\mathbf{r}_1, \mathbf{r}_2)\chi(\sigma_1, \sigma_2), \tag{7.1}$$

where $\mathbf{r}_1$ and $\mathbf{r}_2$ are the position coordinates, and σ_1 and σ_2 the spins of electrons 1 and 2, respectively. The corresponding wavefunctions are ϕ and χ, respectively. The global two-electron wavefunction must be anti-symmetric under exchange of the positions and spins of the two particles, since electrons are fermions and obey the Pauli exclusion principle.

Spin is an intrinsic angular momentum that was introduced to explain why the observed angular momentum of a particle was greater than its orbital momentum by a fixed amount. The name 'spin' was given by analogy with a classical particle spinning around its axis of rotation, but is a purely quantum mechanical quantity. Electrons have spin $\sqrt{3/4}\hbar$, which can be written in terms of their *spin quantum number* $s = 1/2$ as $\sqrt{s(s+1)}\hbar$ (unlike orbital angular momentum states, which always have even quantum numbers, spin quantum numbers may be even or odd). In a medium with axial symmetry (e.g., a material placed in a strong magnetic field or a uniaxial crystal) around the OZ axis, the z-component of the electron spin is quantised and can only take the values $\pm(1/2)\hbar$. The spin components along directions perpendiculat to OZ must at each instant obey the Heisenberg uncertainty relations and their expectation values therefore vanish, but their squares take the well-defined value $(1/4)\hbar^2$, whence the electron spin angular momentum is, in units of $\hbar$, $s^2 = s_x^2 + s_y^2 + s_z^2 = 3/4 \Rightarrow |s| = \sqrt{3/4}$.

By convention, the spin state with $s_z = +1/2$ corresponds to the spin wavefunction $\chi_\alpha(\sigma)$, and the spin state with $s_z = -1/2$ corresponds to the spin wavefunction $\chi_\beta(\sigma)$. Under these conditions, and assigning to s_z the spin operator $\hat{s}_z$, we can write the following eigen equations in units of $\hbar$:

$$\hat{s}_z\chi_\alpha(\sigma) = \frac{1}{2}\chi_\alpha(\sigma), \tag{7.2}$$

$$\hat{s}_z\chi_\beta(\sigma) = -\frac{1}{2}\chi_\beta(\sigma). \tag{7.3}$$

The spin wavefunction of the two-electron system is a symmetric or anti-symmetric linear combination of the wavefunctions of each individual electron. There are four possibilities:

$$\chi_A(\sigma_1,\sigma_2) = \chi_\alpha(\sigma_1)\chi_\beta(\sigma_2) - \chi_\beta(\sigma_1)\chi_\alpha(\sigma_2) = -\chi_A(\sigma_2,\sigma_1), \tag{7.4}$$

$$\chi_S^I(\sigma_1,\sigma_2) = \chi_\alpha(\sigma_1)\chi_\beta(\sigma_2) + \chi_\beta(\sigma_1)\chi_\alpha(\sigma_2) = \chi_S^I(\sigma_2,\sigma_1), \tag{7.5}$$

$$\chi_S^{II} = \chi_\alpha(\sigma_1)\chi_\alpha(\sigma_2) = \chi_S^{II}(\sigma_2,\sigma_1), \tag{7.6}$$

$$\chi_S^{III} = \chi_\beta(\sigma_1)\chi_\beta(\sigma_2) = \chi_S^{III}(\sigma_2,\sigma_1), \tag{7.7}$$

where the indices A and S denote anti-symmetric or symmetric linear combinations, respectively.

To find the z-component of the total spin for these states, the following operator is used:

$$\hat{S}_z = \hat{s}_{z,1} + \hat{s}_{z,2}. \tag{7.8}$$

For the first symmetric state, Eqs. (7.8) and (7.5)–(7.7), together with Eqs. (7.2) and (7.3), yield

$$
\begin{aligned}
\hat{S}_z \chi_S^I(\sigma_1,\sigma_2) &= \left(\hat{s}_{z,1} + \hat{s}_{z,2}\right)\chi_\alpha(\sigma_1)\chi_\beta(\sigma_2) + \left(\hat{s}_{z,1} + \hat{s}_{z,2}\right)\chi_\beta(\sigma_1)\chi_\alpha(\sigma_2) \\
&= \frac{1}{2}\chi_\alpha(\sigma_1)\chi_\beta(\sigma_2) + \left(-\frac{1}{2}\right)\chi_\alpha(\sigma_1)\chi_\beta(\sigma_2) \\
&\quad + \left(-\frac{1}{2}\right)\chi_\beta(\sigma_1)\chi_\alpha(\sigma_2) + \frac{1}{2}\chi_\beta(\sigma_1)\chi_\alpha(\sigma_2) \\
&= \frac{1}{2}\chi_S^I(\sigma_1,\sigma_2) - \frac{1}{2}\chi_S^I(\sigma_1,\sigma_2) = 0,
\end{aligned}
\tag{7.9}
$$

whence the eigenvalue associated with this state is $S_z = 0$. For the second symmetric state we have

$$
\begin{aligned}
\hat{S}_z \chi_S^{II}(\sigma_1,\sigma_2) &= \left(\hat{s}_{z,1} + \hat{s}_{z,2}\right)\chi_\alpha(\sigma_1)\chi_\alpha(\sigma_2) \\
&= \frac{1}{2}\chi_\alpha(\sigma_1)\chi_\alpha(\sigma_2) + \frac{1}{2}\chi_\alpha(\sigma_1)\chi_\alpha(\sigma_2) \\
&= 1\chi_S^{II}(\sigma_1,\sigma_2),
\end{aligned}
\tag{7.10}
$$

whence in this state $S_z = +1$. We likewise obtain the eigenvalues $S_z = -1$ for the third symmetric state and $S_z = 0$ for the anti-symmetric state.

To find the total spin we use the operator:

$$
\begin{aligned}
\hat{\mathbf{S}}^2 &= \hat{S}_x^2 + \hat{S}_y^2 + \hat{S}_z^2 \\
&= \left(\hat{s}_{x,1} + \hat{s}_{x,2}\right)^2 + \left(\hat{s}_{y,1} + \hat{s}_{y,2}\right)^2 + \left(\hat{s}_{z,1} + \hat{s}_{z,2}\right)^2 \\
&= \hat{s}_{x,1}^2 + \hat{s}_{x,2}^2 + \hat{s}_{y,1}^2 + \hat{s}_{y,2}^2 + \hat{s}_{z,1}^2 + \hat{s}_{z,2}^2 + 2\hat{s}_{x,1}\hat{s}_{x,2} + 2\hat{s}_{y,1}\hat{s}_{y,2} + 2\hat{s}_{z,1}\hat{s}_{z,2} \\
&= \hat{\mathbf{s}}_1^2 + \hat{\mathbf{s}}_2^2 + 2\left(\hat{\mathbf{s}}_1 \cdot \hat{\mathbf{s}}_1\right).
\end{aligned}
\tag{7.11}
$$

It can be shown that this operator has eigenvalues $S^2 = 0$ for the anti-symmetric state, and $S^2 = 2\hbar^2$ for each of the symmetric states (Schiff, 1968, chap. 10, sec. 41). In terms of the total spin quantum number s, we write $S^2 = s(s+1)\hbar^2$; hence $s = 0$, for χ_A and $s = 1$ for χ_S^I, χ_S^{II} and χ_S^{III}. The three symmetric spin states form a *triplet*, for each of which we have

$$S = \frac{1}{2} + \frac{1}{2} = 1,$$

i.e., the quantum number of the the total spin is the sum of the quantum numbers of the spins of the individual electrons. For this reason, this spin configuration is called 'parallel spins', from the old quantum theory. This term is also used to describe states with total spin z-component

$$S_z = \left(\frac{1}{2} + \frac{1}{2} \right) \hbar.$$

This leaves out the anti-symmetric state, Eq. (7.10), for which $S_z = 0$. The anti-symmetric state is a *singlet* and its spin configuration is called 'anti-parallel spins' because

$$S = \frac{1}{2} - \frac{1}{2} = 0.$$

We construct the spatially symmetric and anti-symmetric wavefunctions of the two-electron system using linear combinations of the spatial wavefunctions of each of the electrons, $\phi_a(\mathbf{r}_1)$ and $\phi_b(\mathbf{r}_2)$, which gives respectively

$$\phi_+(\mathbf{r}_1, \mathbf{r}_2) = \phi_a(\mathbf{r}_1)\phi_b(\mathbf{r}_2) + \phi_a(\mathbf{r}_2)\phi_b(\mathbf{r}_1) = \phi_+(\mathbf{r}_2, \mathbf{r}_1), \qquad (7.12)$$

$$\phi_-(\mathbf{r}_1, \mathbf{r}_2) = \phi_a(\mathbf{r}_1)\phi_b(\mathbf{r}_2) - \phi_a(\mathbf{r}_2)\phi_b(\mathbf{r}_1) = -\phi_-(\mathbf{r}_2, \mathbf{r}_1). \qquad (7.13)$$

Consequently, the global normalised wavefunction, Eq. (7.1), is

$$\psi_A(1, 2) = \frac{1}{\sqrt{2}} \begin{cases} \phi_+(\mathbf{r}_1, \mathbf{r}_2)\chi_A(\sigma_1, \sigma_2) \\ \phi_-(\mathbf{r}_1, \mathbf{r}_2)\chi_S(\sigma_1, \sigma_2) \end{cases}. \qquad (7.14)$$

Therefore when $\mathbf{r}_1 = \mathbf{r}_2$ (i.e., when the two electrons are in the same position), the total wavefunction vanishes for 'parallel' spins, but is non-zero for 'anti-parallel' spins.

We shall now calculate the Coulomb energy of the two-electron system. Defining $r_{12} = \|\mathbf{r}_1 - \mathbf{r}_2\|$, the Hamiltonian operator of the electrostatic interaction in vacuum is

$$\hat{U}(r_{12}) = \frac{1}{4\pi\epsilon_0} \frac{e^2}{r_{12}}. \qquad (7.15)$$

Using first-order perturbation theory (Schiff, 1968, chap. 8, sec. 31), we obtain the quantum expectation value of the Coulomb energy of the two-electron system:

$$E_{\pm} = \langle \psi_{\pm} | \hat{U}(r_{12}) | \psi_{\pm} \rangle = \langle \phi_{\pm} | \hat{U}(r_{12}) | \phi_{\pm} \rangle$$

$$= \underbrace{\frac{1}{2} \int d^3 r_1 \int d^3 r_2 |\phi_a(\mathbf{r}_1)|^2 |\phi_b(\mathbf{r}_2)|^2 \frac{1}{4\pi\epsilon_0} \frac{e^2}{r_{12}}}_{I_0}$$

$$+ \underbrace{\frac{1}{2} \int d^3 r_1 \int d^3 r_2 |\phi_a(\mathbf{r}_2)|^2 |\phi_b(\mathbf{r}_1)|^2 \frac{1}{4\pi\epsilon_0} \frac{e^2}{r_{12}}}_{I_0}$$

$$\pm \underbrace{\int d^3 r_1 \int d^3 r_2 \phi_a^*(\mathbf{r}_1) \phi_b^*(\mathbf{r}_2) \phi_a(\mathbf{r}_2) \phi_b(\mathbf{r}_1) \frac{1}{4\pi\epsilon_0} \frac{e^2}{r_{12}}}_{J/2}$$

$$= I_0 \pm \frac{J}{2}, \tag{7.16}$$

where I_0 is known as the *Coulomb integral*, and $J/2$ as the *exchange integral* (the latter name refers to the exchange of the position coordinates in the arguments of ϕ_a and ϕ_b in the integrand). In Eq. (7.16), the second equality follows from the fact that the Coulomb operator is spin-independent, hence integration over the (orthonormal) spin wavefunctions gives unity. The energy difference between state ϕ_+ ('antiparallel' spins) and state ϕ_- ('parallel' spins) is thus

$$\Delta E \equiv E_+ - E_- = J. \tag{7.17}$$

Hence from Eqs. (7.14) and (7.17) it follows that for $J > 0$ a state with 'parallel' spins has a lower energy than a state with 'antiparallel' spins, whereas for $J < 0$ the reverse applies. Consequently, according to this model, the state with 'parallel' spins is favoured for materials with $J > 0$, and the state with 'antiparallel' spins is favoured for materials with $J < 0$. Note that J has a *short-range character*, because the product $\phi_a(\mathbf{r}_1)\phi_b(\mathbf{r}_2)$ in the integrand of Eq. (7.16) is non-zero only when both individual wavefunctions are non-zero, i.e., only where $\phi_a(\mathbf{r}_1)$ and $\phi_b(\mathbf{r}_2)$ overlap. Thus J is a rapidly decreasing function of the interatomic separation, and therefore each electron will only significantly interact with the electrons of nearest-neighbour atoms.

Note that the exchange integral J *has no direct connection with the magnetic properties of the electrons*: the spins are taken into consideration only to endow the wavefunction with the correct symmetry, and the energy difference between the two states of the two-electron system results from its electrostatic interaction. The connection between the exchange energy, of electrostatic nature, and the spin configuration is therefore indirect, but fundamental, since it originates in the symmetry of the wavefunction. *The exchange energy can thus be regarded as if resulting from direct interactions between the spins.* In fact, it can be shown that the spin-dependent part of the energy of a two-electron system can be written as

$$\varepsilon_{ij} = -J_{ij} \left(\frac{1}{2} + \frac{2\langle \hat{\mathbf{s}}_i \cdot \hat{\mathbf{s}}_j \rangle}{\hbar^2} \right), \tag{7.18}$$

where J_{ij} is the *exchange coefficient*. To derive this, start by calculating $\langle \hat{\mathbf{s}}_i \cdot \hat{\mathbf{s}}_j \rangle$. From Eq. (7.11) we have

$$2 \left(\hat{\mathbf{s}}_i \cdot \hat{\mathbf{s}}_j \right) = \hat{\mathbf{S}}^2 - \hat{\mathbf{s}}_i^2 - \hat{\mathbf{s}}_j^2, \tag{7.19}$$

whence for the triplet state ($S = 1$) we get, in terms of the spin quantum numbers,

$$2 \langle \hat{\mathbf{s}}_i \cdot \hat{\mathbf{s}}_j \rangle = 1(1+1)\hbar^2 - \frac{1}{2} \left(\frac{1}{2} + 1 \right) \hbar^2 - \frac{1}{2} \left(\frac{1}{2} + 1 \right) \hbar^2 = \frac{1}{2} \hbar^2,$$

and for the singlet state ($S = 0$),

$$2 \langle \hat{\mathbf{s}}_i \cdot \hat{\mathbf{s}}_j \rangle = 0(0+1)\hbar^2 - \frac{1}{2} \left(\frac{1}{2} + 1 \right) \hbar^2 - \frac{1}{2} \left(\frac{1}{2} + 1 \right) \hbar^2 = -\frac{3}{2} \hbar^2.$$

Substituting these results into Eq. (7.18) we obtain

$$\varepsilon_{ij} = \begin{cases} +J_{ij}, & \text{singlet} \\ -J_{ij}, & \text{triplet} \end{cases}. \tag{7.20}$$

Identifying the quantity J_{ij} with the exchange integral for two electrons, $J/2$, Eq. (7.16), yields Eq. (7.17).

The above analysis allowed us to describe states with 'parallel' or 'anti-parallel' spins in terms of the sign of J. However, this is an incomplete picture, since it neglects the contribution of the electron–ion interactions to the potential energy. This contribution is important whenever the two interacting electrons belong to different ions in the solid. In fact, identifying J_{ij} as the exchange integral for two electrons, $J/2$, given by Eq. (7.16), is valid only when both electrons belong to the same ion, as it is assumed that the wavefunctions of each electron are orthogonal, which is only true in this case. Therefore, the contribution of the electron–ion electrostatic interaction energy to the exchange coefficient vanishes in this case. This is no longer true for electrons on neighbouring ions of the crystal lattice. We write the Hamiltonian of the electrostatic interaction of the system of electrons and ions as

$$\hat{U}(\mathbf{r}_1, \mathbf{r}_2) = \hat{U}_{\text{ei}}(\mathbf{r}_1) + \hat{U}_{\text{ei}}(\mathbf{r}_2) + \hat{U}_{\text{ee}}(r_{12}), \tag{7.21}$$

where the first two terms are the potential energies of the electron–ion interactions and the last term is the Hamiltonian of the electron–electron interaction, Eq. (7.15). Proceeding as before, the energy difference between the triplet and the singlet states is given by the electron–electron contribution, Eq. (7.17), plus the sum of the electron–ion contributions (which are identical):

$$\Delta E_{\text{ei}} = 4 \int \phi_a^*(\mathbf{r}_2) \phi_b(\mathbf{r}_2) d^3 r_2 \int \phi_a^*(\mathbf{r}_1) \hat{U}_{\text{ei}}(\mathbf{r}_1) \phi_b(\mathbf{r}_1) d^3 r_1. \tag{7.22}$$

If ϕ_a and ϕ_b represent two atomic orbitals centred on the same atom or ion, the first integral in Eq. (7.22) vanishes; hence the exchange integral is the only contribution to the electrostatic energy of the system. However, the most interesting case for the study of magnetism in solids is when the two wavefunctions ϕ_a and ϕ_b are centred at different atoms. In this case, the exchange coefficient is a sum of two terms, corresponding to the electron–electron interaction and the electron–ion interaction, which have opposite signs. The sign of J_{ij} will depend on the relative magnitude of these two terms. Consequently, when J_{ij} is positive, ferromagnetism is favoured; when it is negative, antiferromagnetism is favoured (see e.g., Christman (1988, chap. 11) or Ibach and Lüth (2003, chap. 8), for details).

7.2.2 The Heisenberg Hamiltonian

The effective spin–spin interaction is taken into account by including an additional term in the Hamiltonian of the two-electron system:

$$\mathcal{H}_{ij} = -J_{ij} \left(\frac{1}{2} + \frac{2\hat{\mathbf{s}}_i \cdot \hat{\mathbf{s}}_j}{\hbar^2} \right). \tag{7.23}$$

We generalise this expression by assuming that the same functional form holds for the exchange Hamiltonian of two atoms with spins $\mathbf{S}_i$ and $\mathbf{S}_j$:

$$\mathcal{H}_{ij} = -J_{ij} \left(\frac{1}{2} + \frac{2\hat{\mathbf{S}}_i \cdot \hat{\mathbf{S}}_j}{\hbar^2} \right). \tag{7.24}$$

The exchange coefficient for two atoms is more complicated than for two electrons. Its magnitude is a function of the interatomic separation; in general it is negative for small separations (nearest neighbours) and positive for larger separations. This change of sign explains why some transition metals, like Fe, Co and Ni, are ferromagnetic, and others, like Mn, are antiferromagnetic (Christman, 1988, chap. 11). On the other hand, an explanation is in order as to why the exchange Hamiltonian is written as a function of the spin atomic angular momenta $\mathbf{S}_i$ only, rather than the total atomic angular momenta $\mathbf{J}_i = \mathbf{L}_i + \mathbf{S}_i$. It is shown in quantum mechanics that, in a central field, one component of the total angular momentum, usually taken as L_z, and the square of the total angular momentum, L^2, are constant. In a crystal, each ion is subjected to a non-central electric field generated by its neighbouring ions, called the crystal field. In this case, the plane of the electron orbits changes with time, so L_z is no longer constant and its average value can vanish, as it does for the transition metals. When $\langle L_z \rangle \approx 0$, the orbital angular momentum of the atom is said to be *quenched*. In this case, $J \simeq S$, and S is determined by the first Hund rule, just as for an isolated atom (for more on the Hund rules see any of the solid-state physics books listed in the References of this chapter).

The total Hamiltonian in the presence of an applied magnetic field is written

$$\mathcal{H} = \mathcal{H}_0 + \mathcal{H}_{\text{exch}}, \tag{7.25}$$

where $\mathcal{H}_0$ is the part of the Hamiltonian that describes the interaction of the atomic magnetic moments with the local field, which equals the sum of the applied field plus contributions from the surrounding atoms. If the orbital angular momentum of the atom is quenched (Kittel, 1995, chap. 14), this part of the Hamiltonian is

$$\mathcal{H}_0 = -\sum_{i=1}^{N} \hat{\boldsymbol{\mu}}_i \cdot \mathbf{B} = -\frac{g\mu_B}{\hbar} \sum_{i=1}^{N} \hat{\mathbf{S}}_i \cdot \mathbf{B}, \tag{7.26}$$

where g is the Landé factor. This was already used in Chapter 4 when studying the ideal system of N spins, for which the local field was assumed to coincide with the applied field. The exchange Hamiltonian is

$$\mathcal{H}_{\text{exch}} = \sum_{\{i,j\}}^{N} \mathcal{H}_{ij}, \tag{7.27}$$

with $\mathcal{H}_{ij}$ given by (7.24) and where $\{i,j\}$ means that the sum is over all pairs of nearest-neighbour atoms. The Hamiltonian (7.25), with Eqs. (7.26), (7.27) and (7.24) is called the *Heisenberg model*, and is the starting point for several calculations in magnetism. In the remainder of this book we shall make several simplifications to this model, such as replacing the sum in Eq. (7.27) by an average (mean-field Weiss theory), or considering a uniaxial environment (Ising model).

In addition to the exchange interaction, there is the direct magnetic interaction between the atomic spin magnetic moments. This is a classical dipolar interaction, whose Hamiltonian is of the form

$$\mathcal{H}_{\text{mag}} \propto \frac{\hat{\boldsymbol{\mu}}_i \cdot \hat{\boldsymbol{\mu}}_j}{r_{ij}^3} - \frac{3\left(\hat{\boldsymbol{\mu}}_i \cdot \mathbf{r}_{ij}\right)\left(\hat{\boldsymbol{\mu}}_j \cdot \mathbf{r}_{ij}\right)}{r_{ij}^5}, \tag{7.28}$$

where r_{ij} is the separation between each pair of electrons. For nearest-neighbour atoms the exchange energy can be several orders of magnitude larger than the direct magnetic energy. However, because the exchange interaction is short-ranged, the direct magnetic interaction dominates at larger distances and should be taken into consideration in a more detailed analysis of magnetism.

7.3 Weiss's mean-field theory of magnetism or the Weiss model

In the mean-field approximation, the sum over spin operators of the ν nearest-neighbour atoms in the Heisenberg Hamiltonian (7.27) is replaced by its average,

$$\sum_{j=1}^{\nu \text{ nearest neighbours}} \hat{\mathbf{S}}_j \simeq \nu \overline{\mathbf{S}}, \tag{7.29}$$

where $\overline{\mathbf{S}}$ is the statistical mean of the quantum expectation value $\langle \hat{\mathbf{S}}_j \rangle$. Under approximation (7.29), called the *mean-field* or *molecular field approximation*, the Heisenberg Hamiltonian, Eq. (7.25), becomes an *effective* Hamiltonian of the form

$$\hat{\mathcal{H}}_{\textit{eff}} = \underbrace{-\frac{g\mu_B}{\hbar}\sum_i \hat{\mathbf{S}}_i \cdot \mathbf{B}}_{\hat{\mathcal{H}}_0} \underbrace{-\frac{2\nu J}{\hbar^2}\sum_i \hat{\mathbf{S}}_i \cdot \overline{\mathbf{S}}}_{\hat{\mathcal{H}}_1}, \tag{7.30}$$

where we have dropped the constant term in Eq. (7.24), and $J_{ij} = J$ if i and j are nearest-neighbour electrons and zero otherwise. $\hat{\mathcal{H}}_0$ is the part of the Hamiltonian that describes the interaction of the spins with the applied field, and $\hat{\mathcal{H}}_1$ is the exchange interaction part. Equation (7.30) can be rewritten as

$$\hat{\mathcal{H}}_{\mathit{eff}} = -\frac{g\mu_B}{\hbar} \sum_i \hat{\mathbf{S}}_i \cdot \underbrace{\left(\mathbf{B} + \frac{2vJ}{\hbar g\mu_B}\overline{\mathbf{S}} \right)}_{\mathbf{B}_{\mathit{eff}}}, \tag{7.31}$$

where the effective field $\mathbf{B}_{\mathit{eff}}$ formally replaces the local field in Hamiltonian $\hat{\mathcal{H}}_0$. This effective field is itself a sum of two terms, the second of which is called the 'molecular field' or 'internal field'. The mean-field approximation thus consists in replacing, in the Hamiltonian of the system, the true exchange interaction between spins of pairs of electrons by an interaction between the spin of each electron and an 'average spin' due to all other neighbouring electrons. This term can be written in terms of the magnetisation, as we shall now show. Start by writing

$$\mathbf{B}_{\mathit{eff}} = \mathbf{B} + \lambda\mu_0\boldsymbol{\mathcal{M}}, \tag{7.32}$$

where $\boldsymbol{\mathcal{M}}$ is the magnetisation, μ_0 is the magnetic permeability of free space and λ is a non-dimensional parameter to be determined later. Neglecting the contribution of the direct dipolar magnetic interaction (7.28) to the local field (see discussion of the magnetic free energy), we approximate the local field $\mathbf{B}$ by the applied field $\mathbf{B}_0$ and write

$$\mathbf{B}_{\mathit{eff}} = \mathbf{B}_0 + \mathbf{B}_M, \quad \mathbf{B}_M = \lambda\mu_0\boldsymbol{\mathcal{M}}, \tag{7.33}$$

where $\mathbf{B}_M$ is the field due to the magnetisation of the material. In terms of the magnetic field $\mathbf{H}$, with $\mathbf{B}_0 = \mu_0\mathbf{H}$, we get

$$\mathbf{B}_{\mathit{eff}} \simeq \mu_0 \left(\mathbf{H} + \lambda\boldsymbol{\mathcal{M}} \right) = \mu_0\mathbf{H}_{\mathit{eff}}. \tag{7.34}$$

The magnetisation is defined as the mean total magnetic moment per unit volume, which can be written in terms of the atomic spin magnetic moment $\boldsymbol{\mu}$:

$$\boldsymbol{\mathcal{M}} = \frac{\overline{\mathbf{M}}}{V} = \frac{N}{V}\overline{\boldsymbol{\mu}} = n\overline{\boldsymbol{\mu}}, \quad \overline{\boldsymbol{\mu}} = -g\mu_B\,\mathbf{S}, \quad n = \frac{N}{V}. \tag{7.35}$$

Comparing the definition of $\mathbf{B}_{\mathit{eff}}$ in Eq. (7.31) with Eqs. (7.32) and (7.35), we conclude that

$$\lambda = \frac{2vJ}{ng^2\mu_B^2\mu_0}. \tag{7.36}$$

The *Weiss parameter* λ was initially introduced to explain the ferromagnetic behaviour of a material in the context of a phenomenological approach. It is a non-dimensional parameter that measures the ratio of the exchange interaction to the direct dipolar magnetic interaction between nearest-neighbour electrons and that is much larger than unity, as will be shown later.

A crystal is an anisotropic medium. In what follows, for simplicity we shall consider a uniaxial environment along *OZ*, which will also be taken as the direction of the applied field. Consequently, we want to calculate the mean value of the magnetisation along *OZ*. We shall consider the case of materials such as the iron group, in which the orbital contribution to the atomic magnetic moments of the atoms is quenched (Kittel, 1995, chap. 14),

We shall further simplify our study of magnetism by considering the specific case $S = 1/2$. Under these conditions, $g = 2$. From Hamiltonian (7.31) we can find the eigenvalues of the energy of the system of N spins 1/2, formally replacing the spin operators $\hat{\mathbf{S}}_{zi}$ by $m_i\hbar$, $i = 1, \ldots, N$. The energy levels of the ith atom are then:

$$\varepsilon_i = -2\mu_B B_{eff} m_i, \quad m_i = \pm 1/2,$$
$$= \mu_B B_{eff}, \tag{7.37}$$

which is formally analogous to Eq. (3.36), upon substitution of the field B by the effective field B_{eff}. To proceed with the calculation of the magnetisation, we now assume the system of N spins to be in equilibrium with a heat reservoir at temperature T (which is the sample itself). Consequently, we can use the results obtained in Chapter 4 for the ideal system of N spins 1/2, by formally replacing B by B_{eff}. Hence from Eq. (4.52) it follows, for the mean magnitude of the total magnetic moment along OZ:

$$\overline{M} = N\mu_B \tanh\left(\frac{\mu_B B_{eff}}{k_B T}\right). \tag{7.38}$$

The interaction between spins is thus taken into account through the effective field. Using Eqs. (7.34) and (7.35), we can write Eq. (7.38) in terms of the magnetisation along OZ:

$$\mathcal{M} = n\mu_B \tanh\left[\frac{\mu_B \mu_0}{k_B T}(H + \lambda \mathcal{M})\right]. \tag{7.39}$$

This is the key result of the *Weiss model* for spin 1/2. It is an implicit equation for the magnetisation, to be solved graphically (or numerically). Before doing so, however, we shall derive the behaviour of the magnetisation in the limiting cases of low temperatures (L.T.) and high temperatures (H.T.). Defining

$$x = \frac{\mu_B \mu_0}{k_B T}(H + \lambda \mathcal{M}), \tag{7.40}$$

we get, at low temperatures:

$$\text{L.T.} \quad x \gg 1 \Rightarrow \tanh x \simeq 1 \Rightarrow \mathcal{M} \simeq n\mu_B = \mathcal{M}_0, \tag{7.41}$$

where $\mathcal{M}_0$ is the *saturation magnetisation*: the maximum value the magnetisation can achieve. On the other hand, at high temperatures,

$$\text{H.T.} \quad x \ll 1 \Rightarrow \tanh x \simeq x \Rightarrow \mathcal{M} \simeq n\frac{\mu_B^2 \mu_0}{k_B T}(H + \lambda \mathcal{M}). \tag{7.42}$$

Defining

$$T_c = \frac{n\mu_B^2 \lambda \mu_0}{k_B}, \tag{7.43}$$

which, as we shall see later, is the *critical temperature* of the paramagnetic–ferromagnetic phase transition, Eq. (7.42) is written

$$\mathcal{M} = \frac{T_c}{T - T_c}\frac{H}{\lambda}. \tag{7.44}$$

This is an approximate form of Eq. (7.39) valid for $T > T_c$, i.e., in the paramagnetic phase. Under this approximation, the magnetisation is proportional to the magnetic field. The thermodynamic parameter that describes the magnetic response of the system to the applied magnetic field is the magnetic susceptibility, defined by Eq. (2.76):

$$\chi_T = \frac{1}{\mu_0}\left(\frac{\partial \mathcal{M}}{\partial H}\right)_T.$$

From Eq. (7.44) it follows that

$$\chi_T = \frac{C}{\mu_0(T - T_c)}, \quad C = \frac{T_c}{\lambda}. \tag{7.45}$$

This result is the *Curie–Weiss law*, where C is the *Curie constant*, which was initially introduced to explain the behaviour of the paramagnetic susceptibility in the context of a phenomenological approach; T_c is also called the *Curie temperature*. From Eqs. (7.36) and (7.43) we obtain for the Curie constant:

$$C = \frac{n\mu_B^2 \mu_0}{k_B}. \tag{7.46}$$

The Curie law, Eq. (7.45), describes the behaviour of the magnetic susceptibility when the temperature approaches the critical temperature T_c from above:

$$\chi_T \propto (T - T_c)^{-\gamma}, \quad \gamma = 1, \tag{7.47}$$

where γ is the *susceptibility critical exponent*, which characterises the behaviour of the susceptibility near the phase transition. From Eq. (7.47) we see that the susceptibility diverges when $T \rightarrow T_c^+$; according to its definition, this means that an arbitrarily weak field generates a finite variation of the magnetisation. As we shall see in Chapter 10, this is a consequence of the large spontaneous fluctuations of the magnetisation occurring near the critical temperature. For $T < T_c$, because the susceptibility is very large, the system can exhibit a non-zero magnetisation in zero applied magnetic field; this is the *spontaneous magnetisation*, which is a feature of the ferromagnetic phase, as we shall see shortly.

7.3.1 Zero-field, or spontaneous, magnetisation

The magnitude of the spontaneous (i.e., zero-field) magnetisation $\mathcal{M}_S$ is obtained from Eq. (7.39) with definitions (7.41) and (7.43) and setting $H = 0$:

$$\frac{\mathcal{M}_S}{\mathcal{M}_0} = \tanh\left(\frac{T_c}{T}\frac{\mathcal{M}_S}{\mathcal{M}_0}\right), \tag{7.48}$$

It is convenient to work with the non-dimensional *reduced spontaneous magnetisation*:

$$m_S = \frac{\mathcal{M}_S}{\mathcal{M}_0}. \tag{7.49}$$

The state with $m_S = 1$ is the state of maximum magnetisation, characterised by a perfect alignment of the N spins, i.e., a state of maximum order. On the other hand, a state with

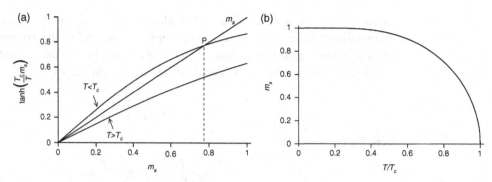

(a) The point of intersection P is the solution of Eq. (7.50) for a given T/T_c. As T approaches T_c, point P moves from $m_S = 1$ (for $T = 0$) to $m_S = 0$ (for $T = T_c$). The two curves shown are for $T/T_c = 3/4$ and $T/T_c = 4/3$. (b) Solution of Eq. (7.50) vs T/T_c, obtained using the method described in (a).

$m_S = 0$ is a state of minimum order, or maximum disorder. Hence m_S is the order parameter of the system. Equation (7.48) can be rewritten in terms of m_s:

$$m_S = \tanh\left(\frac{T_c}{T}m_S\right). \tag{7.50}$$

As shown in Figure 7.2b, and in Figure 7.3 for nickel, the behaviour of the system changes qualitatively at the critical temperature: below T_c, the order parameter $m_S \neq 0$, whereas above T_c, $m_S = 0$. Since m_S decreases continuously to zero when $T \to T_c^-$ (i.e., there is no finite jump of m_S to zero at $T = T_c$), the transition from the ferromagnetic phase to the paramagnetic phase is *continuous* or *second-order*. In summary, from the Weiss model in zero applied magnetic field, we get

$$T < T_c \;\Rightarrow m_S \neq 0 \quad \text{(ferromagnetic order)}$$

$$T > T_c \Rightarrow m_S = 0 \quad \text{(paramagnetic disorder)}.$$

This phase transition can be studied analytically in the vicinity of T_c. Because $m_S \ll 1$ when $T \lesssim T_c$, the hyperbolic tangent in Eq. (7.50) can be expanded in powers of m_S up to third order:

$$\tanh\left(\frac{T_c}{T}m_S\right) \simeq \frac{T_c}{T}m_S - \left(\frac{T_c}{T}\right)^3 \frac{m_S^3}{3},$$

whereupon Eq. (7.50) yields

$$\frac{m_S^2}{3} = \left(\frac{T}{T_c}\right)^3\left(\frac{T_c}{T} - 1\right) = \left(\frac{T}{T_c}\right)^2\left(1 - \frac{T}{T_c}\right) \Rightarrow m_S \simeq \sqrt{3}\left(\frac{T}{T_c}\right)\left(1 - \frac{T}{T_c}\right)^{1/2}. \tag{7.51}$$

Thus when $T \to T_c^-$, m_S goes to zero as

$$m_S \propto (T - T_c)^\beta, \quad \beta = \frac{1}{2}. \tag{7.52}$$

Here β is the *critical exponent of the order parameter* (the reduced spontaneous magnetisation in this problem). This value of β and that of γ in Eq. (7.47) are mean-field results and do not agree with experiment (see the discussion at the end of this section).

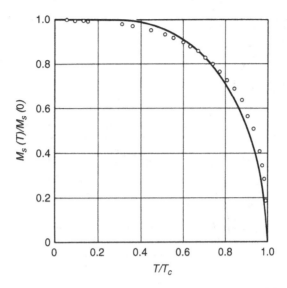

Figure 7.3   Reduced spontaneous magnetisation of nickel vs reduced temperature. Dots: experimental data points. Line: Weiss model for $S = 1/2$. (From Kittel (1995, chap. 15). Data from Weiss, P., and and Forrer, R. 1926. Aimantation et phénomène magnetocalorique du nickel. *Ann. Phys. (Paris)* **5**, 153–213. http://www.annphys.org.)

7.3.2 Non-zero-field magnetisation

To study the behaviour of the system in an applied magnetic field, we start from the full equation for the magnetisation (7.39). Using definitions (7.41) and (7.43), and introducing the non-dimensional variables

$$m = \frac{\mathcal{M}}{\mathcal{M}_0}, \tag{7.53}$$

$$h = \frac{\mu_B \mu_0 H}{k_B T}, \tag{7.54}$$

the reduced magnetisation is written

$$m = \tanh\left(h + \frac{T_c}{T} m\right). \tag{7.55}$$

It is convenient to solve for the inverse function:

$$h = \operatorname{arctanh} m - \frac{T_c}{T} m. \tag{7.56}$$

The behaviour of $m(h)$, as shown in Figure 7.4b, is qualitatively different for $T > T_c$ and for $T < T_c$. In the special case of zero field, we recover the results of the preceding section: zero magnetisation for $T > T_c$, while for $T < T_c$ a spontaneous (reduced) magnetisation $\pm m_S$ arises. As T approaches T_c from below, m_S continuously approaches zero. For $h \neq 0$, the dashed branches for $T < T_c$ are unphysical, i.e., the (reduced) magnetisation jumps from $-m$ to $+m$ when the field is inverted. In zero applied field, the reduced magnetisation can be either $+m_S$ or $-m_S$ at $T = Tc$.

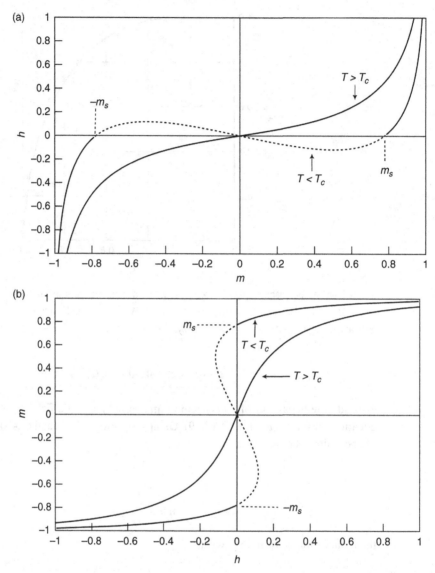

Figure 7.4   (a) Plots of $h(m)$, Eq. (7.56), for $T/T_c = 4/3$ and $T/T_c = 3/4$. (b) By flipping the axes in (a), we obtain $m(h)$. The dashed branches for T_iT_c are unphysical.

7.3.3 Ferromagnetic domains. Hysteresis

A ferromagnetic sample is composed of regions called *Weiss domains*, such that the easy direction of magnetisation in neighbouring domains is different, as illustrated in Figure 7.5a. Domain formation lowers the magnetic energy of the sample. This energy is given by the integral of the energy density associated with the field **B** inside the sample over the volume of the sample: $U_{\mathrm{mag}} \propto \int B^2 dV$. When domains with different orientations of

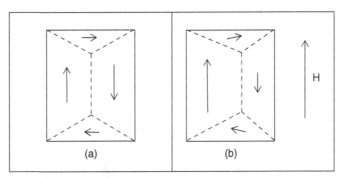

Figure 7.5   Sketch of magnetic domains in a ferromagnetic sample: (a) sample with zero total magnetisation; (b) sample is magnetised through the application of a magnetic field.

the magnetisation are formed inside the sample, the magnitude of the mean magnetic field generated by the magnetisation decreases, hence the total magnetic energy decreases. On the other hand, there is an energy associated with the boundaries between domains: spins on opposite sides of the boundary are not parallel; consequently, their exchange energy is greater than that for parallel spins. Globally, the magnetic energy is minimised when the transition from one orientation to another is gradual, across what is called a *Bloch wall* (Kittel, 1995, chap. 15). Note that the analyses presented previously are valid for *ferromagnetic monodomains* only.

When a magnetic field is applied to a ferromagnetic sample, the boundaries between domains move in such a way that domains whose magnetisation has direction close to that of the applied field tend to grow, whereas those whose magnetisation is in very different directions to the applied field tend to shrink (see Figure 7.5b for a sketch). In a sample without defects this motion is reversible. Defects, however, hinder the motion of domain boundaries, and so in real samples domain reorientation induced by the application of strong magnetic fields becomes irreversible. As a result, the plot of the magnetisation or the magnetic induction versus the applied magnetic field strength H is the *hysteresis loop*, shown schematically in Figure 7.6. If the applied magnetic field is switched off after a small increase from zero, B returns approximately to zero, because domain rotation is negligible in this case. If, however, H attains large enough values, B will remain non-zero even after H has been switched off. Since the samples under these conditions show irreversible behaviour, the reversible thermodynamic theory that follows is not valid for these systems.

7.4 Landau theory of magnetism

The Landau theory of the paramagnetic–ferromagnetic phase transition is a paradigm in the physics of phase transitions. It is a thermodynamic theory. In the thermodynamic study of a (monodomain) ferromagnet, it is convenient to take the magnetisation $\mathcal{M}$ as the independent variable, and not the magnetic field, as was done in the paramagnetic case in Chapter 4, This is because the magnetisation is due mainly to the exchange interaction and not to the

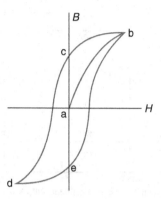

Figure 7.6 When a magnetic field of strength H is applied to a non-magnetised ferromagnetic sample (i.e., with zero total magnetisation), the magnitude of the magnetic induction, B, follows path a–b. When H is decreased to zero, B follows path b–c. For oscillatory applied fields, B follows a closed path b–c–d–e–b, called a *hysteresis loop*, shown schematically here.

reorientation of permanent magnetic moments in an applied magnetic field (Landau and Lifshitz, 1960, chap. V). More precisely, the mean total magnetic moment of the system, $\overline{\mathbf{M}} = V\mathcal{M}$ should be used, which is the extensive thermodynamic variable conjugate to the intensive variable magnetic field. Under these conditions, we start from the fundamental thermodynamic relation, Eq. (2.70):

$$dU = TdS + \mu_0 H d\overline{M}.$$

Helmholtz free energy

Using the magnetic analogue of the Helmholtz free energy, the fundamental thermodynamic relation is

$$F = U - TS \Rightarrow dF = -SdT + \mu_0 H d\overline{M} = dF_0 + \mu_0 H d\overline{M}, \qquad (7.57)$$

where F_0 is a non-magnetic contribution. The change in free energy when the magnetisation increases from zero to $\mathcal{M}$ is

$$F = \int dF = F_0 + V\mu_0 \int_0^{\mathcal{M}} H d\mathcal{M}'. \qquad (7.58)$$

We use the Weiss model to calculate this change in free energy in the vicinity of the Curie temperature. This is done by expanding Eq. (7.56) to third order in powers of the reduced magnetisation m: $m \ll 1 \Rightarrow \operatorname{arctanh} m \simeq m + m^3/3$, whence

$$h(m) \simeq m \left(1 - \frac{T_c}{T} \right) + \frac{m^3}{3}. \qquad (7.59)$$

The *critical isotherm* is the h vs m curve for $T = T_c$. From Eq. (7.59) this is

$$h \propto m^\delta, \quad \delta = 3, \qquad (7.60)$$

where δ is the critical exponent of the magnetic field (see Chapter 10).

Replacing in Eq. (7.59) the non-dimensional variables h and m by the physical variables $\mu_0 H$ and $\mathcal{M}$, and using relations (7.54), (7.53) and (7.43), we find

$$\mu_0 H(\mathcal{M}) = a\mathcal{M} + b\mathcal{M}^3, \tag{7.61}$$

with

$$a = \mu_o \lambda \left(\frac{T}{T_c} - 1 \right), \tag{7.62}$$

$$b = \frac{k_B T}{3n^3 \mu_B^4}, \tag{7.63}$$

and λ given by Eq. (7.36).

If we now insert the magnetic field given by (Eq. (7.61) into Eq. (7.58) and calculate the integral, the magnetic contribution to the free energy per unit volume is

$$f_{\text{mag}} \equiv \frac{F - F_0}{V} = \frac{1}{2} a\mathcal{M}^2 + \frac{1}{4} b\mathcal{M}^4. \tag{7.64}$$

The equilibrium state is found by minimising f_{mag} with respect to the magnetisation $\mathcal{M}$, from which follows

$$a\mathcal{M} + b\mathcal{M}^3 = 0, \tag{7.65}$$

with solutions

$$\mathcal{M} = 0, \tag{7.66}$$

$$\mathcal{M} = \pm \left(-\frac{a}{b} \right)^{1/2}, \tag{7.67}$$

which correspond to the following physical situations:

$$T < T_c \Rightarrow a < 0 \Rightarrow \mathcal{M} = \pm \left(\frac{|a|}{b} \right)^{1/2} = \pm \mathcal{M}_S, \tag{7.68}$$

$$T > T_c \Rightarrow a > 0 \Rightarrow \mathcal{M} = 0. \tag{7.69}$$

Note that the zero-magnetisation solution still exists below the critical temperature, but as we will see shortly, the corresponding free energy is higher than that of the solution with non-zero magnetisation.

In what follows it is convenient to write the free energy density (7.64) in non-dimensional form as a function of the reduced magnetisation m given by Eq. (7.53). After some simple algebra we get

$$f'_{\text{mag}} \equiv \frac{f_{\text{mag}}}{\mu_0 n^2 \mu_B^2} = \frac{1}{2} a'm^2 + \frac{1}{4} b'm^4, \tag{7.70}$$

where the non-dimensional coefficients a' and b' are given by

$$a' = \lambda \left(\frac{T}{T_c} - 1 \right), \tag{7.71}$$

$$b' = \frac{k_B T}{3\mu_0 n \mu_B^2}. \tag{7.72}$$

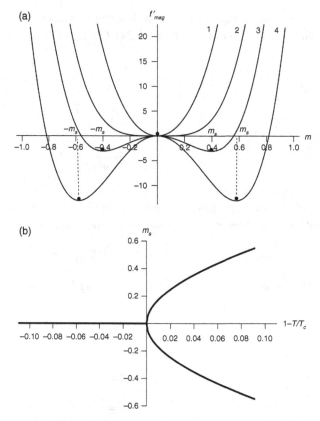

(a) Plot of the reduced Helmholtz free energy, Eq. (7.70), vs the reduced magnetisation (7.53), for (1) $T/T_c = 1.11$; (2) $T/T_c = 1$; (3) $T/T_c = 0.95$; (4) $T/T_c = 0.90$; using parameters appropriate for nickel (see the text). The black circles mark the minima of the free energy, which move from $m = 0$ to $m = \pm m_S$ as the temperature is lowered through T_c. (b) Order parameter m vs $1 - T/T_c$ for the curves in (a), showing a bifurcation at $T = T_c$, with the corresponding symmetry breaking for $T < T_c$: when the order parameter becomes non-zero, it can be either positive or negative with equal probability.

Equation (7.70) is the ratio of the free energy density to the magnitude of the magnetic dipolar interaction energy density, Eq. (7.28). This equation clearly displays the three contributions to the magnetisation of a material (within the approximations made): in coefficient a', which with λ given by Eq. (7.36) is the ratio of exchange energy to magnetic dipolar energy; and in coefficient b', which is the ratio of thermal energy to magnetic dipolar energy. Equation (7.70) with parameters appropriate for nickel is plotted in Figure 7.7a at four different values of T/T_c. Note that, while for $T > T_c$ the minimum is at $\mathcal{M} = 0$, for $T < T_c$ there are two symmetric minima at $\pm m_S$. These values of the reduced spontaneous magnetisation are in good agreement with those obtained graphically from Figure 7.2, in the same reduced temperature range.

At $T = T_c$, f_{mag} is flat around the origin. This change in the shape of the thermodynamic potential as the control parameter (in this case the reduced temperature T/T_c) is varied, and the resulting continuous variation of the order parameter (in this case m) at the transition is the hallmark of the *Landau theory of second-order (or continuous) phase transitions*.

According to Landau theory, near the transition the appropriate thermodynamic potential can be expanded in a power series of the order parameter. Depending on the symmetry of the problem, this expansion may or may not contain odd-order terms. Here we have truncated the expansion at fourth order (Landau and Lifshitz, 1960, chap. XIV), Chap. XIV], hence the transition is discontinuous, i.e., first-order, if a third power of the order parameter is present, and continuous, or second-order, when it is not (Goldenfeld, 1992, chap. 5 and sec. 9.4).

The free energy (7.15) plotted in Figure 7.7 was calculated with parameters appropriate for nickel, which were determined as follows: using $n = 9.14 \times 10^{28}$ atoms/m^3 (Kittel, 1995, Table 4), $k_B = 1.3807 \times 10^{-23}$ J/K, $\mu_0 = 4\pi \times 10^{-7}$ H/m, and $\mu_B = 9.2741 \times 10^{-24}$ A m^2, from Eq. (7.46) we get for the Curie constant $C = 0.715$ K. The Curie temperature of nickel is $T_c = 631$ K, so from Eq. (7.45) we find the Weiss parameter, $\lambda = T_c/C = 882$. Once C and λ are known, we can calculate the free energy coefficients a and b, Eqs. (7.62) and (7.63), as functions of the temperature. We can also find the exchange coefficient of nickel: using Eq. (7.36) and noting that the number of nearest neighbours is $\nu = 12$ (Kittel, 1995, chap. 1), we get $J = 1.45 \times 10^{-21}$ J.

There are discrepancies between the predictions of the Weiss model for nickel and experimental results. In particular, it appears that the experimental values of the magnetisation fit to non-integer numbers of electrons per atom. This is explained by the band model of ferromagnetism: in transition metals, such as nickel or iron, partially itinerant electrons contribute to the magnetisation. We can still, however, describe the magnetic behaviour of these materials with the localised electron model provided an *effective* number of electrons per atom is used, which is calculated by band theory (see e.g., Kittel (1995, chap. 15) or Ibach and Lüth (2003, chap. 8)). In this case, to get quantitative agreement between the calculated and experimental values of the magnetisation, gS must be replaced by the effective number of Bohr magnetons n_B in Eq. (7.41) for the saturation magnetisation, which thus becomes $\mathcal{M}_0 = n n_B \mu_B$. For nickel, $n_B = 0.6$ and $\mathcal{M}_0 = 0.51 \times 10^6$ A m^{-1} (Kittel 1995, chap. 15). Using this and $\lambda = 882$ in Eq. (7.33), the exchange field corresponding to the saturation magnetisation is $B_{\mathcal{M}_0} \sim 10^3$ T. On the other hand, for the dipolar magnetic field $B_{dip} \sim \mu_0 \mu_B / a^3$ between nearest-neighbour ions we obtain, taking $a = 2.5$ Å for nickel (Kittel, 1995, Table 4), $B_{dip} \sim 1$ T. This gives $B_{\mathcal{M}_0}/B_{dip} \sim 10^3$, whence the dipolar magnetic field contribution is negligible compared to the exchange field for nearest-neighbour ions, which justifies neglecting the dipolar magnetic contribution to the local field in Eq. (7.33).

Gibbs free energy

The Gibbs energy (2.75) is the appropriate thermodynamic potential with which to study a system in an applied magnetic field:

$$G = F - \mu_0 H \overline{M},$$

where $-\mu_0 H\overline{M}$ is the potential energy of the spin system in the magnetic field. From the Helmholtz potential (7.64), we obtain the following magnetic contribution to the Gibbs free energy per unit volume:

$$g_{mag} = \frac{1}{2}a\mathcal{M}^2 + \frac{1}{4}b\mathcal{M}^4 - \mu_0 H\mathcal{M}. \tag{7.73}$$

When no field is applied, the Helmholtz and Gibbs free energies are identical, and therefore give the same equilibrium result, Eqs. (7.66) and (7.67). Proceeding as for the reduced Helmholtz energy, we obtain the reduced Gibbs free energy as a function of the reduced magnetisation m given by Eq. (7.53):

$$g'_{mag} = \frac{g_{mag}}{\mu_0 n^2 \mu_B^2} = \frac{1}{2}a'm^2 + \frac{1}{4}b'm^4 - h'm, \tag{7.74}$$

with a' given by Eq. (7.71), b' given by Eq. (7.72), and h' given by

$$h' = \frac{H}{n\mu_B}. \tag{7.75}$$

As can be seen in Figures 7.8 and 7.9, for $T < T_c$ a jump is observed from negative to positive values of the order parameter when h' goes through zero; this is a *discontinuous transition* or *first-order transition*, in which the order parameter changes discontinuously as a function of the magnetic field for $T < T_c$.

Heat capacity

The Helmholtz and Gibbs free energies of the system follow from multiplying Eq. (7.64) or Eq. (7.73), respectively, by the volume, from which we can obtain the heat capacity using Eq. (2.72):

$$C_M = -T\left(\frac{\partial^2 F}{\partial T^2}\right)_M,$$

or Eq. (2.73):

$$C_H = -T\left(\frac{\partial^2 G}{\partial T^2}\right)_H.$$

We leave it as an exercise to show that in the Weiss model the heat capacity has a finite discontinuity at the transition temperature $T = T_c$:

$$\Delta C = \frac{3}{2}Nk_B. \tag{7.76}$$

This jump to lower values of the heat capacity at the paramagnetic–ferromagnetic phase transition is not observed experimentally and is a consequence of the mean-field approximation. Experimentally, $C(T)$ exhibits a peak at the transition. This disagreement between the Weiss model and experiment for the behaviour of the heat capacity, as well as the

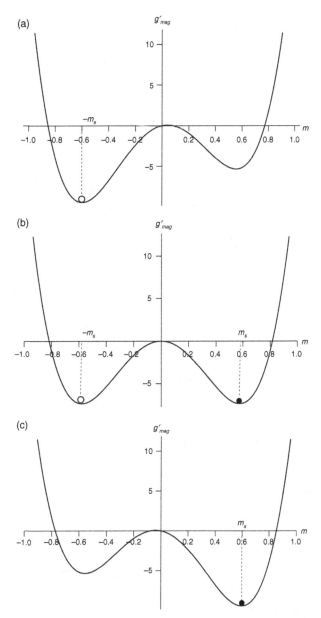

Figure 7.8 Reduced Gibbs free energy (7.74) vs reduced magnetic field (7.75), calculated with the parameters of nickel, for $T/T_c = 0.9$ and (a) $H = -3 \times 10^6$ A m^{-1} $\Rightarrow$ $h' = -3.8$; (b) $H = 0$; (c) $H = 3 \times 10^6$ A m^{-1} $\Rightarrow$ $h' = 3.8$. The white and the black circles mark the absolute minima of the free energy for negative and positive values of the order parameter, respectively.

incorrect values it predicts for the critical exponents of the magnetic susceptibility, γ (Eq. (7.47)), the magnetisation, β (Eq. (7.52)) and magnetic field, δ (Eq. (7.60)), is typical of mean-field theories. A more detailed analysis is therefore necessary near the critical temperature. This is discussed in Chapter 10.

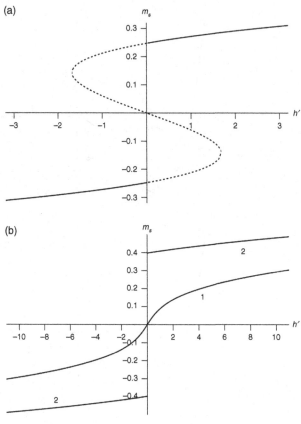

Figure 7.9 Order parameter m_s vs reduced field (7.75), as obtained by minimisation of the reduced Gibbs free energy (7.74) with respect to m, for (a) $T/T_c = 0.98$ and (b) (1) $T/T_c = 1.01$, (2) $T/T_c = 0.95$ and calculated with the parameters of nickel. In (a) the dashed line is unphysical. In (b) only the physical solutions are plotted. This is the same qualitative behaviour as shown in Figure 7.4b with a different reduced field variable.

Advanced topic 7.1

In order to construct the Weiss model for materials composed of ions whose permanent magnetic moments result from a generic spin angular momentum, one should start instead from the Brillouin magnetisation, Eqs. (4.85)–(4.87):

$$\mathcal{M} = ng\mu_B J B_J(y), \quad y = \frac{g\mu_B B}{k_B T}, \tag{7.77}$$

where $B_J(y)$ is the Brillouin function, defined in Eq. (4.84). Previously we only considered the contribution of the electron spin to the total angular moment of the ion, and from the general result (7.77) with $g=2$ and $S=1/2$ we obtain (on multiplying by the volume) the Weiss model for spin 1/2, Eq. (7.38). Following the same procedure as before, we replace in the variable y of Eq. (7.77) the field B by the effective field B_{eff} given by Eq. (7.32). The corresponding expression for the Weiss magnetisation is thus more general than Eq. (7.39). The saturation magnetisation is given by

$$\mathcal{M}_0 = ng\mu_B J,$$ (7.78)

and the Curie constant by

$$C = \frac{ng^2\mu_B^2\mu_0 J(J+1)}{3k_B}.$$ (7.79)

For example, the Weiss model for iron has $J = S \simeq 1$ and is developed along the same lines as for nickel with $S = 1/2$ (see e.g., Christman (1988, chap. 11)).

7.5 Ferrimagnetism and antiferromagnetism

In our discussion of ferromagnetism we assumed that all atoms are identical, as they clearly are for the ferromagnetic elements listed in Table 7.1. However, most materials with interesting magnetic properties are compounds. It is often the case that, in such materials, atoms of different chemical species are located on different sub-lattices. The spins of atoms on the same sub-lattice tend to be parallel ($J_{ij} > 0$ if i and j belong to the same sub-lattice), while the spins of atoms on different sub-lattices tend to be anti-parallel ($J_{ij} < 0$ if i and j belong to different sub-lattices). An example of great relevance to applications (e.g., in magnetic storage systems) is ferrites, of generic composition Fe_2MO_4, where M may be copper, lead, magnesium, manganese, cobalt, nickel or iron, occurring as a divalent ion. One such ferrite is magnetite, the magnetic constituent of lodestone, for which M is Fe^{2+}. Typically, one of the sub-lattices is occupied by the Fe^{3+} ions and the other by the M^{2+} and O^{2-} ions.

In these materials the total magnetisation is the sum of the magnetisations of the two sub-lattices, which have the same direction but different senses. If the total spontaneous magnetisation vanishes at all temperatures, the material is said to be *antiferromagnetic*; otherwise, it is called *ferrimagnetic*.

Clearly, antiferromagnetism may be seen as a particular case of ferrimagnetism, so we shall treat ferrimagnetism first. A ferrimagnetic material is qualitatively similar to a ferromagnet: it shows spontaneous magnetisation, magnetic domains and hysteresis at temperatures lower than the Curie temperature T_c, and paramagnetic behaviour for $T > T_c$. However, the spontaneous magnetisation and the susceptibility behave differently from those of a ferromagnetic material, as described below. Our analysis will focus on the $S = 1/2$ case for consistency with the study of ferromagnetism presented before; the general case is discussed in Christman (1988, chap. 11.4).

The simplest Weiss model of a ferrimagnet consists in assuming that the total local magnetic induction acting on the atoms of each sub-lattice (hereafter labelled 1 and 2) is the sum of the applied magnetic induction plus the induction due to the magnetisation of the other sub-lattice. We thus generalise Eq. (7.33) and (7.34) by writing

$$\mathbf{B}_{eff1} = \mathbf{B}_0 + \lambda_2 \mu_0 \mathcal{M}_2 = \mu_0 \left(H + \lambda_2 \mathcal{M}_2 \right), \tag{7.80}$$

$$\mathbf{B}_{eff2} = \mathbf{B}_0 + \lambda_1 \mu_0 \mathcal{M}_1 = \mu_0 \left(H + \lambda_1 \mathcal{M}_1 \right), \tag{7.81}$$

which, by analogy with Eq. (7.36), allows us to find the constants λ_1 and λ_2 for each of the sub-lattices:

$$\lambda_1 = \frac{2v_1 J}{n_1 g_1^2 \mu_B^2 \mu_0} < 0, \tag{7.82}$$

$$\lambda_2 = \frac{2v_2 J}{n_2 g_2^2 \mu_B^2 \mu_0} < 0 \tag{7.83}$$

(recall that $J < 0$). Clearly, the total magnetisation is the sum of the magnetisations of the two sub-lattices:

$$\mathcal{M} = \mathcal{M}_1 + \mathcal{M}_2. \tag{7.84}$$

Without loss of generality, we may take $\mathbf{H}$, $\mathcal{M}_1$ and $\mathcal{M}_2$ along the OZ axis; Eq. (7.39) thus gives, for the magnitudes of magnetisations of each of the two sub-lattices:

$$\mathcal{M}_1 = n_1 \mu_B \tanh \left[\frac{\mu_B \mu_0}{k_B T} \left(H - |\lambda_2| \mathcal{M}_2 \right) \right], \tag{7.85}$$

$$\mathcal{M}_2 = n_2 \mu_B \tanh \left[\frac{\mu_B \mu_0}{k_B T} \left(H - |\lambda_1| \mathcal{M}_1 \right) \right]. \tag{7.86}$$

Since the hyperbolic tangent has the same sign as its argument, setting $H = 0$ clearly gives $\mathcal{M}_1 = -\mathcal{M}_2$, i.e., the magnetisations of the two sub-lattices indeed have opposite senses. Henceforth, we shall take $\mathcal{M}_1 > 0$ and $\mathcal{M}_2 < 0$.

As in the ferromagnetic case, we start our analysis by deriving the behaviour of the magnetisation in the low- (L.T.) and high- (H.T.) temperature limits. By analogy with Eq. (7.40), we define:

$$x_1 = \frac{\mu_B \mu_0}{k_B T} \left(H - |\lambda_1| \mathcal{M}_1 \right), \quad x_2 = \frac{\mu_B \mu_0}{k_B T} \left(H - |\lambda_2| \mathcal{M}_2 \right), \tag{7.87}$$

from which results

$$\text{L.T.} \quad \begin{cases} x_1 < 0, & |x_1| \gg 1 \\ x_2 > 0, & |x_2| \gg 1 \end{cases} \Rightarrow \begin{cases} \tanh x_1 \simeq -1 \\ \tanh x_2 \simeq 1 \end{cases} \Rightarrow \begin{cases} \mathcal{M}_1 = n_1 \mu_B \\ \mathcal{M}_2 = -n_2 \mu_B \end{cases} \tag{7.88}$$

$$\text{L.T.} \quad \mathcal{M} = \mathcal{M}_1 + \mathcal{M}_2 = (n_1 - n_2) \mu_B, \tag{7.89}$$

and

$$\text{H.T.} \begin{cases} x_1 < 0, & |x_1| \ll 1 \\ x_2 > 0, & |x_2| \ll 1 \end{cases} \Rightarrow \begin{cases} \tanh x_1 \simeq x_1 \\ \tanh x_2 \simeq x_2 \end{cases}$$

$$\Rightarrow \begin{cases} \mathcal{M}_1 = \frac{n_1 \mu_B^2 \mu_0}{k_B T} \left(H - |\lambda_2| \mathcal{M}_2 \right) \\ \mathcal{M}_2 = \frac{n_2 \mu_B^2 \mu_0}{k_B T} \left(H - |\lambda_1| \mathcal{M}_1 \right) \end{cases}. \tag{7.90}$$

The solutions of the set of Eqs. (7.90) are thus the magnetisations of the two sub-lattices at (fairly) high temperatures. To investigate whether there might be a finite magnetisation in zero applied field, i.e., a ferrimagnetic phase, we set $H = 0$ and equate the determinant of the coefficients to zero, whence

$$T_c = \sqrt{T_1 T_2}, \quad T_i = \frac{n_i \mu_B^2 \mu_0 |\lambda_i|}{k_B}. \tag{7.91}$$

The system is thus ferrimagnetic below T_c, and paramagnetic above T_c.[1] To study the behaviour of the paramagnetic phase in the vicinity of the transition, we solve Eqs. (7.90), for $H \neq 0$ and $T < T_c$, with the result:

$$\begin{cases} \mathcal{M}_1 = n_1 \left(1 - \frac{T_2}{T}\right)\left(1 - \frac{T_c^2}{T^2}\right)^{-1} \frac{\mu_B^2 \mu_0}{k_B T} H \\ \mathcal{M}_2 = n_2 \left(1 - \frac{T_1}{T}\right)\left(1 - \frac{T_c^2}{T^2}\right)^{-1} \frac{\mu_B^2 \mu_0}{k_B T} H \end{cases} \Rightarrow \begin{cases} \mathcal{M}_1 = n_1 \left(\frac{T - T_2}{T^2 - T_c^2}\right) \frac{\mu_B^2 \mu_0}{k_B} H \\ \mathcal{M}_2 = n_2 \left(\frac{T - T_1}{T^2 - T_c^2}\right) \frac{\mu_B^2 \mu_0}{k_B} H \end{cases} . \tag{7.92}$$

For the magnitude of the total magnetisation, we get

$$\begin{aligned} \mathcal{M} = \mathcal{M}_1 + \mathcal{M}_2 &= \frac{(n_1 + n_2)\frac{\mu_B^2 \mu_0}{k_B} T - \frac{n_1 n_2 \mu_B^4 \mu_0^2}{k_B^2}(|\lambda_1| + |\lambda_2|)}{T^2 - T_c^2} H \\ &= \frac{aT + b}{T^2 - T_c^2} \mu_0 H, \end{aligned} \tag{7.93}$$

where we have defined the constants a and b. The susceptibility of the paramagnetic phase is then

$$\chi_T = \frac{1}{\mu_0}\left(\frac{\partial \mathcal{M}}{\partial H}\right)_T = \frac{aT + b}{T^2 - T_c^2}, \tag{7.94}$$

which is indeed different from the paramagnetic susceptibility of a Weiss ferromagnet, Eq. (7.45).

Antiferromagnetism is a special case of ferrimagnetism, in which the magnetisations of the two sub-lattices have the same magnitude, but opposite signs, so that the total magnetisation always vanishes: the two sub-lattices have equal numbers of atoms, $n_1 = n_2 = n/2$, with equal magnetic moments, $\lambda_1 = \lambda_2 \equiv \lambda$. Examples of antiferromagnetic solids are elements like cerium, chromium and neodymium, and compounds like CuO, FeS and MnO_2. In this case, Eqs. (7.85) and (7.86) reduce to

$$\mathcal{M}_1 = \frac{n}{2} \mu_B \tanh\left[\frac{\mu_B \mu_0}{k_B T}\left(H - |\lambda|\mathcal{M}_2\right)\right], \tag{7.95}$$

$$\mathcal{M}_2 = \frac{n}{2} \mu_B \tanh\left[\frac{\mu_B \mu_0}{k_B T}\left(H - |\lambda|\mathcal{M}_1\right)\right]; \tag{7.96}$$

whence it straightforwardly follows that $\mathcal{M}_1 = -\mathcal{M}_2$ if $H = 0$. Moreover,

$$\mathcal{M}_1 = \frac{n}{2} \mu_B \tanh\left(\frac{\mu_B \mu_0 |\lambda|}{k_B T}\mathcal{M}_1\right), \tag{7.97}$$

which, except for the factor $1/2$, is identical to Eq. (7.39) for the ferromagnetic solid in zero applied field.

---

[1] We are assuming that the temperature is high, but not so high that the spins are necessariliy disordered.

In the low- and high-temperature limits we have, for an antiferromagnet:

$$\text{L.T.} \quad \mathcal{M}_1 \simeq \frac{n}{2}\mu_B = -\mathcal{M}_2, \tag{7.98}$$

$$\text{H.T.} \quad \begin{cases} \mathcal{M}_1 \simeq \frac{n\mu_B^2\mu_0}{2k_BT}\left(H - |\lambda|\mathcal{M}_2\right) \\ \mathcal{M}_2 \simeq \frac{n\mu_B^2\mu_0}{2k_BT}\left(H - |\lambda|\mathcal{M}_1\right) \end{cases}. \tag{7.99}$$

Again requiring a solution of these equations with non-zero sub-lattice magnetisations at high temperatures and no applied field, we find the *Néel temperature*:

$$T_N = \frac{n\mu_B^2\mu_0|\lambda|}{2k_B}, \tag{7.100}$$

which is the transition temperature between the paramagnetic state ($T > T_N$) and the antiferromagnetic state ($T < T_N$). The magnetisation and the susceptibility of the paramagnetic phase in zero applied field are found as before:

$$\mathcal{M}_1 = \frac{\frac{n\mu_B^2\mu_0}{2k_B}\left(T - \frac{n\mu_B^2\mu_0|\lambda|}{2k_B}\right)}{T^2 - T_N^2}H = \frac{\frac{T_N}{|\lambda|}(T - T_N)}{T^2 - T_N^2}H$$

$$= \frac{T_N}{|\lambda|}\frac{1}{T + T_N}H = -\mathcal{M}_2, \tag{7.101}$$

$$\mathcal{M} = \mathcal{M}_1 + \mathcal{M}_2, \tag{7.102}$$

$$\chi_T = \frac{1}{\mu_0}\left(\frac{\partial\mathcal{M}}{\partial H}\right)_T = \frac{2T_N}{\mu_0|\lambda|}\frac{1}{T + T_N}. \tag{7.103}$$

Note the similarity of Eqs. (7.103) and (7.45), the latter obtained for the Weiss ferromagnet. In both cases, $1/\chi_T$ is linear at the temperature in the vicinity of the transition. A fundamental difference lies in the sign of the critical temperature: the Néel temperature appears with a plus sign.

7.6 Spin waves and magnons

As seen in the preceding sections, the ground state of a system of N spins described by the Heisenberg Hamiltonian with a positive exchange coefficient $J_{ij} > 0$ is the state with all spins parallel to each other, i.e., all the vectors pointing in the same direction: taking this direction (the direction of spontaneous magnetisation) as the OZ axis, then $S_{iz} = S$ for all $i = 1, \ldots, N$. We can ask what the excited states of such a system are. We shall deal with the first excited state, that is, the state with energy immediately above that of the ground state, but lower than those of all the other excited states. We might think that such a state would consist of just one misaligned spin, $S_{jz} = S - \hbar$, and $N - 1$ spins aligned as in the ground state, $S_{iz} = S$, $i \neq j$. In fact this would correspond to a state with higher energy than that allowed by an additional degree of freedom of the spins, which is the possibility of precessing around the direction of spontaneous magnetisation. This precession deviates the

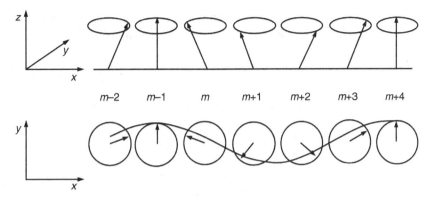

Figure 7.10 Propagating spin wave: the spins precess around the magnetisation direction, Top panel: side view; bottom panel: top view. The position of a lattice spin is indexed by m; consecutive spins are a distance a (the lattice constant) apart, hence the m-th spin has position coordinate $x = ma$.

spins away from their preferred orientation, and consequently decreases the magnetisation, and increases the energy, relative to the ground state. The description that follows is mostly classical, in the sense of non-quantum; see, e.g., Christman (1988, chap. 11.5) or Kittel (1995, chap. 15) for more details. This will later be adapted to the quantum case, in order to study the thermodynamics of elementary magnetic excitations (magnon gas).

We consider a one-dimensional lattice – a row of spins – shown schematically in Figure 7.10. This will allow an easier derivation of results that are of general relevance.

We focus on the mth spin in the chain, $\mathbf{S}_m$: its position coordinate is $x = ma$, where a is the distance between consecutive spins (the lattice constant). According to the Heisenberg model, this mth spin interacts with its nearest neighbours only: the $(m-1)$th spin, $\mathbf{S}_{m-1}$, on its left, and the $(m+1)$th spin, $\mathbf{S}_{m+1}$, on its right. Because the spin is an angular momentum, its time evolution is driven by the total torque τ_m acting on it, which is generated by its nearest neighbours:

$$\boldsymbol{\tau}_m = \boldsymbol{\mu}_m \times \mathbf{B}_{eff}, \quad \boldsymbol{\mu}_m = -\frac{g\mu_B}{\hbar}\mathbf{S}_m. \tag{7.104}$$

The effective field is obtained by analogy with the energy of a magnetic moment in a magnetic field, $\varepsilon = -\boldsymbol{\mu} \cdot \mathbf{B}$. Neglecting the constant term in the exchange energy (7.18) and writing $\mathbf{S}_i = -(\hbar/g\mu_B)\boldsymbol{\mu}_i$, we get[2]

$$\varepsilon_{ij} = -2J_{ij}\frac{\mathbf{S}_i \cdot \mathbf{S}_j}{\hbar^2} = -\boldsymbol{\mu}_i \cdot \left(-\frac{2J_{ij}}{g\mu_B\hbar}\mathbf{S}_j\right) \Rightarrow \mathbf{B}_{eff} = -\frac{2J_{ij}}{g\mu_B\hbar}\mathbf{S}_j. \tag{7.105}$$

We can thus write the effective field acting on magnetic moment $\boldsymbol{\mu}_m$ as

$$\mathbf{B}_{eff} = -\frac{J}{\hbar g\mu_B}(\mathbf{S}_{m-1} + \mathbf{S}_{m+1}), \tag{7.106}$$

---

[2] We drop the hats, as these are 'classical' spins.

where $J = J_{ij}$ if i and j are nearest neighbours and zero otherwise, and a factor of 1/2 was introduced to avoid double counting. The equation of motion for the mth spin is then

$$\frac{d\mathbf{S}_m}{dt} = \boldsymbol{\tau}_m = \frac{J}{\hbar^2}\left(\mathbf{S}_m \times \mathbf{S}_{m-1} + \mathbf{S}_m \times \mathbf{S}_{m+1}\right). \tag{7.107}$$

In terms of the Cartesian components of $\mathbf{S}_m$, this is written

$$\frac{dS_{mx}}{dt} = \frac{J}{\hbar^2}\left(S_{my}S_{m-1z} - S_{mz}S_{m-1y} + S_{my}S_{m+1z} - S_{mz}S_{m+1y}\right), \tag{7.108}$$

$$\frac{dS_{my}}{dt} = -\frac{J}{\hbar^2}\left(S_{mx}S_{m-1z} - S_{mz}S_{m-1x} + S_{mx}S_{m+1z} - S_{mz}S_{m+1x}\right), \tag{7.109}$$

$$\frac{dS_{mz}}{dt} = \frac{J}{\hbar^2}\left(S_{mx}S_{m-1y} - S_{my}S_{m-1x} + S_{mx}S_{m+1y} - S_{my}S_{m+1x}\right). \tag{7.110}$$

The lowest-energy excitations above the ground state correspond to spins deviating only slightly from the OZ axis, that is, whose x- and y-components are much smaller than their z-component: $S_{mx}, S_{my} \ll S_{mz}$. Consequently, we shall neglect in Eqs. (7.108)–(7.110) terms that are products of x and y spin components (e.g., $S_{mx}S_{m+1y}$), leading to

$$\frac{dS_{mx}}{dt} = \frac{J}{\hbar^2}S_z\left(2S_{my} - S_{m-1y} - S_{m+1y}\right), \tag{7.111}$$

$$\frac{dS_{my}}{dt} = -\frac{J}{\hbar^2}S_z\left(2S_{mx} - S_{m-1x} - S_{m+1x}\right), \tag{7.112}$$

$$\frac{dS_{mz}}{dt} \approx 0 \Rightarrow S_{mz} = \text{const.} \equiv S_z, \tag{7.113}$$

where, from Eq. (7.113), S_{mz} was taken to be approximately constant and the same for all spins.

Since the spins are arranged regularly in a very long chain (actually infinite, when periodic boundary conditions are used), we seek periodic solutions of Eqs. (7.111)–(7.113) that we assume to be of the form:

$$S_{mx} = S_0 \cos(kma - \omega t), \tag{7.114}$$

$$S_{my} = S_0 \sin(kma - \omega t), \tag{7.115}$$

$$S_{mz} = S_z = \sqrt{S^2 - S_0^2}, \tag{7.116}$$

where $S = \|\mathbf{S}\|$. These solutions describe precessing spins, i.e., spins that trace out cones of circular cross-section with axes along the direction of magnetisation. The phase of the spins' motion changes along the chain. As can be seen in the bottom panel of Figure 7.10, this corresponds to a propagating *spin wave*. We obtain its angular frequency ω by substituting Eqs. (7.114) and (7.115) into Eqs. (7.111) and (7.112):

$$\omega = \frac{2JS_z}{\hbar^2}(1 - \cos ka) = \frac{4JS_z}{\hbar^2}\sin^2\left(\frac{ka}{2}\right). \tag{7.117}$$

This is the dispersion relation for spin waves. Note that in the limit $ka \ll 1$ (long-wavelength limit, i.e., the wavelength is much larger than the spacing between spins), we get

$$\omega \approx \frac{JS_z}{\hbar^2}(ka)^2. \tag{7.118}$$

Thus the angular frequency is quadratic in the wavevector k.

Finally, we calculate the exchange energy from Eq. (7.24), omitting the constant term:

$$
\begin{aligned}
E_{\text{exch}} &= -\frac{J}{2}\sum_{m=1}^{N}\left[\frac{2\mathbf{S}_m \cdot (\mathbf{S}_{m-1} + \mathbf{S}_{m+1})}{\hbar^2}\right] \\
&= -\frac{2NJ}{\hbar^2}\left[S^2 - 2S_0^2 \sin^2\left(\frac{ka}{2}\right)\right] \\
&= -\frac{2NJS^2}{\hbar^2} + \frac{NS_0^2}{S_z}\omega \approx -\frac{2NJS^2}{\hbar^2} + \frac{NS_0^2}{S}\omega.
\end{aligned} \tag{7.119}
$$

The first term is the groundstate energy. The last term is the increase in energy due to the excitation of a spin wave with angular frequency ω given by Eq. (7.117).

As stated before, our description of spin waves has been entirely classical up to now (except that Planck's constant appears in some formulae). The quantum character is introduced by restricting the possible values of the energy of the wave, which should be multiples of $\hbar\omega$. The quantum of energy of a spin wave is called a *magnon*. Quantisation of magnons is performed by following the same procedure as for phonons in Chapter 6. The energy of a mode with n_k magnons with frequency ω_k is

$$\varepsilon_k = \left(n_k + \frac{1}{2}\right)\hbar\omega_k. \tag{7.120}$$

Like phonons, magnons obey the Planck statistics:

$$\overline{n_k} = \frac{1}{e^{\beta\hbar\omega_k} - 1}, \tag{7.121}$$

where we have dropped the zero-point energy. Hence the Planck statistics shows up whenever we deal with a system of quantum harmonic oscillators. The mean total number of magnons that are excited at a temperature T is

$$\overline{N_M} = \int_0^\infty \overline{n}(\omega)g(\omega)d\omega. \tag{7.122}$$

At low enough temperatures, the limits of the integral can be taken as 0 and ∞, because $\overline{n}(\omega) \to 0$ exponentially when $\omega \to 0$. Now we have to calculate the density of states $g(\omega)$ for magnons. Magnons have only one polarisation for each value of the wavevector k (Kittel, 1995, chap. 15). Using the density of states (6.116),

$$g(k)dk = \frac{Vk^2}{2\pi^2}dk,$$

and the approximate dispersion relation (7.118), the density of states per unit volume for magnons is

$$g(\omega)d\omega = \frac{1}{4\pi^2}\left(\frac{\hbar^2}{JSa^2}\right)^{3/2}\omega^{1/2}d\omega, \tag{7.123}$$

whence

$$\overline{N_M} = \frac{1}{4\pi^2}\left(\frac{\hbar^2}{JSa^2}\right)^{3/2}\int_0^\infty \frac{\omega^{1/2}}{e^{\beta\hbar\omega}-1}d\omega = \frac{1}{4\pi^2}\left(\frac{\hbar k_B T}{JSa^2}\right)^{3/2}\int_0^\infty \frac{x^{1/2}}{e^x-1}dx, \tag{7.124}$$

where we have made the change of variables $x = \hbar\omega/k_B T$, and the integral is a constant:

$$\int_0^\infty \frac{x^{1/2}}{e^x-1}dx = 0.0587\left(4\pi^2\right).$$

The number N of atoms per unit volume is Q/a^3, where $Q = 1$, 2 or 4 for the simple cubic, body-centred cubic and face-centred cubic lattices, respectively (Kittel, 1995, chap. 15). Noting that $\overline{N_M}/NS$ equals the relative change of the magnetisation $\Delta\mathcal{M}/\mathcal{M}_0$ (where $\Delta\mathcal{M} = \mathcal{M}_0 - \mathcal{M}(T)$ is the deviation away from saturation), we find

$$\frac{\Delta\mathcal{M}}{\mathcal{M}_0} = \frac{0.0587}{SQ}\left(\frac{\hbar k_B T}{JS}\right)^{3/2}. \tag{7.125}$$

This result is known as *Bloch's law* and explains the $T^{3/2}$ dependence of the magnetisation at low temperatures, as found experimentally (Kittel, 1995, chap. 15), as opposed to the exponential temperature dependence predicted by mean-field theory (see Problem 7.1).

Problems

7.1 Show that for the Weiss model the relative change of the magnetisation away from saturation, $\Delta\mathcal{M} \equiv \mathcal{M}_0 - \mathcal{M}(T)$, is given by

$$\frac{\Delta\mathcal{M}}{\mathcal{M}_0} \simeq 2e^{-2T_c/T}.$$

7.2 Consider a hypothetical magnetic ion whose magnetic moment m can take any value between $-\infty$ and $+\infty$. The corresponding density of states is $g(m) \propto \exp\left(-\alpha m^2\right)$, where α is a constant with the appropriate dimensions. In a magnetic field H, the energy of one such ion is $-mH$.

a. Calculate the partition function, the Helmholtz free energy and the specific heat at constant magnetic field per magnetic ion.

b. Calculate the mean magnetic moment $m(T,H)$.

c. A material is now made from N such ions, placed on the sites of a crystal lattice. Assume that, in a mean-field approximation, each ion feels the effective interaction $B_{eff} = \eta m^{1/2}$ (where η is a constant) due to all the other ions.

Show that the material is spontaneously magnetised at all temperatures with mean magnetisation

$$\mathcal{M}(T) = \left(\frac{\eta}{2\alpha T}\right)^2.$$

References

Christman, J. R. 1988. *Fundamentals of Solid State Physics*. Wiley.

Goldenfeld, N. 1992. *Lectures on Phase Transitions and the Renormalization Group*. Addison-Wesley.

Ibach, H., and Lüth, H. 2003. *Solid State Physics*, 3rd edition. Springer Verlag.

Kittel, C. 1995. *Introduction to Solid State Physics*, 7th edition. Wiley.

Landau, L. D., and Lifshitz, E. M. 1960. *Electrodynamics of Continuous Media*. Pergamon Press.

Landau, L. D., Lifshitz, E. M., and Pitaevskii, L. P. 1980. *Statistical Physics, Part 1*, 3rd edition. Pergamon Press.

Schiff, L. I. 1968. *Quantum Mechanic*s, 3rd edition. McGraw-Hill.

The Ising model

8.1 Introduction

The Ising model[1] was invented by Wilhelm Lenz in 1920, who gave it as a problem to his student Ernst Ising (Brush, 1962). Originally intended as a simplified model of magnetism, it has found application to many other problems, of which we shall see a few examples later in this chapter.

The Ising model can be derived from the spin-1/2 Heisenberg model for a uniaxial sample, i.e., when there is a single preferential direction around which the system exhibits continuous rotational symmetry. We assume that there is only one unpaired electron per atom, that OZ is along the crystal axis, and that the N spins are quantised according to $s_i = \pm 1/2\hbar$, $i = 1,\ldots,N$. Equation (7.18) for the interaction between a pair of spins is now replaced by

$$\varepsilon_{ij} = -J_{ij}\sigma_i\sigma_j, \tag{8.1}$$

where σ_i and σ_j are the non-dimensionalised spin variables:

$$\sigma_i = \pm 1; \quad \sigma_j = \pm 1. \tag{8.2}$$

We thus have

$$\varepsilon_{ij} = \begin{cases} +J_{ij} & \text{if } (\sigma_i,\sigma_j) = (+,-) \text{ or } (-,+) \\ -J_{ij} & \text{if } (\sigma_i,\sigma_j) = (+,+) \text{ or } (-,-) \end{cases}, \tag{8.3}$$

whence we recover Eq. (7.20). Equations (8.1) and (8.2) thus mimic the exchange interaction between pairs of spins 1/2: if $J_{ij} > 0$, then $\varepsilon_{ij} < 0$ for two parallel spins and $\varepsilon_{ij} > 0$ for two antiparallel spins, which favours ferromagnetic order. Conversely, if $J_{ij} < 0$, then $\varepsilon_{ij} > 0$ for two parallel spins and $\varepsilon_{ij} < 0$ for two antiparallel spins, whereby antiferromagnetic order is favoured instead.

Here we shall restrict ourselves to a simulation study of the two-dimensional Ising model It can be shown that the one-dimensional Ising model has zero spontaneous magnetisation at all finite temperatures. The proof is a simple application of the powerful transfer matrix method; we give it in the next section, following Binney *et al.* (1993, chap. 3.2).

---

[1] Pronounced 'easing' and not 'eye-sing'.

8.2 Exact solution of the one-dimensional Ising model

Consider a linear chain composed of N spins. Each spin interacts only with its nearest neighbours, i.e., the ith spin interacts only with the $(i-1)$th and $(i+1)$th spins. Let the interaction strength be $J > 0$. To eliminate boundary effects – which, in any case, would be unimportant in the thermodynamic limit, $N \to \infty$ – we wrap the chain up into a ring: thus spin $i = 1$ interacts with spins $i = 2$ and $i = N$, while spin $i = N$ interacts with spins $i = N - 1$ and $i = 1$. The partition function is then

$$Z = \sum_{\{\sigma_i\}} \exp\left(-\beta \sum_{i=1}^{N} \varepsilon_{ii+1}\right) = \sum_{\{\sigma_i\}} \exp\left(\beta J \sum_{i=1}^{N} \sigma_i \sigma_{i+1}\right) = \sum_{\{\sigma_i\}} \prod_{i=1}^{N} \exp\left(\beta J \sigma_i \sigma_{i+1}\right), \quad (8.4)$$

where $\{\sigma_i\}$ denotes a sum over all (2) values of each of the (N) spins. Note that the product of exponentials in the rightmost side of Eq. (8.4) is of the form

$$\exp\left(\beta J \sigma_1 \sigma_2\right) \exp\left(\beta J \sigma_2 \sigma_3\right) \cdots \exp\left(\beta J \sigma_{N-1} \sigma_N\right) \exp\left(\beta J \sigma_N \sigma_1\right).$$

In other words, Z is the sum over $\sigma_1, \sigma_2, \ldots, \sigma_N$ of the product of Boltzmann factors of the interactions between pairs of consecutive spins. Because we have imposed periodic boundary condition, the first and last spin in each of these products is σ_1.

We now define the *transfer matrix* $\mathbf{T}$ as the matrix whose entries are the Boltzmann factors for all possible interactions between pairs of spins:[2]

$$\mathbf{T} = \begin{bmatrix} \exp(\beta J) & \exp(-\beta J) \\ \exp(-\beta J) & \exp(\beta J) \end{bmatrix}. \quad (8.5)$$

On the basis of this definition and Eq. (8.4), it is not difficult to convince ourselves that

$$Z = \sum_{i=1}^{2} \left(\sum_{k_1=1}^{2} \sum_{k_2=1}^{2} \cdots \sum_{k_{N-1}=1}^{2} T_{ik_1} T_{k_1 k_2} \cdots T_{k_{N-1} i} \right) = \mathrm{Tr}\, \mathbf{T}^N, \quad (8.6)$$

where Tr denotes the operation trace of a matrix.[3] That Z equals the trace of a matrix is a consequence of the imposed periodic boundary condition. The problem of finding the partition function thus reduces to finding the trace of the Nth power of the transfer matrix, which according to any linear algebra textbook is

$$\mathrm{Tr}\, \mathbf{T}^N = \lambda_>^N + \lambda_<^N, \quad (8.7)$$

where $\lambda_>$ and $\lambda_<$ are, respectively, the largest and the smallest eigenvalue of $\mathbf{T}$.

Because $\mathbf{T}$ is a real symmetric matrix, it has real eigenvalues. They are

$$\lambda_> = 2\cosh(\beta J), \quad \lambda_< = 2\sinh(\beta J). \quad (8.8)$$

---

[2] In the Ising model each spin can only take one of two different values, so the transfer matrix is a 2×2 matrix. In the more general case of a model where each spin can take one of S different values, the corresponding transfer matrix will be an $S \times S$ matrix.

[3] Readers are invited to check this by explicit calculation of Z, e.g., for $N = 3$ or $N = 4$.

In the limit $N \to \infty$, the partition function will be dominated by the largest eigenvalue:

$$Z = \mathrm{Tr}\,\mathbf{T}^N \approx \lambda_>^N = 2^N \cosh^N(\beta J) \Rightarrow F = -\frac{N}{\beta}\ln\left[2\cosh(\beta J)\right]. \tag{8.9}$$

We see that the Helmholtz free energy, F, does not exhibit any discontinuities or singularities. Therefore, by the results of Chapter 10, the system does not exhibit any phase transition.

Consider now the same model in an external magnetic field h (in reduced units, to be defined later) directed along OZ. The interaction energy between a pair of spins is now:

$$\varepsilon_{ii+1} = -J\sigma_i\sigma_{i+1} - \frac{1}{2}h(\sigma_i + \sigma_{i+1}). \tag{8.10}$$

The transfer matrix is now written:[4]

$$\mathbf{T} = \left[\begin{array}{cc} \exp[\beta(J+h)] & \exp(-\beta J) \\ \exp(-\beta J) & \exp[\beta(J-h)] \end{array} \right], \tag{8.11}$$

with eigenvalues

$$\lambda_> = e^{\beta J}\left[\cosh(\beta h) + \sqrt{\cosh^2(\beta h) - \left(1 - e^{-4\beta J}\right)}\right], \tag{8.12}$$

$$\lambda_< = e^{\beta J}\left[\cosh(\beta h) - \sqrt{\cosh^2(\beta h) - \left(1 - e^{-4\beta J}\right)}\right], \tag{8.13}$$

whence we conclude that

$$Z \simeq e^{N\beta J}\left[\cosh(\beta h) + \sqrt{\cosh^2(\beta h) - \left(1 - e^{-4\beta J}\right)}\right]^N, \tag{8.14}$$

$$F = -\frac{1}{\beta}\ln Z = -N\left\{J + \frac{1}{\beta}\ln\left[\cosh(\beta h) + \sqrt{\cosh^2(\beta h) - \left(1 - e^{-4\beta J}\right)}\right]\right\}. \tag{8.15}$$

Again, the free energy does not exhibit any discontinuity or singularity. The mean value of the spin is

$$\overline{\sigma} = \frac{1}{N}\sum_{\{\sigma_i\}}\sigma_i p_i = \frac{1}{NZ}\sum_{\{\sigma_i\}}\sigma_i \exp\left(\beta J\sum_{i=1}^{N}\sigma_i\sigma_{i+1} + \beta h\sum_{i=1}^{N}\sigma_i\right)$$

$$= \frac{1}{N\beta}\frac{\partial \ln Z}{\partial h} = \frac{\sinh(\beta h)}{\sqrt{\cosh^2(\beta h) - \left(1 - e^{-4\beta J}\right)}}, \tag{8.16}$$

which vanishes when $h \to 0$, i.e., there is no spontaneous magnetisation of the one-dimensional Ising model at any finite temperature.

The above result, which we found by explicit calculation of the partition function, can be cast in more qualitative terms through the *Landau–Peierls argument*, which states that there is no ordered phase of any one-dimensional model with short-range interactions. To prove this, imagine that there is a very low-temperature ordered state where all spins take

---

[4] In Eq. (8.10) we have symmetrised the field term with respect to σ_i and σ_{i+1} in order to preserve the symmetry of the transfer matrix.

the value $+1$. Because the temperature is finite (albeit very low), there are still fluctuations – some spins will flip to -1. Assume that all spins in a block of $n \ll N$ contiguous spins are now -1. Now two 'walls' have been created, one on either side of this block, at which there is a pair of antiparallel spins, where previously there was a pair of parallel spins. The energy cost of creating these walls is then $2 \times [J - (-J)] = 4J$. That is, creating a block of spins of the 'wrong' sign raises the energy, as we would have expected.

On the other hand, there are approximately N^2 different ways of placing 2 walls in a chain of N spins, so the entropy associated with wall creation is approximately $k_B \ln N^2 = 2k_B \ln N$. The Helmholtz free energy change associated with creating a block of n contiguous spins with the 'wrong' orientation, relative to the perfectly ordered state, is then

$$\Delta F = 4J - 2k_B T \ln N. \tag{8.17}$$

However low the temperature, for sufficiently large N we will always have $\Delta F < 0$, i.e., inverting the orientation of a block of n contiguous spins will lower the Helmholtz free energy, and is, therefore, favourable. It follows hat the ordered state is unstable *but only because the energy cost does not depend on n*: if this were not so – if, e.g., the energy cost of inverting the orientations of a block of n spins were an increasing function of n – then it might be advantageous to invert the orientations of small blocks, but not of large blocks. A state with a small, but non-vanishing, degree of order might then survive. And that is indeed so in higher dimensions.

In dimensions $d \geq 2$, it can be shown that the Ising model exhibits a ferromagnetic phase (see chap. 10) for $d = 2$. Consider a square lattice of N sites, with a spin variable $\sigma_i = \pm 1$ at each site. The total energy – the Hamiltonian – of the Ising model is

$$E = -\frac{1}{2} \sum_{i,j} J_{ij} \sigma_i \sigma_j - H \sum_i \sigma_i, \tag{8.18}$$

where

$$\begin{cases} J_{ij} = J & \text{if } i,j \text{ are nearest neighbours} \\ J_{ij} = 0 & \text{otherwise} \end{cases}. \tag{8.19}$$

The factor $1/2$ is introduced to correct for double counting of the same pair of spins (i,j). The second term on the right-hand side of Eq. (8.18) is the interaction energy of the spins with a magnetic field directed along OZ:

$$H = \mu_B B. \tag{8.20}$$

The two-dimensional Ising model on a square lattice was solved analytically by Onsager (Onsager, 1944); we discuss it in more detail in Chapter 10. Here we just collect the basic results in zero applied field:

Reduced critical temperature: $\quad t_c \equiv \dfrac{k_B T_c}{J} = 2.269,$ $\qquad$ (8.21)

Magnetic moment per lattice site: $\quad m \equiv \dfrac{M}{N} \begin{cases} \propto (T_c - T)^{1/8} & \text{for } T \to T_c^- \\ \simeq 1 & \text{for } T \to 0 \end{cases}.$ $\quad$ (8.22)

8.3 Monte Carlo simulations of the Ising model

The basics of Monte Carlo simulation have been outlined in Chapter 1. In summary, in a Monte Carlo simulation a subset of configurations of the configuration space of a given system is generated, according to a predefined probability distribution. This procedure is known as 'sampling of configuration space'. Any relevant physical quantities for the system under study will then be computed as statistical averages over this subset, or sample. When studying the random walk in Chapter 1 we used the simplest unbiased sampling method, in which the configuration space of random walks was sampled homogeneously. In our Ising model simulations we shall employ a biased sampling method which selects such regions of configuration space as make the most significant contributions to the quantities we wish to compute. This is known as *importance sampling*.

8.3.1 Importance sampling

When performing an unbiased sampling of the configuration space of some system, it is often the case that most configurations contribute very little to the quantities that we wish to compute as statistical averages. We therefore need to generate a huge number of configurations in order to obtain a representative sample that will yield statistically significant results. Let us see this in more detail for the two-dimensional Ising model. A given configuration of the N-spin lattice is specified by the set of all σ_i, $i = 1, 2, \ldots N$, e.g.,

$$\sigma_1 = +1; \quad \sigma_2 = -1; \quad \sigma_3 = -1; \quad \ldots; \quad \sigma_N = +1. \tag{8.23}$$

The total number of configurations is huge, even for a very small system. If L is the linear dimension of the square lattice, i.e., the number of spins in a row or in a column, then $N = L^2$. If, e.g., $L = 20$, then $N = 400$ and the number of possible configurations (remember that each spin has two possible states) is $2^N \approx 2.58 \times 10^{120}$. A representative sample of these will still be enormous, which negates the core assumption of the Monte Carlo method – namely, that it is enough randomly to generate a relatively small number of configurations.

The *Metropolis method* or *Metropolis algorithm* offers a way to perform importance sampling of the configuration space. In this method a Markov chain of configurations is generated in which each configuration x_{l+1} is obtained from the previous one x_l with a suitably chosen transition probability $w(l \to l+1)$. The time evolution of this stochastic process is governed by the master equation

$$\frac{dP_l(t)}{dt} = -\sum_{l \neq m} \left[P_l(t)w_{(l \to m)} - P_m(t)w_{(m \to l)} \right], \tag{8.24}$$

where $P_l(t)$ is the probability that the system is in state l (i.e., in configuration x_l) at time t. At equilibrium, $dP_l(t)/dt = 0$, whence from Eq. (8.24),

$$P_l(t)w_{(l \to m)} = P_m(t)w_{(m \to l)}, \tag{8.25}$$

i.e., at equilibrium the system must 'enter' and 'exit' configuration x_l (for all l) with equal probability. This condition that the transition probability $w_{(l \to m)}$ must satisfy is known as *detailed balance*. Furthermore, $w_{(l \to m)}$ must be chosen so that, at equilibrium, the configurations $x_0, x_1, x_2, \ldots, x_l, \ldots$ are generated according to the Boltzmann distribution:

$$P_{eq}(x_l) = \frac{1}{Z} \exp\left(-\frac{E(x_l)}{k_B T}\right), \tag{8.26}$$

where Z is the partition function and $E(x_l)$ is the energy of configuration x_l. This ensures that we only sample regions of configuration space where the Boltzmann factor is not too small, i.e., that we select only a few configurations with a low probability of occurence. From Eqs. (8.25) and (8.26) we conclude that the ratio of probabilities for the transition $l \to m$ and the reverse transition $m \to l$ depends only on the energy difference between the initial and final states:

$$\frac{w_{(l \to m)}}{w_{(m \to l)}} = \exp\left(-\frac{\Delta E}{k_B T}\right), \quad \Delta E = E(x_m) - E(x_l). \tag{8.27}$$

This does not, however, uniquely determine $w_{(l \to m)}$. Here we shall use the Metropolis function:

$$w_{(l \to m)} = \min\left[1, \exp\left(-\frac{\Delta E}{k_B T}\right)\right], \tag{8.28}$$

i.e.,

$$w(l \to m) = \exp\left(-\frac{\Delta E}{k_B T}\right), \text{if } \Delta E > 0, \tag{8.29}$$

$$w(l \to m) = 1, \text{if } \Delta E < 0. \tag{8.30}$$

What is the meaning of this transition function? A new configuration is obtained from an old one by attempting to flip a spin; i.e., in terms of variables σ_i, by attempting the moves $+ \to -$ or $- \to +$. Suppose that the energy goes down on flipping a spin. Then by Eq. (8.26) the new configuration is more probable, so this transition should be accepted with probability one: hence when $\Delta E < 0$, $w = 1$. On the other hand, we do not want the algorithm to be trapped in a local minimum, so we need to allow moves that increase the energy, albeit with a low probability if they increase the energy by a large amount; hence when $\Delta E > 0$, w is given by the Boltzmann factor of the energy variation.

The problem thus reduces to generating a reasonably large number X of statistically independent configurations distributed according to Eq. (8.26). The law of large numbers (see Chapter 1) allows us to replace the statistical average by the arithmetic mean, yielding for the mean total magnetic moment and mean total energy:

$$\overline{M} = \frac{1}{X} \sum_{i=1}^{X} M_i \quad, \quad M_i \equiv M\left(\sigma_1^i, \sigma_2^i, \ldots, \sigma_N^i\right), \tag{8.31}$$

$$\overline{E} = \frac{1}{X} \sum_{i=1}^{X} E_i \quad, \quad E_i \equiv E\left(\sigma_1^i, \sigma_2^i, \ldots, \sigma_N^i\right). \tag{8.32}$$

8.3.2 Computational method

Computationally the Metropolis method is applied as follows: we start by selecting the lattice geometry and size, e.g., a square lattice of linear size L, as illustrated in Figure 8.1; and the boundary conditions, e.g., periodic along both dimensions. We also need to specify the initial configuration, e.g., all spins 'point down' (i.e., $\sigma_i = -1$, for all i) at $t = 0$. Next, we sweep through the lattice, performing the following six steps at each site:

1. Select one lattice site i at random where a spin flip ($\sigma_i \to -\sigma_i$) will be attempted.
2. Compute the associated energy change δE.
3. Compute the transition probability w.
4. Generate a random number r uniformly distributed between zero and one.
5. If $r < w$, flip the spin; otherwise, do not flip. In either case, the resulting configuration is counted as a new configuration for the purpose of computing the statistical averages.
6. Save the new configuration and compute the desired averages.

Steps 1–5 above are repeated for all lattice sites, i.e., L^2 times. This constitutes one *Monte Carlo step* (MCS), which may be taken as our unit of computational 'time'. Because each configuration is obtained from the previous one by flipping a single spin (or even by flipping no spin at all, if the move is not accepted), consecutive configurations are strongly correlated. For this reason, statistical averages should not be computed at every step 5, but only at longer 'time intervals', to ensure that the configurations used are statistically independent. It is therefore advisable to implement step 6 above *only once every MCS*.

On the other hand, starting from some initial configuration and performing steps 1–5 yields a configuration that differs from the initial one by at most one spin. It is then necessary to go through steps 1–5 many times in order to lose memory of the initial configuration and attain the equilibrium distribution at the temperature under study. It is recommended to discard (at least) the first L^2 configurations (Binder and Heermann, 1992, sec. 2.3.5), i.e., *the first configuration to be saved for the purpose of computing statistical averages should be that at the end of the first MCS (or later)*. Near the critical point, it will be necessary to discard even more configurations on account of the *critical slowing down* that occurs in the vicinity of a continuous phase transition. Basically, as we approach the transition the system takes longer and longer to relax to its equilibrium state, this relaxation time diverging

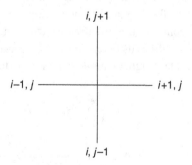

Figure 8.1   Nearest neighbours of site (i,j) on a square lattice.

at the transition. As a rule of thumb, a time $\tau = n_1 L^2$, where n_1 is an integer greater than or equal to one. should be allowed to elapse before storing the first valid configuration.

As we are primarily interested in the equilibrium properties of our system, we may take as *initial condition* a perfectly ordered state, say $\sigma_i = -1$, for all i. At sub-critical temperatures, this gives much faster relaxation to equilibrium than if we started from a randomly disordered state. It also circumvents the problem of domain formation (Binder and Heermann, 1992).

8.3.3 Metropolis algorithm

The following algorithm is from Binder and Heermann (1992).

```
(*Monte Carlo step*)
count = 0
do mcs = 1 to mcsmax
   begin
   (*Sweep through lattice*)
   do i = 1 to L
      do j = 1 to L
         generate a random number r between zero and one
         find energy change ΔE from flipping a spin
         if r < w(ΔE) then
         flip spin
         if mcs ≥ n0 then
            begin
            count = count + 1
            if count = n1 then
               begin
               perform averages
               count = 0
               end
            end
   end
```

where n_0 is the number of configurations that must be discarded at the beginning, because they are non-equilibrium, and afterwards averages are only computed every n_1 configurations.

8.3.4 Transition probabilities

In practice it is too time-consuming to compute $w(\Delta E)$ every time we want to flip a spin. Considerable savings can be made by tabulating all possible values of this function and then looking up the table whenever necessary. Of course, this is only possible because the spin is a discrete variable taking a finite number of values. In zero applied field there are

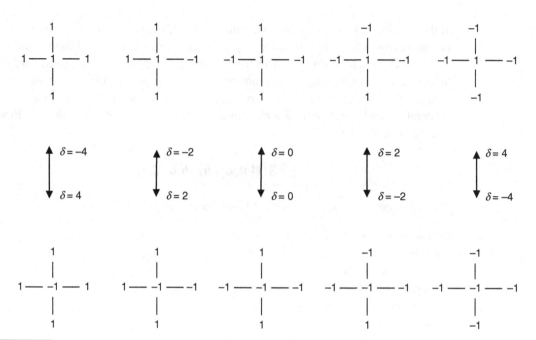

Figure 8.2 Possible energy changes (in units of 2J) of a spin and its nearest neighbours when the central spin is flipped.

only five different transition probabilities (see Figure 8.2). If a given spin σ flips from $\sigma_{initial} = \pm 1$ to $\sigma_{final} \mp 1$, we have

$$\Delta E = J\left(\sigma_{final}\sum_j \sigma_j - \sigma_{initial}\sum_j \sigma_j\right) = J\left(\pm 2\sum_j \sigma_j\right) \equiv 2J\delta$$

$$\Rightarrow \delta = \pm\sum_j \sigma_j = -4, -2, 0, 2, 4. \tag{8.33}$$

The following algorithm tabulates all possible transition probabilities.

(*Tabulate the transition probabilities in zero applied field; here $t = k_B/J*$)
do $i = -4, -2, 0, 2, 4$
   $w(i) = 1$
   if $i > 0$ then
   $w(i) = \exp(-2i/t)$

In non-zero applied field, there is an additional contribution $\pm\mu_B B$ when a spin is flipped; hence there are now 10 different transition probabilities, which can be tabulated thus:

(*Tabulate the transition probabilities in non-zero applied field; here $h = \mu_B B/J*$)
do $i = -1, 1$
   do $j = -4, -2, 0, 3, 4$

```
          begin
          delta(i,j) = j + ih
          end
do i = −1, 1
   do j = −4, −2, 0, 2, 4
      begin
      w(i,j) = 1
      if delta(i,j) > 0 then
      w(i,j) = exp(−2delta(i,j)/t)
      end
```

For sites on the boundary we need to specify the boundary conditions. We choose periodic boundary conditions, a popular choice for mimicking an infinite system. To save computer time, these are also tabulated, as follows:

```
(*Ising model*)
(*Tabulate periodic boundary conditions*)
do i = 1 to L
   begin
   iplus(i) = i + 1
   iminus(i) = i − 1
   end
iplus(L) = 1
iminus(1) = L
(*Tabulate transition probabilities in non-zero applied field*)
do i = −1, 1
   do j = −4, −2, 0, 2, 4
      begin
      delta(i,j) = j + ih
      end
do i = −1, 1
   do j = −4, −2, 0, 2, 4
      begin
      w(i,j) = 1
      if delta(i,j) > 0 then
      w(i,j) = exp(−2delta(i,j)/t)
      end
(*Initialise sample*)
do i = 1 to L
   do j = 1 to L
      (*Set all spins to −1*)
      lattice(i,j) = −1
(*Sampling loop*)
   count = 0
   do mcs = 1 to mcsmax
```

```
(*Monte Carlo loop*)
  begin
    do i = 1 to L
      do j = 1 to L
      begin
      site = lattice(i,j)
      sum = lattice(iplus(i),j)+lattice(iminus(i),j)
        + lattice(i,iplus(j))+lattice(i,iminus(j))
      (*Find the energy change from flipping a spin and retrieve
      transition probability for that energy change*)
      k1=site×sum
      k2=site
      (*Generate random number r between 0 and 1*)
      if r < w(k1,k2) then
      (*Flip spin*)
      lattice(i,j) = −site
      end
    if mcs ≥ n0 then
      begin
      count = count + 1
      if count = n1 then
        begin
        (*Perform analysis*)
        count = 0
        end
      end
  end
```

8.3.5 Computation of physical quantities

Magnetisation

For a two-dimensional square lattice, the magnetisation (or magnetic moment per unit volume) is written

$$\mathcal{M} = \frac{\overline{M}}{V} = \frac{\overline{M}}{L^2},$$ (8.34)

where $\overline{M}$ is given by Eq. (8.31). Note that, in this model, both M and $\mathcal{M}$ are non-dimensional quantities, where the latter is the mean total magnetic moment per lattice site. Now a finite sample always has zero spontaneous magnetisation (Binder and Heermann, 1992, sec. 2.3) because states with M and $-M$ are equivalent and the system alternates between them, so the total magnetic moment averaged over a large enough number of configurations will always vanish. Figure 8.3 illustrates this behaviour, as well as the *fluctuations* in $\mathcal{M}$ which, as we shall see later, are a measure of the *magnetic susceptibility* of the system.

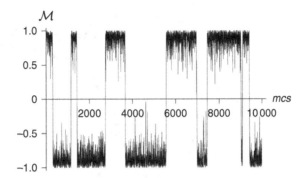

Figure 8.3 Spontaneous magnetisation, Eq. (8.34), vs number of mcs, for a two-dimensional Ising model of linear size $L = 10$, at $t = 2$.

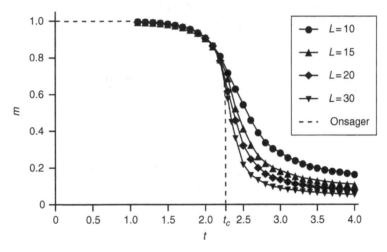

Figure 8.4 Order parameter m, Eq. (8.35), vs reduced temperature t, Eq. (8.37), for a number of system sizes. Averages are taken over 10 000 configurations. m does not go to zero at high temperatures because of finite-size effects, but becomes smaller the larger the system. The dashed line is the exact Onsager result, Eq. (10.83).

So instead of $\mathcal{M}$ we shall use as our order parameter *the mean of the absolute value of the magnetic moment per lattice site*:

$$m = \frac{|\overline{M}|}{L^2}. \tag{8.35}$$

Now $m = 1$ corresponds to perfect order, and $m = 0$ to perfect disorder. This is plotted in Figure 8.4, alongside the exact Onsager result, Eq. (10.83): for the small systems simulated there are pronounced finite-size effects in the high-temperature (paramagnetic) phase. Figure 8.5 shows m vs the applied field strength at two temperatures, above and below the transition.

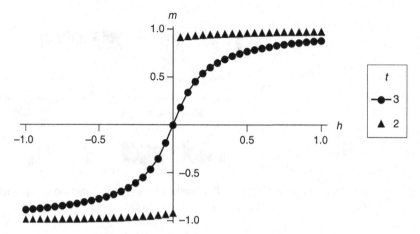

Figure 8.5 Order parameter m, Eq. (8.35), vs reduced field, Eq. (8.38). Averages are taken over 10 000 configurations. For this system size (L = 20) the reduced transition temperature is $t_c \simeq 2.3$, cf. Figures 8.7 and 8.8.

Energy

It is convenient to compute the mean energy per lattice site, defined as

$$\varepsilon = \frac{\overline{E}}{L^2}, \tag{8.36}$$

where $\overline{E}$ is given by Eq. (8.32). As for the (redefined) magnetisation m, Eq. (8.35), this allows us to compare results for different system sizes.

Note that, if the reduced variables

$$t = k_B T/J, \tag{8.37}$$

$$h = \mu_B B/J, \tag{8.38}$$

are used in the simulation algorithm, the mean energy computed from Eq. (8.36) will be in units of J, i.e., it will actually be the non-dimensional quantity $\overline{E}/J$. This is plotted in Figure 8.6.

Critical temperature

We find the (reduced) transition temperature $t_c(L)$ by locating the maxima of the thermodynamic derivatives of the magnetisation and the energy calculated as described earlier. These are the magnetic susceptibility, given by Eq. (2.76),

$$\chi_T = \frac{1}{\mu_0} \left(\frac{\partial \mathcal{M}}{\partial H} \right)_T,$$

and the heat capacity at constant total magnetic moment, given by Eq. (2.72),

$$C_M = \left(\frac{\partial U}{\partial T} \right)_M.$$

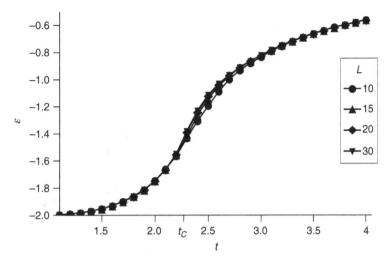

Mean energy per lattice site ε, Eq. (8.36), vs reduced temperature t, Eq. (8.37), for a number of system sizes at zero field, $h = 0$. Averages are taken over 10 000 configurations.

However, the actual computation of these quantities uses the fluctuation-dissipation relations for the susceptibility, Eq. (4.65),

$$\chi_T = \frac{\sigma^2(M)}{Vk_BT}, \tag{8.39}$$

where $\sigma^2(M)$ is the variance of the total magnetic moment; and for the heat capacity, Eq. (4.60),

$$C_H = \frac{\sigma^2(E)}{Vk_BT^2}, \tag{8.40}$$

where $\sigma^2(E)$ is the variance of the energy and we have identified the internal energy U with the mean total energy $\overline{E}$. Notice that we are now computing C_H and not C_M: this is because Eq. (2.72) was derived on the assumption that the internal energy comprises both the work performed on setting up a magnetic field of strength H and on increasing the total magnetic moment of the sample to its final value M, whereas Eq. (4.60) was derived in the context of paramagnetic systems, where we defined the internal energy as comprising only the work performed on magnetising the sample. As discussed in Section 2.2.4, C_M and C_H thus defined are equivalent.

In zero applied field, however, the isothermal susceptibility of a finite system does not exhibit a peak at the transition, because the magnetic moment alternates between M and $-M$ in the ferromagnetic phase, as mentioned before. Moreover, in a finite system, m defined by Eq. (8.35) is finite even in the paramagnetic phase, i.e., above t_c (see Figure 8.4). Equation (8.39) is therefore not usable, and instead the *zero-field susceptibility* is computed from the dimensionless expression

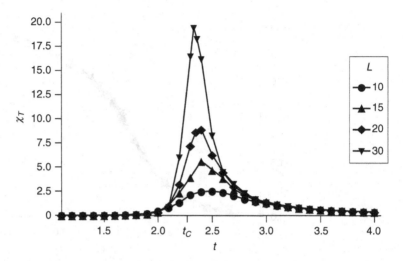

Figure 8.7 Magnetic susceptibility, Eq. (8.41), vs reduced temperature t, Eq. (8.37), for a number of system sizes at zero field, $h = 0$. Averages are taken over 10 000 configurations. In small systems the order parameter is finite even in the high-temperature phase, see Figure 8.4, so the transition is washed out, i.e., the peak in the susceptibility is very small.

$$\chi_T = \frac{\left[\overline{(M/L^2)^2} - \overline{(|M|/L^2)}^2\right]}{t}L^2. \tag{8.41}$$

In non-zero field the states M and $-M$ are no longer equivalent and the true variance of M may be used in Eq. (8.41); see Figure 8.7.

Finally, the non-dimensional *specific heat* per lattice site is computed from Eq. (8.40) for the heat capacity and the definition of reduced temperature, Eq. (8.42); see Figure 8.8:

$$\frac{C_H}{Nk_B} = \frac{C_H}{L^2 k_B} = \frac{\sigma^2(E/J)}{L^2 t^2}. \tag{8.42}$$

The transition temperature of the infinite system, t_c, can be found from $t_c(L)$ for different system sizes L by an appropriate finite-size scaling analysis. This, however, requires very good-quality data: different thermodynamic functions scale differently with system size and therefore peak at different temperatures (Binder and Heermann, 1992, sec. 4.2.3).

8.4 Other Ising models

The Ising model, Eq. (8.18), is equivalent to the simple lattice gas model for the liquid–vapour transition, as well as to models of binary alloys used in metallurgy. Furthermore, it can be used to describe antiferromagnetism in zero applied field; see Figure 8.9. Here we follow Ma (1985).

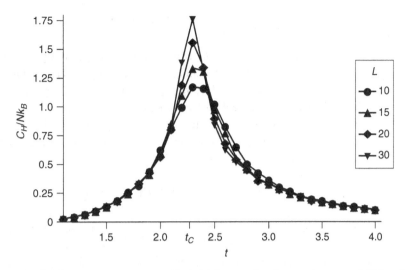

Figure 8.8   Specific heat, Eq. (8.42), vs reduced temperature t, Eq. (8.37), for a number of system sizes at zero field, $h = 0$. Averages are taken over 10 000 configurations.

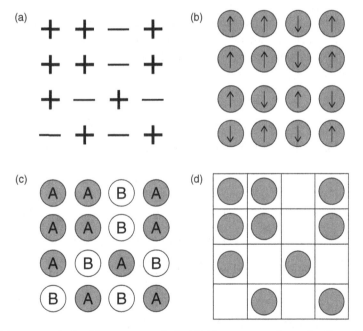

Figure 8.9   (a) Ising model on a square lattice; (b) arrangement of spins 1/2; (c) arrangement of atoms in a binary alloy; (d) two-dimensional lattice gas with hole (0) and particle (1) states; see the text for details.

8.4.1 Antiferromagnetism

If we replace J by $-J$ in Eq. (8.18), the lowest-energy configuration of a pair of nearest-neighbour spins will be antiparallel, which yields an antiferromagnetic state at sufficiently low temperatures. Each product $\sigma_i\sigma_{i+1}$ will carry a minus sign, which cancels that of $-J$; hence in zero applied field the total energy is the same as that of the ferromagnetic Ising model – provided the lattice topology is such as to allow all pairs of nearest neighbours to have antiparallel spins, i.e., if there is no *frustration*.[5] In non-zero applied field, the ferromagnetic and antiferromagnetic Ising models are no longer equivalent.

8.4.2 The lattice gas

A *lattice gas* is a model – perhaps the simplest model – for liquid–vapour phase transitions. This can be mapped onto the Ising model by assuming that $\sigma_i = +1$ corresponds to site i being occupied by a fluid particle, whereas $\sigma_i = -1$ corresponds to site i being unoccupied (i.e., site i is occupied by a 'hole', or the absence of a particle). It is further assumed that no lattice site may be occupied by more than one particle (or more than one hole), which implements a short-range repulsion between particles (or holes). The interaction energy of a pair of lattice sites is written

$$\varepsilon_{ij} = -J\sigma_i\sigma_j - \frac{1}{z}H\left(\sigma_i + \sigma_j\right) + K, \tag{8.43}$$

where z is the number of nearest neighbours, and J, H and K are as yet undetermined parameters. If we now consider an L^d-site lattice, where L is the linear size, i.e., the number of sites along a single dimension, and $d = 2$ or 3 is the lattice dimension, then the total energy will be

$$E = \frac{1}{2}\sum_{i,j=1}^{L^d}\varepsilon_{ij}. \tag{8.44}$$

Let $-\varepsilon$ be the interaction strength between two particles. From Eq. (8.43) we have

$$\sigma_i = \sigma_j = 1 \Rightarrow -J - \frac{2}{z}H + K = -\varepsilon \quad \text{(interaction between two particles)},$$

$$\sigma_i = \sigma_j = -1 \Rightarrow -J + \frac{2}{z}H + K = 0 \quad \text{(interaction between two holes)},$$

$$\sigma_i = 1, \quad \sigma_j = -1 \Rightarrow J + K = 0 \quad \text{(interaction between a particle and a hole)}.$$

Solving with respect to J, H and K yields

$$J = \frac{\varepsilon}{4}, \quad H = \frac{z\varepsilon}{4}, \quad K = -J. \tag{8.45}$$

Substituting Eqs. (8.45) into Eq. (8.43), we find

$$\varepsilon_{ij} = -\frac{\varepsilon}{4}\left(\sigma_i\sigma_j + \sigma_i + \sigma_j\right) + \text{constant}, \tag{8.46}$$

---

[5] In two dimensions, this is possible on a square lattice, but not on a triangular lattice. Exercise: check.

and for the total energy, Eq. (8.44):

$$E = -\frac{\varepsilon}{8} \sum_{i,j} \sigma_i \sigma_j - \frac{z\varepsilon}{4} \sum_i \sigma_i + \text{constant}. \tag{8.47}$$

Equation (8.47) with Eq. (8.45) and $h = H = z\varepsilon/4$ is, to within an unimportant additive constant, identical to Eq. (8.18) for the total energy of the ferromagnetic Ising model. It follows that the ferromagnetic Ising model is also a model for the liquid–vapour transition where the interparticle interaction has strength $-\varepsilon = -4J$. Recalling that the magnetic moment per lattice site is

$$m \equiv \frac{M}{N} = \frac{1}{L^d} \sum_{i=1}^{L^d} \sigma_i, \tag{8.48}$$

we conclude that a spin configuration with $m > 0$ is equivalent to a high-density state of a lattice gas, i.e., to the liquid phase, whereas a configuration with $m < 0$ is equivalent to a low-density state, i.e., to the vapour phase.

Defining site occupation variables as

$$\alpha_i = \frac{1+\sigma_i}{2}, \tag{8.49}$$

we see that $\sigma_i = 1 \Rightarrow \alpha_i = 1$, if the ith site is occupied; and $\sigma_i = -1 \Rightarrow \alpha_i = 0$, if it is vacant. It is now important to note that, if the Ising model is studied in the canonical ensemble, i.e., at constant number of spins, then any spin flip will cause a change in occupation number, i.e., in the number of lattice-gas particles. In other words, the Ising model in the canonical ensemble is equivalent to the lattice-gas model in the grand canonical ensemble. This enables us to find the number of particles N of the gas, as well as the number of particles per lattice site, i.e., the gas density, for a given spin configuration, using the occupation variables α_i defined in Eq. (8.49):

$$n = \frac{N}{L^d} = \frac{1}{L^d} \sum_{i=1}^{L^d} \alpha_i, \tag{8.50}$$

whence it follows that n is always greater than zero and less than one. Equation (8.49) allows us to relate the lattice-gas density n, given by Eq. (8.50), to the magnetic moment per lattice site m, given by Eq. (8.48), with the result

$$n = \frac{1+m}{2}. \tag{8.51}$$

Thus m found from a Monte Carlo simulation of the Ising model immediately gives n for the lattice gas.

Problems

8.1 Simulate the two- and three-dimensional ferromagnetic Ising model, using the Metropolis algorithm.

    a. In zero applied field, compute the order parameter, mean energy per lattice site, specific heat and zero-field magnetic susceptibility, as functions of the reduced temperature of a system of N spins 1/2. Vary the linear size L of the lattice, the initial conditions (for instance, start from a random or a perfectly ordered configuration) and the boundary conditions (for instance, simulate an *interface* by fixing $\sigma_i = +1$ at the top edge of the lattice, $\sigma_i = -1$ at the bottom edge, with periodic boundary conditions in the left–right direction.

    b. Repeat the above simulations in an applied field. Compute the order parameter vs strength of applied field, above and below the critical temperature.

8.2 Simulate the lattice gas in two and three dimensions, using the Metropolis algorithm. Compute the following quantities as functions of the reduced temperature $t = k_B T/\varepsilon$:

    a. The mean and the variance of the total number of particles N, and of the density n, the latter given by Eq. (8.50).

    b. The isothermal compressibility K_T, from the fluctuation-dissipation relation, Eq. (4.131):

$$\sigma^2(N) = k_B T \frac{\overline{N}^2}{V} K_T.$$

8.3 Consider a simple model of a binary alloy: $\sigma_i = +1$ corresponds to an atom of species A occupying site i, and $\sigma_i = -1$ corresponds to an atom of species B at site i (see Figure 8.9c). Let $-\varepsilon_{AA}$, $-\varepsilon_{BB}$ and $-\varepsilon_{AB}$ be the interaction energies between AA, BB and AB pairs of nearest neighbours, respectively. Use Eqs. (8.43) and (8.44) and follow the same steps as for the lattice gas to derive the expression for the energy. What do you expect the ground state will be when $J > 0$ or when $J < 0$?

References

Binney, J. J., Dowrick, N. J., Fisher, A. J., and Newman, M. E. J. 1993. *The Theory of Critical Phenomena*. Oxford University Press.

Binder, K., and Heermann, D. W. 1992. *Monte Carlo Simulation in Statistical Physics*, 2nd (corrected) edition. Springer Verlag.

Brush, S. G. 1962. History of the Lenz–Ising Model *Rev. Mod. Phys.* **39**, 883–893.

Ma, S. K. 1984. *Statistical Mechanics*. World Scientific.

Onsager, L. 1944. Crystal statistics. I. A two-dimensional model with an order–disorder transition. *Phys. Rev.* **65**, 117–149.

9 Liquid crystals

9.1 Generalities

Most solids are positionally ordered in three dimensions: if the position of one of their constituent particles is known, then the positions of all the other particles will also be known. Particle positions are said to be *correlated*, and solids possessing this property are termed *crystalline*. On the other hand, most liquids are positionally disordered: knowledge of the position of one particular particle tells us nothing about the positions of any other particles that are not its near neighbours, as these positions continually change. In this case particle positions are said to be *uncorrelated*, and liquids possessing this property are termed *isotropic*. If, however, the constituent particles are markedly non-spherical, they may be orientationally ordered, if their orientations, rather than their positions, are correlated. Materials made up of non-spherical particles may therefore exhibit *liquid crystalline phases* or *mesophases*: states of matter in which particles are on average parallel to one another, but where their positions are uncorrelated in one, two or three dimensions. These phases are intermediate between the crystalline solid (which is positionally as well as orientationally ordered) and the isotropic liquid (which is positionally as well as orientationally disordered). A material or substance that exhibits mesophases is called a *mesogen*.

Liquid crystals were first discovered by the Austrian biochemist Friedrich Reinitzer (1858–1927). In 1888, while studying cholesterol derivatives extracted from carrots, he was surprised to observe that cholesteryl benzoate appeared to have two melting points. Between the solid and the liquid there was a turbid liquid phase. Reinitzer sent some of his samples to the German physicist Otto Lehmann (1855–1922), who examined them using his microscope fitted with light polarisers and a hot stage. Lehmann found that the turbid liquid phase was homogeneous, but had the optical properties of a crystal when illuminated with polarised light. This is how the name 'liquid crystal' eventually came to be applied to this class of materials. To know more about the history of liquid crystals, see Chandrasekhar (1992), de Gennes and Prost (1992), Dunmur and Sluckin (2011).

Some substances exhibit more than one mesophase. If the transitions between these phases are driven by temperature changes, the substance is said to be *thermotropic*. If, on the other hand, the transitions are driven by changing the concentration of the substance in some solvent, then the substance is said to be *lyotropic*. More recently, liquid crystals have been synthesised that exhibit both thermotropic and lyotropic behaviours; these are termed *amphotropic*. On heating a thermotropic liquid crystal with only one mesophase, the solid will melt into a turbid liquid at the melting point; this turbid liquid has anisotropic physical properties. On further heating, a second transition occurs whereupon

(a)   (b)   (c)   (d)

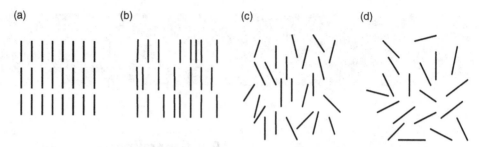

Figure 9.1 Sketch of the molecular orientations in the different phases of a thermotropic liquid crystal, as the temperature is raised from left to right: (a) crystal; (b) smectic-A; (c) nematic; (d) isotropic.

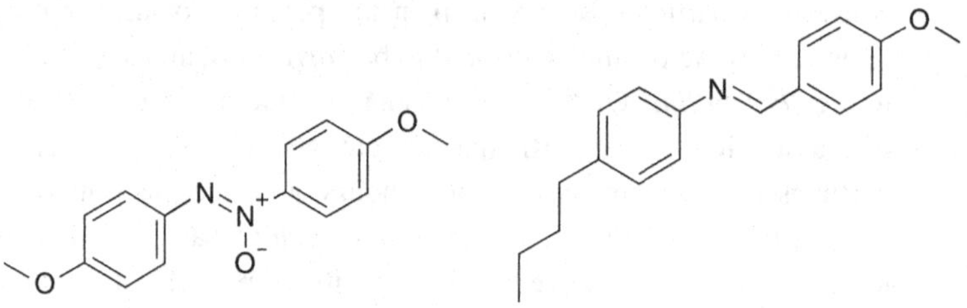

Figure 9.2 Two examples of common calamitic nematic molecules. *Left*: 1-methoxy-4-[(4-methoxyphenyl)-*NNO*-azoxy]benzene, or para-azoxyanisole (PAA); $T_{KN} = 118°C$, $T_{NI} = 135.5°C$. *Right*: *N*-(4-Methoxybenzylidene)-4-butylaniline (MBBA); $T_{KN} = 22°C$, $T_{NI} = 47°C$. Here T_{KN} is the melting (i.e., crystal–nematic transition) temperature, and T_{NI} is the isotropic–nematic transition temperature (Papon *et al.* 2002, chap. 8.1)

the turbid liquid becomes clear; this is known as the *clearing point*. Figure 9.1 illustrates a possible phase sequence for a thermotropic substance on varying the temperature: see http://plc.cwru.edu/tutorial/enhanced/lab/lab.htm for an interactive computer simulation of liquid crystal phase transitions.

All liquid crystalline substances are made up of non-spherical, fairly stiff building blocks. Most liquid crystals studied so far consist of rigid or semi-flexible rod-shaped molecules; they are known as *calamitics*. Substances composed of disc- or plate-shaped molecules can also exhibit mesophases; these materials are called *discotics*. Other meso-genic molecules may be shaped like planks, pyramids or – in the case of polymer liquid crystals – linear or branched chains. In what follows we shall restrict ourselves to low-molecular-weight liquid crystals, i.e., non-polymeric. Figure 9.2 shows two examples of calamitic nematic mesogens.

Mesophases can be classified as *nematic*, *smectic* and *columnar*. In the nematic phase, there is a macroscopic direction of long-range preferential alignment, given by a unit vec-tor, the *director*, denoted by **n**. This axis of preferential orientation is non-polar, i.e., **n** and −**n** are physically equivalent. The degree of alignment along the director is measured by the *nematic order parameter*, to be defined below. In this phase the positions of the molecular centres of mass are not correlated over long distances and the molecules may translate freely, hence the phase is fluid. In calamitic nematics, it is the long axes of the molecules that are, on average, parallel, whereas in discotic nematics it is the axes perpendicular to the discs that align. In a *uniaxial* nematic phase there is full rotational symmetry about the director. The symmetry of the isotropic phase, where all spatial directions are equivalent, is therefore broken at the isotropic–nematic transition, as the nematic phase has a lower symmetry. Recently, *biaxial* nematic phases have been reported in the literature: on lowering the temperature from the uniaxial nematic phase, there is a second symmetry-breaking transition at which two additional preferential directions of alignment – *secondary directors* – appear, which are perpendicular to the principal director of the parent uniaxial phase. Symmetry breaking is actually a common feature of many phase transitions, see, e.g., Goldenfeld (1992) or Papon *et al.* (2002). Here we shall be primarily concerned with the uniaxial nematic phase of calamitic mesogens.

In the absence of any constraints, a nematic would consist of a single domain (mono-domain sample) with uniform **n** pointing along some arbitrary direction, except for thermal fluctuations. In practice, however, a nematic sample breaks up into domains similar to those of a ferromagnetic material with **n** approximately uniform within each domain. The direction of **n** is determined by the boundaries – this is known as *anchoring* of the liquid crystal – and/or by applied fields; it is these effects that are exploited in liquid crystal displays. A monodomain sample is an optically uniaxial medium with its optical axis lying along **n**. Its *birefringence*, defined as the difference between indices of refraction for light polarised parallel and perpendicular to **n**, may be substantial. Other technologically relevant properties of liquid crystals include the anisotropies of their electrical and magnetic susceptibilities, which enable their alignment by applied fields. Moreover, because these fluids possess a small, but finite, orientational elastic energy, the effect of surface anchoring may propagate into the bulk of a sample. Local director structure inside a sample at equilibrium is thus determined by the balance of elastic and electric or magnetic torques. It is this local director structure that in turn determines the optical response of a liquid crystal device.

A calamitic liquid crystal made up of optically active molecules may exhibit a chiral nematic phase – i.e., one where there is no mirror symmetry. In such a phase, known as *cholesteric*, the director rotates from point to point in a helical pattern.

On further lowering of the temperature, some calamitic liquid crystals may exhibit one of more *smectic* phases before freezing into a solid. In a smectic, the molecules are arranged into a stack of layers. Within each layer there is no positional order and the molecules are free to move, as in an isotropic liquid. There is, however, long-range positional order along the direction perpendicular to the layers, hence this phase has a lower symmetry than the nematic. There are many different types of smectic, characterised by different degrees of orientational and positional order. In the simplest of these, the smectic-A, the director is perpendicular to the layers. In other smectic phases the director may be tilted (smectic-C), or there may be some positional order within the layers (smectic-B, etc.).

Finally, in discotic liquid crystals molecules may stack themselves into columns, yielding *columnar* phases. Columns in turn may arrange themselves into hexagonal, rectangular or oblique patterns. These phases are ordered positionally in two dimensions: within each column the molecules are free to move, as in an isotropic liquid, but the positions of the columns are correlated.

Readers wishing to know more about the fascinating world of liquid crystals and their applications are referred to Dunmur and Sluckin's informative and entertaining book (Dunmur and Sluckin, 2011). The professional liquid crystal scientist's standard reference is P. G. de Gennes' classic monograph (de Gennes and Prost, 1992).

As stated above, here we shall concentrate on thermotropic calamitics, which are the simplest and commonest type of liquid crystal. We shall study the isotropic-to-uniaxial-nematic transition by means of a mean-field statistical mechanical theory – the theory of Maier and Saupe – and of Monte Carlo computer simulations. Both approaches are closely related to those used to model ferromagnetism in Chapter 7. An alternative approach is Onsager theory, which is often used to describe phase transitions in lyotropic liquid crystals – discotic as well as calamitic. We shall discuss it briefly and point out its formal relationship to the Maier–Saupe theory.

9.2 Maier–Saupe theory

We start by introducing the long-range order parameter of the nematic phase. We assume that the liquid crystal is composed of rigid rod-shaped molecules. If we choose the OZ axis of our coordinate frame to point along the director $\mathbf{n}$, then the orientation of each rod is given by its polar and azimuthal angles, θ and ϕ respectively – see Figure 9.3.

Because the phase is uniaxial, the molecular orientational distribution function $f(\theta, \phi)$ depends only on the polar angle, i.e., $f(\theta, \phi) = f(\theta)$. Furthermore, because $\mathbf{n}$ and $-\mathbf{n}$ are equivalent, we have $f(\theta) = f(\pi - \theta)$. It then follows that the simplest, non-trivial measure of the degree of alignment is the statistical average of the second Legendre polynomial of the molecular orientations (de Gennes and Prost, 1992) – see Figure 9.4. We thus define the *nematic order parameter* Q as

$$Q = \frac{1}{2}\langle 3\cos^2\theta - 1\rangle = \frac{1}{2}\int_0^\pi \left(3\cos^2\theta - 1\right)f(\theta)\sin\theta\,d\theta, \tag{9.1}$$

where $\langle\rangle$ denotes a statistical average.

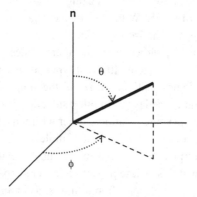

Figure 9.3 Cartesian coordinate frame with OZ axis along the director $\mathbf{n}$. The orientation of a rod-shaped molecule is given by the polar angle θ and the azimuthal angle ϕ.

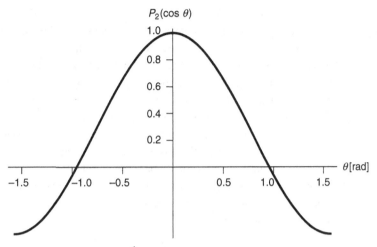

Figure 9.4 Second Legendre polynomial $P_2(\cos\theta) = \frac{1}{2}(3\cos^2\theta - 1)$.

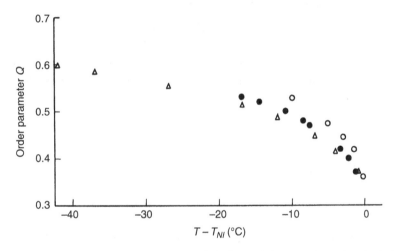

Figure 9.5 Temperature dependence of the nematic order parameter of PAA. The symbols are experimental results obtained by a variety of experimental methods: open circles – NMR; filled circles – diamagnetic anisotropy; triangles – birefringence. (From Chandrasekhar (1992, Chap. 2), by permission of Cambridge University Press.)

If the molecular axes are randomly oriented, then $f(\theta) =$ constant and $Q = 0$: this is the isotropic phase. If, on the other hand, there is perfect orientational order, then $f(\theta)$ will be a sum of two delta functions located at $\theta = 0$ and $\theta = \pi$; hence $Q = 1$. Finally, in the physically unlikely event (for rods) that $f(\theta)$ is a delta function located at $\theta = \pi/2$, then $Q = -\frac{1}{2}$, In the nematic phase, Q is a function of temperature (for a thermotropic) or of composition (for a lyotropic); Figure 9.5 iluustrates this dependence. From the above it also follows that $-\frac{1}{2} \leq Q \leq 1$. We shall now find the temperature dependence of Q from Maier–Saupe theory.

Maier and Saupe proposed the model that now bears their names around 1960 (Maier and Saupe, 1958, 1959, 1960); for its historical background, see Dunmur and Sluckin (2011).

In the spirit of a mean-field theory analogous to the Weiss theory of ferromagnetism, the model assumes that each nematic molecule resides in an anisotropic environment due to its interactions with all other molecules. The effect of this environment is embodied in a one-particle pseudopotential. It is further assumed that molecules are cylindrically symmetric hard rods. Here we shall restrict ourselves to the generic features of the intermolecular forces, as their exact nature is unimportant for our purpose; interested readers are referred to Chandrasekhar (1992, chap. 2.3) for details.

In Maier and Saupe's model, the combined effect of the repulsive (steric) and attractive (induced-dipole) interactions is assumed to impose a separation R between a pair of nearest-neighbour molecules; this separation is independent of the relative orientations of the two molecules. Under this assumption, Maier and Saupe showed that the orientation-dependent induced-dipole contribution to the interaction energy of a molecule with its m nearest neighbours is given by

$$u_i = -\frac{a}{R^6}\sum_{k=1}^{m} s_k s_i, \tag{9.2}$$

where a is a positive constant, and

$$s_i = \frac{1}{2}\left(3\cos^2\theta_i - 1\right), \tag{9.3}$$

where θ_i is the angle between the molecule (assumed rod-like) and the nematic director.

The mean-field approximation replaces the sum over the m nearest neighbours of a molecule by its statistical average:

$$\sum_{k=1}^{m} s_k \rightarrow \langle\sum_{k=1}^{m} s_k\rangle,$$

leading to

$$\langle\sum_{k=1}^{m} s_k\rangle = mQ. \tag{9.4}$$

where $Q = \langle s_k \rangle$ is the order parameter, given by Eq. (9.1). It then follows from Eq. (9.2) that, in the mean-field approximation, we have

$$u_i = -\frac{a}{R^6}mQs_i. \tag{9.5}$$

This can be rewritten in terms of the molar volume V, which is proportional to R^3:

$$u_i = -\frac{A}{V^2}Qs_i, \tag{9.6}$$

where A is a constant, independent of pressure, volume or temperature, and

$$\frac{A}{V^2}Q \equiv \mathcal{N} \tag{9.7}$$

is the 'orientational mean field' acting on s_i, in perfect analogy with the Weiss model of ferromagnetism, Eq. (7.31). Equation (9.6) is the Maier–Saupe pseudopotential: it tells us that, in the mean-field approximation, the interaction energy between any given molecule and its nearest neighbours is a function of: (i) its orientation relative to the average orientation of the uniaxial environment where the molecule resides; (ii) the order parameter describing the degree of order of that environment.

As seen in Section 5.8, the orientational distribution function of molecule i will be of the Boltzmann form:

$$f(\theta) = \frac{e^{-u_i/k_BT}}{Z_1},\tag{9.8}$$

where u_i is given by Eq. (9.6) and Z_1 is the single-particle partition function:

$$Z_1 = \int_0^\pi \sin\theta_i d\theta_i\, e^{-u_i/k_BT} \int_0^{2\pi} d\phi_i.\tag{9.9}$$

In the isotropic phase $u_i = 0$, and Eqs. (9.8) and (9.9) yield

$$f_{\text{iso}}(\theta_i) = \frac{1}{4\pi}.\tag{9.10}$$

Substituting Eq. (9.10) into Eq. (9.1), we consistently obtain $Q = 0$.

Following Papon *et al.* (2002), if we now define a new parameter α as

$$\alpha = \frac{1}{k_BT}\frac{A}{V^2}Q,\tag{9.11}$$

and introduce a new integration variable $x_i = \cos\theta_i$, it follows from Eq. (9.8) that the mean value of the one-particle orientation s_i given by Eq. (9.3) equals

$$\langle s_i \rangle = \frac{\int_0^1 dx_i\, e^{\alpha(3x_i^2-1)/2}(3x_i^2-1)/2}{\int_0^1 dx_i\, e^{\alpha(3x_i^2-1)/2}} = k_BT\frac{\partial \ln Z_1}{\partial \mathcal{N}},\tag{9.12}$$

where $\mathcal{N}$ is the Maier–Saupe orientational mean field, Eq. (9.7). The last equality in Eq. (9.12) aptly illustrates the analogy with magnetism: if we replace $\mathcal{N}$ by the applied field $\mu_0 H$, and s_i by the magnetic moment μ_i, we recover Eq. (4.57) for the mean magnetic moment in the paramagnetic phase. If, on the other hand, we replace $\mathcal{N}$ by the effective Weiss field, Eq. (7.34), we recover the mean magnetic moment of the ferromagnetic phase.

Inserting Maier and Saupe's orientational distribution function, Eq. (9.8), into Eq. (9.1) for the order parameter, the first equality in Eq. (9.12) yields the following consistency relation:

$$Q = \langle s_i \rangle.\tag{9.13}$$

From Eqs. (9.13), (9.12) and (9.11) we then obtain

$$Q = k_BT\frac{\partial \ln Z_1}{\partial \mathcal{N}} = \frac{\partial \ln Z_1}{\partial \alpha}.\tag{9.14}$$

On the other hand, combining Eqs. (9.9) and (9.11) yields for the single-particle partition function:

$$\frac{Z_1}{2\pi} = 2e^{-\frac{1}{2}\alpha}J(\alpha); \qquad J(\alpha) = \int_0^1 dx_i\, e^{\frac{3}{2}\alpha x_i^2},\tag{9.15}$$

whence

$$\ln\left(\frac{Z_1}{4\pi}\right) = -\frac{\alpha}{2} + \ln J(\alpha), \tag{9.16}$$

and, using Eq. (9.14) we get

$$Q = -\frac{1}{2} + \frac{\partial \ln J(\alpha)}{\partial \alpha} = -\frac{1}{2} + \frac{3}{2}\frac{\int_0^1 dx_i\, x_i^2 e^{\frac{3}{2}\alpha x^2}}{J(\alpha)} \equiv I(\alpha). \tag{9.17}$$

Because α is a function of Q, Eq. (9.17) is an implicit equation for the order parameter, just as in the Weiss model, Eq. (7.39). As before, we can find the order parameter by solving Eq. (9.17) together with Eq. (9.11) graphically. The latter can be rewritten in the more convenient form

$$\frac{I(\alpha)}{\alpha} = k_B T \frac{V^2}{A}. \tag{9.18}$$

The non-zero solutions of this equation are the abscissae of the points where $I(\alpha)/\alpha$ intersects the straight line $y = k_B T V^2/A$, as shown in figure 9.6a. There is no non-zero solution for $y > 0.2228$. For $y = 0.2228$, the non-zero solution is $\alpha = 1.453$. For $\alpha < 1.453$ ($0.2 < y < 0.2228$), there is a second non-zero solution, which, however, is unphysical.

We now check the thermodynamic stability of the solution plotted in figure 9.6b by computing the thermodynamic functions of the system.

1. *Internal energy*: by Eqs. (4.16) and (9.9), the orientation-dependent part of the interaction of each molecule with its environment contributes to the internal energy of the nematic phase an amount equal to

$$\bar{\epsilon} = -\frac{\partial \ln Z_1}{\partial \beta} = -\frac{A}{V^2}Q^2. \tag{9.19}$$

As mentioned earlier, this comes from pair interactions. In a mole of molecules there are mN_A nearest-neighbour pairs (where N_A is Avogadro's number), so the orientation-dependent internal energy per mole of a nematic is

$$U = \frac{N_A}{2}\bar{\epsilon} = -\frac{N_A}{2}\frac{A}{V^2}Q^2 = -\frac{1}{2}N_A k_B T\alpha Q, \tag{9.20}$$

where the factor $1/2$ corrects for double counting of pairs.

2. *Entropy*: using Eq. (4.27) and the Maier–Saupe distribution, Eq. (9.8), we get

$$S = -N_A k_B \langle \ln f_i \rangle = N_A k_B \ln Z_1 + \frac{N_A}{T}\langle u_i \rangle. \tag{9.21}$$

From Eq. (9.19), we have $\langle u_i \rangle = -(A/V^2)Q^2$. Equation (9.16) then yields, for the entropy per mole (to within an additive constant):

$$S = N_A k_B \left[\ln J(\alpha) - \frac{1}{2}\alpha(2Q+1)\right]. \tag{9.22}$$

3. *Helmholtz free energy*:

$$F = U - TS = N_A k_B T\left[\frac{1}{2}\alpha(Q+1) - \ln J(\alpha)\right]. \tag{9.23}$$

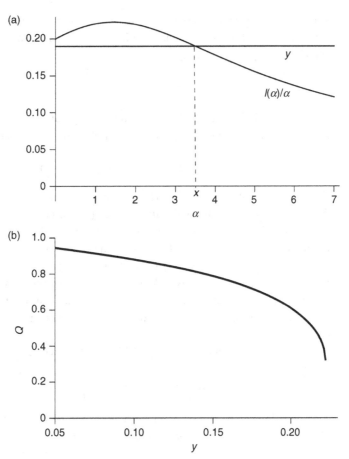

Figure 9.6 (a) x is the value of α that solves Eq. (9.18) for $y = k_B T V^2 / A = 0.19$; it follows from the definitions of y and α that the order parameter Q equals xy. (b) Plot of the order parameter Q vs y, obtained by implementing the method described in part (a) for $\alpha \le 20$. When $\alpha \to \infty$, $I(\alpha)/\alpha$ goes to zero asymptotically; in this limit, the order parameter Q approaches 1 as $y \to 0$.

4. *Gibbs free energy*:

$$G = F - PV, \tag{9.24}$$

where V is the molar volume and P is the pressure.

At the transition temperature T_{NI}, the isotropic and nematic phases are in equilibrium, therefore they must have the same chemical potential; see section 3.4.7. To find T_{NI}, we compute the change in the Gibbs free energy when the order parameter goes from zero (in the isotropic phase) to Q (in the nematic phase). By Eq. (3.81) this is given in terms of the change in chemical potential:

$$\Delta G = N_A \Delta \mu = \Delta F - P \Delta V, \tag{9.25}$$

where ΔV is the change in molar volume. However, in most situations of practical interest ΔV is very small (Chandrasekhar, 1992, Chap. 2.1), so if the pressure is not too high we can usually neglect $P\Delta V$ and write

$$\Delta\mu \simeq \frac{\Delta F}{N_A}. \tag{9.26}$$

Now ΔF is the excess Helmholtz free energy of the nematic phase relative to the isotropic phase, given by Eq. (9.23), At equilibrium, we then have

$$\Delta\mu = 0 \Rightarrow \frac{1}{2}\alpha\,(Q+1) = \ln J(\alpha). \tag{9.27}$$

It is now apparent from Eq. (9.23) how the nematic phase arises from competition between the energetic, Eq. (9.20), and entropic, Eq. (9.22), contributions to the Helmhotlz free energy. At high temperatures, entropy dominates and the isotropic phase wins, in which the molecules are orientationally disordered ($\Delta F > 0$). As the temperature is lowered, it becomes easier for the molecules to align parallel to one another, and the energy decreases, i.e., it becomes more negative. At the same time the entropy decreases as well, i.e., it becomes less positive, due to the increased orientational order. At low enough temperatures, the drop in energy is enough to offset the loss of entropy and the nematic phase wins ($\Delta F < 0$).

Using Eqs. (9.11) and (9.17), Eq. (9.27) may be rewritten as

$$\frac{1}{2}\alpha Q = \ln\left(\frac{Z_1}{4\pi}\right) \Rightarrow \frac{k_B T V^2}{A} = \frac{2\ln(Z_1/4\pi)}{\alpha^2}. \tag{9.28}$$

This equation can be solved graphically using Eq. (9.18): we look for the intersection of $I(\alpha)/\alpha$ and $z = \frac{2\ln(Z_1/4\pi)}{\alpha^2}$. The non-zero solution is

$$\begin{cases} \alpha = 1.96 \\ z = 0.220 \end{cases} \Rightarrow \begin{cases} \frac{1}{k_B T_{NI}}\frac{A}{V^2}Q_{NI} = 1.96 \\ \frac{k_B T_{NI} V^2}{A} = 0.220 \end{cases} \Rightarrow \begin{cases} T_{NI} = 0.220\frac{A}{k_B V^2} \\ Q_{NI} = 0.43 \end{cases}. \tag{9.29}$$

We conclude that the Maier–Saupe theory predicts a discontinuous transition, at which the order parameter jumps from 0.43 in the nematic phase to zero in the isotropic phase. The value of the order parameter at the transition is universal, i.e., it is the same for all substances, and does not depend on volume or on pressure. There is some experimental evidence in support of this result: see figure 9.7, although the value found is somewhat lower, $Q_{NI} \approx 0.3$ (de Gennes and Prost, 1992, Chap. 2).

By Eq. (9.29) the nematic–isotropic transition temperature is proportional to A/V^2, which Maier and Saupe assumed constant. From experimental data for T_{NI} of PAA and MBBA we find that $A/V^2 = 2.57 \times 10^{-20}$ J (PAA) and $A/V^2 = 2.01 \times 10^{-20}$ J (MBBA).

If we now define a dimensionless reduced temperature,

$$t \equiv \frac{T}{T_{NI}} = \frac{k_B T}{0.220}\frac{V^2}{A}, \tag{9.30}$$

then it follows from $k_B T_{NI} V^2/A = 0.220$ and the discussion following Eq. (9.18) that, for $0.220 \le k_B T V^2/A \le 0.2228$, the nematic phase is only metastable, and $k_B T V^2/A = 0.2228$

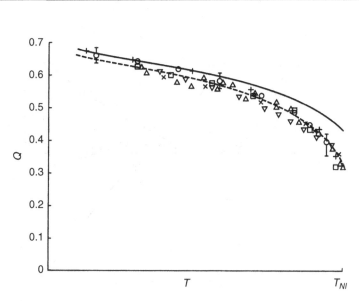

Nematic order parameter of MBBA vs temperature, from experiment (symbols) and theory (lines); here $T_{NI} = 47°$C. Experiment: circles – Raman; squares – NMR; crosses – birefringence; triangles – diamagnetic anisotropy. Theory: solid line – Maier–Saupe model; dashed line – generalisation of Maier–Saupe theory by Humphries, James and Luckhurst. (From de Gennes and Prost (1992, fig. 2.3, p. 52), by permission of Oxford University Press. Data from Deloche *et al.* (1971), Humphries *et al.* (1972).)

is its limit of local stability. This can be verified by direct calculation of the limits of stability of the nematic and isotropic phases, which must satisfy

$$\frac{\partial^2 G}{\partial Q^2} = 0. \tag{9.31}$$

Because $G \approx F$ we obtain from Eqs. (9.31) and (9.23) combined with Eq. (9.16) that

$$\frac{k_B T V^2}{A} = \frac{\partial^2 \ln(Z_1/4\pi)}{\partial \alpha^2}, \tag{9.32}$$

which can be solved graphically in the same way as before (Papon *et al.*, 2002). This yields, for the limit of stability of the nematic ($\alpha \neq 0$) phase,

$$\begin{cases} \alpha = 0.45 \\ \frac{k_B T_N V^2}{A} = 0.2228 \end{cases} \Rightarrow \begin{cases} Q_N = 0.323 \\ T_N = 0.2228 \frac{A}{k_B V^2} \end{cases} \Rightarrow \frac{T_N}{T_{NI}} = 1.013, \tag{9.33}$$

which agrees with our earlier conclusion. For the limit of stability of the isotropic ($\alpha = 0$) phase we get

$$\begin{cases} \alpha = 0 \\ \frac{k_B T_I V^2}{A} = 0.2 \end{cases} \Rightarrow \begin{cases} Q_I = 0 \\ T_I = 0.2 \frac{A}{k_B V^2} \end{cases} \Rightarrow \frac{T_N}{T_{NI}} = 0.91. \tag{9.34}$$

The changing relative stability of the two phases can be visualised by plotting the Helmholtz free energy vs order parameter: see Figure 9.8. For $T < T_{NI}$, the absolute minimum of the free energy occurs at $Q \neq 0$, corresponding to the nematic phase, whereas for

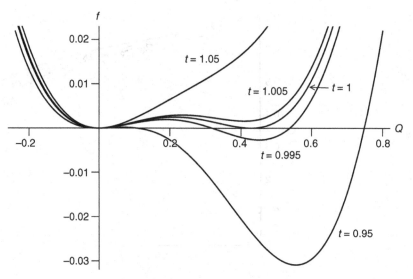

Figure 9.8 Helmholtz free energy, Eq. (9.23), in non-dimensional form, $f = F/N_A k_B T$, vs order parameter Q at five reduced temperatures $t = T/T_{NI}$, straddling the first-order isotropic–nematic transition.

$T > T_{NI}$, the minimum is at $Q = 0$, corresponding to the isotropic phase. At the transition, $T = T_{NI}$, the free energy has two minima of the same depth, one at $Q = 0$ and another at $Q \neq 0$. For T slightly above T_{NI} there are still two minima, but the absolute minimum is the one at $Q = 0$, so the isotropic phase is the stable phase and the nematic phase is only metastable. Conversely, for T slightly below T_{NI} there are again two minima, but the absolute minimum is the one at $Q \neq 0$, so the nematic phase is the stable phase and the isotropic phase is only metastable.

To make further contact with experiment a number of other quantities can be calculated, such as the latent heat L (see Problem 9.1), for which one finds $L/T_{NI} = 0.418R = 3.46$ J mol^{-1} K^{-1}. Maier–Saupe theory therefore predicts universal values of the order parameter jump $\Delta Q = 0.43$ and of the entropy change $\Delta S/R = 0.418$ at the isotropic–nematic transition, which are not in very good quantitative agreement with experimental results; see Table 9.1. Still, the theory provides a good qualitative description of the properties of the nematic phase (Luckhurst and Zannoni, 1977).

There are many possible generalisations of the Maier–Saupe model that lead to improved agreement with experiment, such as the inclusion of higher-order terms in the pseudopotential to account for interactions of higher-than-dipolar order. Other possibilities include the

Table 9.1 Transition temperatures and latent heats for two common nematogens. From Chandrasekhar (1992, Chap. 2.1)

Nematogen	T_{NI} (K)	L (J mol^{-1})	L/T_{NI}
PAA	408.6	570	1.4
MBBA	320.2	582	1.8

treatment of long-range thermal fluctuations, which depress the effective order parameter, or explicit consideration of molecular flexibility. For details see de Gennes and Prost (1992, Chap. 2) or Chandrasekhar (1992, Chap. 2.3).

9.3 Onsager theory

Onsager's theory of the nematic phase, proposed in 1949 (Onsager, 1949), is an alternative theory to Maier and Saupe's, with which it shares a number of formal similarities. It has been applied mostly to lyotropic liquid crystals, though it can be formulated in very general terms. Basically Onsager treated the interactions between liquid crystal molecules at the level of the second virial coefficient, B_2, which was introduced when studying the real gas in Section 5.10.

Our starting point is, therefore, again Eq. (5.103). The only difference relative to the real gas case is that the interaction potential between rigid rod-shaped particles – and consequently also their Mayer function – will now depend on both particle positions and orientations, the latter given by angles θ and ϕ:

$$u \equiv u(\mathbf{r}_i, \theta_i, \phi_i, \mathbf{r}_j, \theta_j, \phi_j) \quad \text{and} \quad \Phi \equiv e^{-\beta u(\mathbf{r}_i, \theta_i, \phi_i, \mathbf{r}_j, \theta_j, \phi_j)} - 1 = \Phi(\mathbf{r}_i, \theta_i, \phi_i, \mathbf{r}_j, \theta_j, \phi_j), \quad (9.35)$$

and all integrals are now taken over not only the positions $\mathbf{r}_i$ and $\mathbf{r}_j$, but also the orientations (θ_i, ϕ_i) and (θ_j, ϕ_j). Moreover, because the particles are now able to align in some preferential direction, the integrand in Eq. (5.104) must be weighted by the particles' orientational distribution functions, leading to

$$B_2[f] = -\frac{1}{2V} \int f(\theta_i) \Phi(\mathbf{r}_i, \theta_i, \phi_i, \mathbf{r}_j, \theta_j, \phi_j) f(\theta_j) d\mathbf{r}_i \sin\theta_i \, d\theta_i d\phi_i d\mathbf{r}_j \sin\theta_j \, d\theta_j d\phi_j. \quad (9.36)$$

This expression may be simplified for convenience. Start by denoting the set of angles (θ, ϕ), by Ω and $\sin\theta \, d\theta d\phi$ by $d\Omega$:

$$B_2[f] = -\frac{1}{2V} \int f(\theta_i) \Phi(\mathbf{r}_i, \Omega_i, \mathbf{r}_j, \Omega_j) f(\theta_j) d\mathbf{r}_i d\Omega_i d\mathbf{r}_j d\Omega_j. \quad (9.37)$$

Next, note that the interparticle interaction, hence also the Mayer function, does not depend on the positions of the individual particles, $\mathbf{r}_i$ and $\mathbf{r}_j$, but only on their relative position, $\mathbf{r}_{ij} = \mathbf{r}_i - \mathbf{r}_j$. Making the following change of variables,

$$\int d\mathbf{r}_i \int d\mathbf{r}_j \rightarrow \int d\mathbf{r}_i \int d\mathbf{r}_{ij}, \quad (9.38)$$

the integral over $\mathbf{r}_i$ in Eq. (9.37) gives just a factor of V, and we obtain

$$B_2[f] = -\frac{1}{2} \int f(\theta_i) \Phi(\mathbf{r}_{ij}, \Omega_i, \Omega_j) f(\theta_j) d\mathbf{r}_{ij} d\Omega_i d\Omega_j. \quad (9.39)$$

This second virial contribution to the Helmholtz free energy has been derived for a fixed orientational distribution function. To it must be added an orientational entropy term given by Eq. (9.21), $-TS_{\text{orient}} = Nk_BT\langle\ln f\rangle,$[1] with the result

$$\frac{F}{Nk_BT} \approx \ln\left(\Lambda^3\rho\right) - 1 + \langle\ln f\rangle + \rho B_2\,[f].\tag{9.40}$$

As in the Maier–Saupe model, Eq. (9.23), it is readily apparent from Eq. (9.40) that the nematic phase is a result of the competition between an entropic term, $\langle\ln f\rangle$, and a term due to particle interactions, $B_2\,[f]$. At high temperatures the entropic term dominates and the phase is isotropic, i.e., fully disordered orientationally ($\Delta F > 0$). As the temperature is lowered, the molecules are able to align more easily along a preferential direction and the energy term dominates, i.e., becomes more negative: the decrease in energy offsets the loss of orientational entropy and the nematic phase prevails ($\Delta F < 0$).

The formalism developed above is quite general: thus far we have said nothing about the nature of the interparticle interaction – except that it is pairwise. This kind of approach is, however, particularly useful when $u(\mathbf{r}_{ij}, \Omega_i, \Omega_j)$ is purely repulsive and short-ranged, as is the case with, e.g., charged colloidal particles in a counterion-rich solvent – the electrostatic interactions are strongly screened and decay exponentially with particle separation. One convenient approximation is then to assume that the interaction is infinitely strongly repulsive and has zero range:

$$u(\mathbf{r}_{ij}, \Omega_i, \Omega_j) = \begin{cases} \infty. & \text{if particles overlap} \\ 0, & \text{if particles do not overlap} \end{cases},\tag{9.41}$$

leading straightforwardly to the Mayer function

$$\Phi(\mathbf{r}_{ij}, \Omega_i, \Omega_j) = \begin{cases} -1. & \text{if particles overlap} \\ 0. & \text{if particles do not overlap} \end{cases}.\tag{9.42}$$

The integral of the Mayer function over the relative position of a pair of such particles (called *rigid rods* for obvious reasons),

$$\int \Phi(\mathbf{r}_{ij}, \Omega_i, \Omega_j)d\mathbf{r}_{ij} \equiv -V_{\text{excl}}(\Omega_i, \Omega_j),\tag{9.43}$$

is the excluded volume of the pair of particles: it is the volume excluded to particle i by the presence of particle j; or, equivalently, the volume excluded to particle j by the presence of particle i. In other words, if we take the centre of mass of one of the particles to be the origin of our coordinate frame, then the excluded volume is the set of points that, if occupied by the centre of mass of the other particle, would lead to overlap of the two particles. Consequently, for an infinitely repulsive interaction, potential $\mathbf{r}_{ij}$ can only take values outside the excluded volume of the pair of particles, which will of course depend on their relative orientations, see figure 9.9.

From Eqs. (9.39) and (9.43) it follows that

$$B_2\,[f] = \frac{1}{2}\int f(\theta_i)V_{\text{excl}}(\Omega_i, \Omega_j)f(\theta_j)d\Omega_i d\Omega_j,\tag{9.44}$$

---

[1] Note that this is the orientational entropy of a sample of N particles. In Eq. (9.21) we took $N = N_A$, i.e., the orientational entropy of one mole.

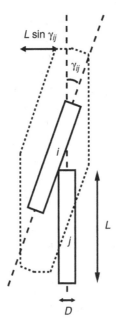

Figure 9.9   Excluded volume of a pair of hard cylinders. The volume bounded by the dashed line is inaccessible to the centre of mass of cylinder j owing to the presence of cylinder i.

which allows us to interpret the second virial coefficient as an average over orientations of the excluded volume of a pair of particles, weighted by their orientational distribution functions.

Now our remarks on Eq. (9.40) above are still valid even if the interactions are purely repulsive, leading us to conclude that a nematic phase may arise even in the absence of any attractions between particles. But why is this so, if there is no attractive interaction energy? Let us consider again the 'entropy' term, $\langle \ln f \rangle$, and the 'energy' term, $B_2[f]$, in Eq. (9.40). The former is clearly the orientational entropy: it is maximal in the isotropic phase. The latter is maximal (in absolute value) or minimal (in signed value, since it is negative) when the rods are parallel to one another – i.e., when their excluded volume is smallest. $B_2[f]$ may then be interpreted as a translational entropy term: when the excluded volume is minimal, the number of positions that are accesible to the centre of mass of a particle is maximal. The physical basis for the isotropic and nematic phases is then the following: in the isotropic phase, the orientational entropy term dominates; whereas in the nematic phase, it is the translational entropy term that dominates. And what about the isotropic–nematic transition? It follows from Eqs. (9.42) and (9.44) that $B_2[f]$ does not depend on temperature, so the only variable that can drive the transition is the density, ρ: Eq. (9.40) then implies that the orientational entropy term dominates at low densities – i.e., in the isotropic phase – whereas the energy term dominates at high densities – i.e., in the nematic phase. This behaviour has been observed in solutions of, *inter alia*, tobacco mosaic virus (TMV), which are, to a good approximation, rigid rods. Systems such as these, where the temperature does not play a role, are called *athermal*.

To find the order parameter and the equilibrium Helmholtz (or Gibbs) free energy of a hard-rod fluid using Onsager's theory, we need to minimise Eq. (9.40) with respect to the orientational distribution function. $f(\theta)$. In general, however, there is no exact analytical expression for the excluded volume of a pair of non-spherical particles, so calculations are usually more involved than in Maier–Saupe theory and often require numerical methods of solution. Consequently, here we shall treat only one of the cases in which it is possible to extract approximate analytical results, following Vroege and Lekkerkerker (1992).

For simplicity, but without much loss of generality, let the hard rods be cylinders of length L and diameter D, such that $L/D \gg 1$. We then have (Vroege and Lekkerkerker, 1992):

$$V_{\text{excl}}(\Omega_i, \Omega_j) \approx 2DL^2 |\sin \gamma_{ij}|, \tag{9.45}$$

where γ_{ij} is the angle between the two rods (see Figure 9.9). From Eqs. (9.44) and (9.45) it then follows that, in the isotropic phase, the second virial coefficient of hard cylinders is[2]

$$B_2^{\text{iso}} \approx \frac{1}{2} \int \frac{1}{4\pi} \left(2DL^2\right) |\sin \gamma_{ij}| \frac{1}{4\pi} d\Omega_i d\Omega_j = \frac{\pi}{4} DL^2, \tag{9.46}$$

and the dimensionless Helmholtz free energy per particle is

$$\frac{F^{\text{iso}}}{Nk_BT} \approx \ln\left(\Lambda^3 \rho\right) - 1 + \rho B_2^{\text{iso}} = \ln\left(\Lambda^3 \rho\right) - 1 + \frac{\pi}{4} DL^2 \rho. \tag{9.47}$$

In the nematic phase, the simplest possible approximation is to assume that the orientational distribution function is Gaussian, i.e.,

$$f(\theta) = \begin{cases} C \exp\left(-\frac{1}{2}\alpha\theta^2\right), & \text{if } 0 \le \theta \le \frac{\pi}{2} \\ C \exp\left(-\frac{1}{2}\alpha(\pi - \theta)^2\right), & \text{if } \frac{\pi}{2} \le \theta \le \pi \end{cases}, \tag{9.48}$$

where α is a parameter that measures the 'sharpness' of $f(\theta)$ ($\alpha = 0$ corresponds to the isotropic phase), and C is a normalisation constant. For large α, i.e., deep in the nematic phase, we have (see Appendix A of Vroege and Lekkerkerker (1992)):

$$C = \frac{\alpha}{4\pi} \left(1 + \frac{1}{3\alpha} + \cdots\right). \tag{9.49}$$

The larger α is, the sharper $f(\theta)$ will be and the larger the order parameter Q

$$Q = \langle P_2(\cos\theta) \rangle \approx 1 - \frac{3}{\alpha}. \tag{9.50}$$

Recall that this approximation is only valid for $\alpha \gg 1$, i.e., away from the isotropic–nematic transition. If we now substitute Eqs. (9.48) and (9.49) into Eq. (9.40) we find, after some algebra (see Appendix A of Vroege and Lekkerkerker (1992)), that

$$\langle \ln f \rangle = \ln \alpha - 1; \qquad B_2[f] = DL^2 \sqrt{\frac{\pi}{\alpha}},$$

---

[2] This is calculated as follows: first, take the OZ axis along the long axis of rod i; then $\gamma_{ij} = \theta_j$ and the integration over Ω_j can be performed. Then integrate over all possible orientations of OZ, i.e., over Ω_i, which gives a factor of 4π. This should be checked as an exercise.

leading to the following approximate expression for the Helmholtz free energy of the nematic phase:

$$\frac{F^{\text{nem}}}{Nk_BT} \approx \ln\left(\Lambda^3\rho\right) - 1 + \ln\alpha - 1 + DL^2\rho\sqrt{\frac{\pi}{\alpha}}. \tag{9.51}$$

Equations (9.47) and (9.51) may be written in a more convenient form by introducing the *volume fraction* ξ, defined as

$$\xi = \rho v_0 = \rho\frac{\pi}{4}D^2L, \tag{9.52}$$

where v_0 is the volume of one rod. We then have (to within an unimportant additive constant):

$$\frac{F^{\text{iso}}}{Nk_BT} \approx \ln\xi - 1 + \frac{L}{D}\xi, \tag{9.53}$$

$$\frac{F^{\text{nem}}}{Nk_BT} \approx \ln\xi - 1 + \ln\alpha - 1 + \frac{L}{D}\xi\frac{4}{\sqrt{\pi\alpha}}. \tag{9.54}$$

It is straightforwardly seen that the orientational entropy term, $\ln\alpha$, dominates at low densities. On raising the density, it is overtaken by the translational entropy term, $(L/D)\xi(4/\sqrt{\pi\alpha})$. The longer the rods – i.e., the larger L/D – the lower the density at which the two terms balance. Thus, longer rods align more easily, as might be intuitively expected.

Figure 9.10 plots the two entropic contributions to the Helmholtz free energy vs the nematic order parameter Q, for a set of values of L/D and volume fractions ξ. At large Q, the loss of translational entropy is less than the loss of orientational entropy and the nematic phase is favoured. The threshold value of Q above which the nematic forms is an increasing function of the elongation and of the volume fraction. Consequently, denser fluids of more elongated particles order more strongly when transitioning from the isotropic phase to the nematic phase. The Helmholtz free energy of the nematic phase is minimised for

$$\alpha = \frac{4}{\pi}\left(\frac{L}{D}\right)^2\xi^2. \tag{9.55}$$

In other words, the more elongated the rods, the sharper the orientational distribution function that minimises the Helmholtz free energy of the nematic phase, and the larger the nematic order parameter.

Let us now consider the isotropic-to-nematic transition in more detail. As we saw in Section 3.4.7, the condition for the coexistence of two phases is the equality of their chemical potentials or, equivalently, of their Gibbs free energies. However, if we assume, as in Maier–Saupe theory (see Section 9.2), that the molar volume changes very little at the transition – i.e., that the isotropic and nematic phases have approximately the same density – then this condition reduces to equality of the Helmholtz free energies of the two phases, cf. Eq. (9.26). From Eqs. (9.53) and (9.54) it then follows that, according to Onsager theory,

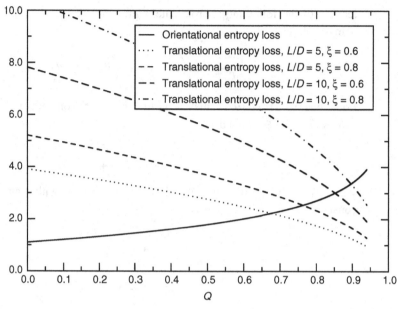

Figure 9.10 Orientational and translational entropy losses (= increases in Helmholtz free energy), $\ln \alpha$ and $(L/D)\xi(4/\sqrt{\pi\alpha})$ respectively, vs nematic order parameter Q, from Eq. (9.40). Because the orientational entropy is a single-particle property, it does not depend on the rod aspect ratio or volume fraction.

the transition between isotropic and nematic phases of hard rods will occur at a volume fraction ξ_{NI} such that

$$\frac{\Delta F}{Nk_BT} = \frac{F^{\text{nem}}}{Nk_BT} - \frac{F^{\text{iso}}}{Nk_BT} \approx \ln \alpha - 1 + \frac{L}{D}\xi\left(\frac{4}{\sqrt{\pi\alpha}} - 1\right) = 0. \qquad (9.56)$$

Figure 9.11 shows $\Delta F/Nk_BT$ vs the nematic order parameter Q, for several volume fractions ξ. As expected, at small ξ (low densities), $\Delta F > 0$) and the isotropic phase is stable, whereas for large ξ (high densities), there is a range of Q for which $\Delta F < 0$, and the equilibrium phase is the nematic. The phase transition occurs when $\Delta F = 0$ and $\partial \Delta F/\partial Q = 0$, i.e., for $(L/D)\xi_{NI} \approx 4.027$, for which one has $Q_{NI} \approx 0.854$.

At this point it should be noted that the density change at the isotropic–nematic transition of hard rods is, however, not small, so the treatment outlined above is a rather rough approximation. In Chapter 10 we shall study phase coexistence more generally.

9.4 Landau–de Gennes theory

Landau–de Gennes theory is a phenomenological theory of the isotropic–nematic transition. It was proposed by de Gennes, who generalised Landau's theory for a generic phase transition, as was applied, e.g., to the paramagnetic–ferromagnetic transition in chapter 7.

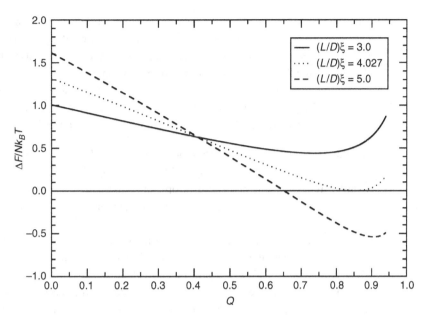

Figure 9.11 Excess Helmholtz free energy of the nematic phase relative to the isotropic phase vs the nematic order parameter Q, for volume fractions ξ below, at, and above, the isotropic–nematic transition of hard rods.

In Landau–de Gennes theory the Helmholtz free energy of a nematogen is expanded in a power series of the nematic order parameter in the vicinity of the phase transition temperature T_{NI}. In the original formulation of the theory the coefficients in this expansion are unknown, and need to be determined from experimental data. However it is possible to relate them to microscopic quantities, as we did in our study of ferromagnetism, by expanding the free energies of the Maier–Saupe or Onsager models and identifying terms with the same powers of Q. Still, the detailed procedure is subtle and beyond the scope of this book, for which reason only the final result is presented at the end of this section. The interested reader is referred to Katriel *et al.* (1986).

In generic terms, Landau theory postulates that there exists a function of the thermodynamic parameters of the system and of its order parameter Q, called the Landau free energy, $\mathcal{L}$, such that the equilibrium state of the system is given by the minimum of $\mathcal{L}$ with respect to Q. For $T > T_{NI}$, the global minimum is at $Q = 0$, which corresponds to the disordered phase; whereas for $T < T_{NI}$, it is at $Q \neq 0$, corresponding to the orientationally ordered phase. If we assume that Q is small in the vicinity of T_{NI}, then $\mathcal{L}$ may be approximated by the first few terms of its expansion in powers of Q. Hence in this approximation, $\mathcal{L}$ is an analytic function of Q. Because a nematic sample under standard experimental conditions is a system at fixed temperature and pressure, the appropriate Landau free energy would be the Gibbs energy. However, we saw in the preceding section that we can usually replace this with the Helmholtz free energy F, whose power series expansion to fourth order in Q is

$$F = A(T)Q^2 + BQ^3 + CQ^4 \equiv \mathcal{L}, \tag{9.57}$$

where only the coefficient of the quadratic term is a function of temperature, assumed linear: $A(T) = a(T - T^*)$. This is not the only possible choice but, as we shall see below, it is the simplest physically reasonable choice that produces the desired result. We shall see shortly that T^* is actually the limit of stability of the isotropic phase, denoted T_I in section 9.2.

The requirement that $\mathcal{L}$ must respect the symmetry of the problem imposes constraints on the coefficients in the Taylor expansion for F, Eq. (9.57). We shall now discuss what these constraints are; for details see Goldenfeld (1992, Chap. 5.3).

Linear term This must vanish, otherwise we would not have $\partial F / \partial Q = 0$ if $Q = 0$, i.e., the isotropic phase ($Q = 0$) would not be stable for $T > T^*$, as it must be.

Quadratic term The nematic–isotropic phase transition occurs in the vicinity of (but, as we shall see below, not exactly at) the temperature T^* at which this term changes sign.

Cubic term In the absence of this term the nematic–isotropic transition would be continuous, just like the paramagnetic–ferromagnetic transition, and states of order parameter Q and $-Q$ would be equivalent, which – unlike in ferromagnetism – is unphysical. Indeed, $Q > 0$ describes a state of preferential alignment of the molecular long axes along the director, whereas $Q < 0$ describes a state of preferential alignment perpendicular to the director; see figure 9.1. Q and $-Q$ are thus associated with different molecular arrangements and there is no reason why they should have the same free energy. Therefore one must have $B \neq 0$ and a first-order transition. Moreover, as can be seen by solving the minimisation equation for $\mathcal{L}$, B must be negative in order to favour the $Q > 0$ solution, which is the one most often realised in nature.

Quartic term C must be positive so that $\mathcal{L}$ is bounded below; indeed, if $C < 0$, the system could always lower its free energy by letting $|Q| \to \infty$ and there would be no stable nematic phase for $T < T^*$.

Let us now find the isotropic–nematic transition temperature in Landau–de Gennes theory. Equating the thermodynamic potentials, Eq. (9.57), at $Q = 0$ and $Q \neq 0$, we get, for $Q \neq 0$, $A(T_{NI}) + BQ + CQ^2 = 0$. The transition will occur at the highest temperature for which this equation has a non-zero solution, i.e.,

$$A(T_{NI}) = \frac{B^2}{4C} \Leftrightarrow T_{NI} = \frac{B^2}{4aC} + T^*. \tag{9.58}$$

The order parameter at the transition is then

$$Q_{NI} = Q(T_{NI}) = -\frac{B}{2C} > 0, \tag{9.59}$$

hence the transition predicted by Landau–de Gennes theory is indeed first-order or discontinuous, as anticipated earlier, which agrees with experiment.

The stability conditions are

$$\frac{\partial F}{\partial Q} = 0 \quad \text{and} \quad \frac{\partial^2 F}{\partial Q^2} > 0. \tag{9.60}$$

The first of these requires that F should be extremised, the second that this extremum should be a minimum. If $Q = 0$, it follows from the second condition that $T > T^*$. Hence T^* is the limit of stability of the isotropic phase, i.e., $T^* = T_I$. By considering the case $Q \neq 0$, we can likewise find the limit of stability of the nematic phase, T_N, as was done earlier using Maier–Saupe theory, which provides a better approximation to reality. We leave this calculation as an exercise for the reader.

From the first of Eqs. (9.60) we can also extract the temperature dependence of the order parameter for $T < T^*$; this is the solution of $2A(T) + 3BQ + 4CQ^2 = 0$:

$$Q(T) = \frac{-3B \pm \sqrt{9B^2 - 32A(T)C}}{8C} = -\frac{3B}{8C}\left[1 \mp \sqrt{1 - \frac{32A(T)C}{9B^2}}\right]. \tag{9.61}$$

This equation has two branches, corresponding to the stable (minus sign, lower free energy) and metastable (plus sign, higher free energy) solutions. Substituting $A(T_{NI}) = B^2/4C$ into this equation with the **minus** sign, we recover the earlier result for the order parameter at the transition, $Q(T_{NI}) = -B/2C$. On the other hand, substituting $A(T) = a(T - T^*)$ into the equation with the **plus** sign, it can be straightforwardly shown that $Q(T^*) = 0$, i.e., the order parameter vanishes in the limit of stability of the isotropic phase.

Critique of Landau–de Gennes theory

The order parameter at the transition is too large – unlike in the Landau theory of the paramagnetic–ferromagnetic transition, which is continuous, it is not consistent to truncate the free energy at fourth order near the nematic–isotropic transition. Landau–de Gennes theory is, however, useful for the study of pretransitional effects in the isotropic phase – see de Gennes and Prost (1992, Chap. 3) for more details. Another shortcoming of the theory is that the coefficients in the expansion of the thermodynamic potential are free parameters that need to be determined either by fitting experimental data or by mapping it onto a more microscopic theory. In the latter vein it can be shown that, within Maier–Saupe theory (Katriel *et al.*, 1986), the coefficients in the Landau excess free energy, Eq. (9.57), are given by

$$A(T) = \frac{5}{2}k_B(T - T^*), \quad k_B T^* = 0.2\frac{A}{V^2},$$

$$B = \frac{25}{21}k_B T, \quad C = \frac{425}{196}k_B T,$$

which shows that T^* is the limit of stability of the isotropic phase from Maier–Saupe theory, T_I, Eq. (9.34).

9.5 Monte Carlo simulations of the Lebwohl–Lasher model[*]

In this section we discuss Monte Carlo simulations of the isotropic–nematic transition in thermotropic calamitic liquid crystals, i.e., thermotropics composed of hard rodlike molecules. We shall study the combined effects of an applied electric field and rigid boundary conditions on the direction of preferential alignment of a nematic film.

We shall simulate the Lebwohl–Lasher model. This consists of uniaxial particles placed at the nodes of a lattice. Each particle interacts with the others via the pair potential given by

$$u_{ij} = -\epsilon_{ij}P_2(\cos\gamma_{ij}),\qquad(9.62)$$

where ϵ_{ij} is the interaction strength – a positive constant equal to ϵ if the two particles are nearest neighbours and zero otherwise, γ_{ij} is, as before, the angle between the long axes of particles i and j, and $P_2(x)$ is the second Legendre polynomial. The Lebwohl–Lasher potential, Eq. (9.62), is a simplification of the Maier–Saupe anisotropic interaction, Eq. (9.2): the key advantage of the former relative to the latter is that it does not depend on the nematic director, which is unknown *a priori*.

The Lebwohl–Lasher model is a staple for computer simulations of systems that exhibit an isotropic–nematic phase tansition. We shall perform Monte Carlo simulations of this model using the Metropolis algorithm, just as we did for the Ising model of ferromagnetism in Section 8.3.

It is known from Monte Carlo simulations that the Lebwohl–Lasher model exhibits an isotropic–nematic phase transition at a reduced temperature $t_c \equiv k_B T_c/\epsilon \approx 1.123$ (Fabbri and Zannoni, 1986; Boschi *et al.*, 2001). This transition appears to be weakly first order, as can be seen from Figure 9.12, where we plot the reduced energy per particle U^*, defined as

$$U^* = \frac{\langle U\rangle}{N\epsilon},\qquad \langle U\rangle = -\frac{1}{2}\sum_{i,j=1}^{N}\epsilon_{ij}\langle P_2\left(\cos\gamma_{ij}\right)\rangle.$$

Here $\langle P_2\left(\cos\gamma_{ij}\right)\rangle$ is the short-range order parameter – the average is calculated over nearest neighbours.

One important aspect of simulating the isotropic–nematic transition is finding the (long-range) orientational order parameter. This is, however, far from simple. The main source of difficulty is that a system with periodic boundary conditions and no applied field is invariant under rotations. In the course of a simulation the nematic director may fluctuate from one configuration to the next, and the order parameter needs to be calculated relative to an instantaneous preferential direction of alignment. How this is actually done is beyond the scope of the present text; the interested reader is referred to, e.g., Fabbri and Zannoni (1986) and Boschi *et al.* (2001), where it is shown that the short- and long-range transition temperatures are the same.

Effect of an applied field

We shall now study the combined effects of an applied electric field and rigid boundary conditions on the preferential alignment of a nematic film. Most mesogens are dielectrics, and this particular configuration is relevant to their electro-optic display applications. In our simulations, particles will be represented by three-dimensional vectors, residing in the cells – or at the nodes – of a two-dimensional square lattice. Boundary conditions are taken to be rigid along one direction, and periodic along the perpendicular direction. We next discuss these two choices.

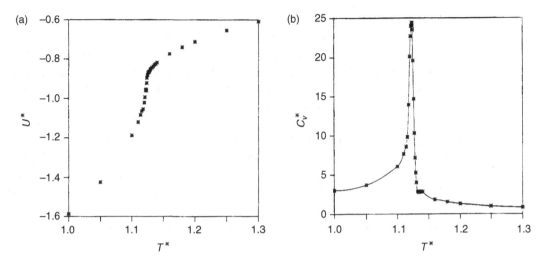

Figure 9.12 Monte Carlo simulation of the Lebwohl–Lasher model on a cubic lattice of size $N = 30 \times 30 \times 30$ and periodic boundary conditions. (a) Reduced energy per particle U^* vs reduced temperature $T^* = k_B T/\epsilon$; a change of slope is readily apparent, which suggests a first-order transition. (b) Solid line: specific heat $C_V^* = \partial U^*/\partial T^*$, obtained by fitting a curve to the simulated data points in (a) and differentiating it analytically. The (short-range) orientational transition temperature is given by the position of the specific heat maximum, as discussed for the Ising model in Chapter 8. For details of the simulations, which were performed on a CRAY supercomputer, see Fabbri and Zannoni (1986), Boschi *et al.* (2001). (From Fabbri and Zannoni (1986), ©Taylor & Francis, www.tandfonline.com.]

Generically, the Lebwohl–Lasher model exhibits an orientational transition if the lattice has dimension $d = 3$ and the molecular axes are vectors in $n = 3$ dimensions. Variants with $d = 2$ or $d = 3$ and $n = 2$ or $n = 3$ have also been studied. There is no phase transition for $(d = 2, n = 2)$, but for $(d = 2, n = 3)$ – the so-called planar Lebwohl–Lasher model – simulations using periodic boundary conditions reveal a pseudo-transition into a low-temperature ordered phase (Chiccoli *et al.*, 1988; Mondal and Roy, 2003). The reduced pseudo-transition temperature is $T^* \approx 0.6$ for a system of size comparable to that which we shall study here. The specific heat peak is, however, insensitive to system size, unlike in the case $d = 3$, where the peak sharpens on increasing the system size, as would be expected of a true phase transition. Nonetheless, the planar Lebwohl–Lasher model is convenient for studying the effect of an external field on the mean particle orientation. It is also computationally cheaper to simulate than the full $d = 3$ model. For these reasons we shall use it in what follows.

We wish to simulate the effect of a field applied perpendicular to the walls confining a nematic thin film, which is the basic geometry of some opto-electronic devices. Molecules located at either wall are assumed to be rigidly anchored parallel to the wall and periodic boundary conditions are imposed in the perpendicular direction. As mentioned in Section 9.1, one of the technologically most relevant properties of liquid crystals is that the electric or magnetic susceptibilities of the ordered phases are anisotropic; hence the nematic director can be oriented by electric or magnetic fields. Most materials used in actual optical devices have positive anisotropy of their electric or magnetic susceptibilities, and consequently the nematic director will align parallel to the applied field. Director orientation within a monodomain is determined by the interplay of boundary conditions, and the direction and magnitude of the applied field. At the molecular level, the alignment due to a field is driven by the interaction between the field and the (permanent or induced) dipole moments along the rod axes. In addition, in an MC simulation the effect of temperature-induced disorder needs to be taken into account; indeed, the competition between this and the ordering effects of fields and boundaries is the central problem of statistical physics. In what follows we shall restrict ourselves to static electric fields. By using the Lebwohl–Lasher model, we focus exclusively on field-induced reorientation and neglect any collective translational motions – hydrodynamic modes – that are known to be obtained for sufficiently large fields.

The interaction energy of molecule i with its nearest neighbour j and an applied electric field is, in non-dimensional form,

$$\frac{u_i}{\epsilon} = -\sum_{j=1}^{m} P\left(\cos\gamma_{ij}\right) - \frac{\chi}{\epsilon}P_2\left(\cos\beta_i\right), \qquad (9.63)$$

where χ is the coupling strength between the induced dipole and the field, and β_i is the angle between the molecule and the field. The reduced electric field $\mathcal{E}^* \equiv \chi/\epsilon$ is a measure of the relative strength of the field–molecule and molecule–molecule interactions. The reduced electric field $\mathcal{E}^*$ and the reduced temperature T^*, defined above, will be the two control parameters in our simulations.

MC simulations of the system shown in figure 9.13 for different values of $\mathcal{E}^*$ and T^* were performed using the Metropolis algorithm. This is the same as described in Chapter 8. The only difference is that consecutive configurations differ in the values of angles Θ_i and Φ_i, which are uniformly distributed continuous random variables. For this reason it is not viable to pre-compute and tabulate the transition probabilities, as we did for the Ising model. These are now computed at every lattice site within each MC cycle, using Eq. (9.63) for the energy.

We measure orientations with respect to the axis parallel to the plates (OX in figure 9.13). The arithmetic mean of the second Legendre polynomial of the angle between the rod axes and the OX axis, taken over all rods except those anchored at the plates, is, for a given configuration,

$$\langle s \rangle = \frac{1}{N}\sum_{i=1}^{N}\frac{1}{2}\left(3\cos^2\Theta_i - 1\right), \qquad N = D(L-2). \qquad (9.64)$$

This will be unity when the rods are uniformly aligned along OX, and $-1/2$ when the rods are uniformly aligned along OZ, i.e., parallel to the applied field, as follows from Figure 7.13. We shall take $\langle s \rangle$ as our measure of rod alignment relative to OX.

Figure 9.14 shows $\langle s \rangle$ vs the reduced electric field $\mathcal{E}^*$ at five reduced temperatures T^*, for a lattice of size $N = 30 \times 40$ ($L = 40$). A transition is apparent, from a state of preferential alignment parallel to the plates, at low fields, to one of preferential alignment parallel to the field, at high fields; this is known as the *saturation transition*. At the lowest temperatures

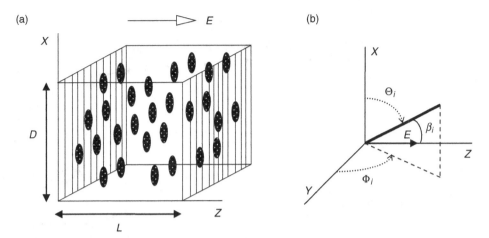

Figure 9.13 (a) Sketch of a nematic film of thickness L between two parallel plates, where a voltage is applied between the plates (shown shaded). L and D are, respectively, the number of lattice points in the directions perpendicular and parallel to the plates. The rod-shaped molecules are anchored parallel to the plates, and an electric field is applied perpendicular to the plates. If the field magnitude is below a value known as the Freedericksz threshold, the elastic energy cost of a deformation away from the anchoring imposed by the plates is larger than the electric energy cost of misalignment with the applied field, and the director remains spatially uniform: $\Theta = 0$ everywhere in the film (Chandrasekhar, 1992; de Gennes and Prost, 1992). (b) Above the Freedericksz threshold, the director will deform, as the molecules rotate towards the field direction: this effect is maximal in the midplane of the film. In the planar Lebwohl–Lasher model ($d = 2, n = 3$) that we simulate, the molecules can rotate out of the (x,z) plane. In the figure; β_i is the angle between the ith molecular rod and the field; the polar and azimuthal angles specifying the orientation of each rod are in the ranges $0 \leq \Theta_i \leq \pi/2$ and $0 \leq \Phi_i \leq \pi$, respectively.

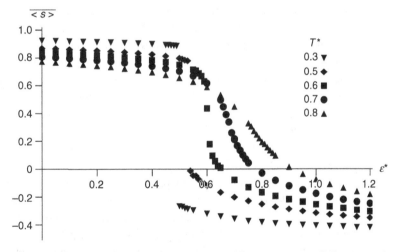

Figure 9.14 Mean rod orientation along OX vs reduced applied field $\mathcal{E}^*$ at five reduced temperatures T^*, for a lattice of size $N = 30 \times 40$ ($L = 40$). The symbols are averages of $\langle s \rangle$ over 10 000 configurations. For all $\mathcal{E}^*$, the following configurations have been omitted from the average: at $T^* = 0.3$, the first 10 000 configurations, or the first 20 000 if in the transition region; at $T^* = 0.5$ and $T^* = 0.6$, the first 5000 configurations, or the first 10 000 if in the transition region; at $T^* = 0.7$ and $T^* = 0.8$, the first 5000 configurations.

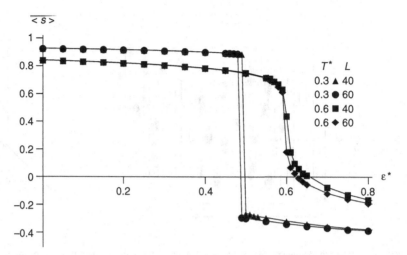

Figure 9.15 Effect of the film thickness L on the critical field, at two different temperatures. The lines are just to guide the eye.

considered, the state of preferential alignment parallel to the plates becomes metastable and there is a discontinuous transition from this to the state of preferential alignment parallel to the applied field, at some temperature-dependent critical field $\mathcal{E}_c^*$. At fixed temperature, $\mathcal{E}_c^*$ decreases as the lattice size L increases, as can be seen in Figure 9.15. Moreover, the peak in the plot of the electric susceptibility vs applied field (not shown) also decreases with increasing L. These effects are easily rationalised: as the film thickness increases, the effect of strong anchoring at the walls is felt less strongly in the bulk of the film, hence the critical field goes down. This suggests that the metastability and the discontinuity at the transition are finite-size effects: at lower temperatures the nematic elastic constants are larger and there is no smooth director deformation if the lattice is small, as the elastic energy cost is too high. Consequently, there is a build-up of elastic energy that is released abruptly at a value of the field. At temperatures above $T^* \sim 0.6$, the transition becomes continuous: the order parameter varies smoothly with the applied field, One possible interpretation is that, at this temperature, the film undergoes a transition into a paranematic state, in which there is (weak) orientational order due exclusively to the strong anchoring at the plates. In other words, in the absence of the bounding plates, a film at this temperature would be isotropic (Chiccoli *et al.* 1988; Mondal and Roy 2003). Figure 9.16 shows the $\langle \Theta \rangle$ profiles across the sample for $T^* = 0.6$.

The saturation transition has been predicted to be continuous (Sluckin and Poniewierski, 1985). It should be noted that this is not the same as the Freedericksz transition, which is a transition between the uniform, undeformed state and a non-uniform, deformed director state, occurring at $\mathcal{E}^*$ smaller than $\mathcal{E}_c^*$. It is not our purpose here to study the Freedericksz transition by MC simulation; the standard treatment using the continuum theory of nematics can be found in, e.g., Chandrasekhar (1992) and de Gennes and Prost (1992). In a thick film only the Freedericksz transition survives.

In our simulations we have discarded the first 5000–20 000 configurations (the latter in the transition region) when calculating averages. This is because a considerable number

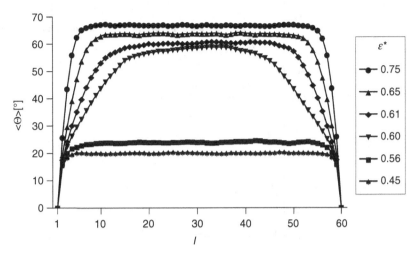

Figure 9.16 Mean angle $\langle\Theta\rangle$ between the rod axis and the *OX* axis (see Figure 9.13) vs position across the film, for $T^* = 0.6$ and a number of applied field strengths. System size is $N = 30 \times 60$ ($L = 60$); the solid lines are just to guide the eye. The corresponding order parameter (of which one example is plotted in figure 9.17) was computed by averaging over 10 000 configurations, after discarding the first 5000 configurations and checking that convergence had been achieved. $\langle\Theta\rangle$ was computed by averaging all spins of each column a distance l from one of the bounding plates. The resulting profile is well described by the continuum theory of nematics.

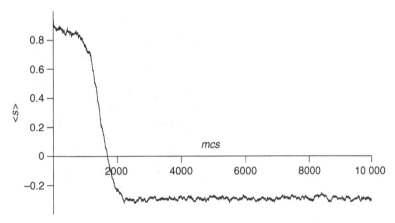

Figure 9.17 $\langle s\rangle$ vs number of MC cycles for $T^* = 0.3$ and $\mathcal{E}^* = 0.55$. Equilibrium is reached only after about 3000 MC cycles. Here $L = 40$.

of MC cycles are necessary to fully equilibrate the system: see Figure 9.17. One should always check that the configurations used in the calculation of averages of any quantities of interest have converged to a stable equilibrium state.

In the nematic phase, there is preferential alignment along the field around the mid-plane of the film. However, the effect of strong anchoring is to create two narrow regions of strong deformation close to either plate: see Figure 9.18, for $T^* = 0.6 < T^*_{NI}$

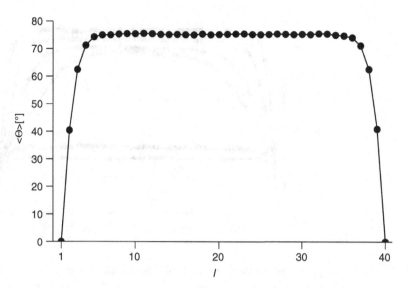

Figure 9.18 Symbols: angle $\langle \Theta \rangle$ between the rod axes and OX (see Figure 9.13), averaged over each lattice column, vs distance l to one of the plates, for $T^* = 0.3 < T^*_{NI}$ and $\mathcal{E}^* = 0.61$. The line is just to guide the eye. System size is $N = 30 \times 40$ ($L = 40$). The global order parameter is $\overline{\langle s \rangle} = -0.321$, computed by averaging over 10 000 configurations, after discarding the first 5000 configurations and checking that convergence had been achieved.

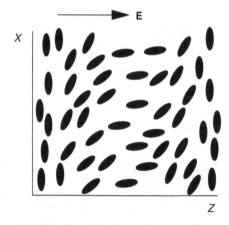

Figure 9.19 Sketch of the splay deformation discussed in the text, for a field just above the Freedericksz threshold.

and $\mathcal{E}^* = 0.61 > \mathcal{E}^*_c$. This is well described by the continuum (Leslie–Ericksen) theory of nematics, if the applied field is much larger than the Freedericksz threshold (Chandrasekhar, 1992; de Gennes and Prost, 1992). These two deformation regions store splay elastic energy, which is one of the nematic modes of deformation; see Figure 9.19. For the present choices of T^* and $\mathcal{E}^*$, $\langle \Phi \rangle \simeq \pi/2$, i.e., the rods reside on average in the plane defined by the applied field and the direction of preferred anchoring at the plates.

Problems

9.1 Find L/T_{NI}, where L is the latent heat of the nematic–isotropic phase transition in the Maier–Saupe model.

9.2 Prove Eq. (9.46).

9.3 Find the limit of thermodynamic stability of the nematic phase from Landau–de Gennes theory.

9.4 *Consider the *Saupe ordering matrix*, defined as

$$\eta_{\alpha\beta} = \frac{1}{2}\left(3u_\alpha u_\beta - \delta_{\alpha\beta}\right),$$

where $\mathbf{u}$ is a unit vector along the molecular long axis. Show that $\eta_{\alpha\beta}$ can be written in the form

$$\eta_{\alpha\beta} = Q\left(3n_\alpha n_\beta - \delta_{\alpha\beta}\right) + P\left(l_\alpha l_\beta - m_\alpha m_\beta\right),$$

where $\mathbf{n}$ is the nematic director, and $(\mathbf{l}, \mathbf{m})$ are unit vectors such that $(\mathbf{l}, \mathbf{m}, \mathbf{n})$ are a right-handed orthogonal basis. Derive expressions for Q and P, which measure, respectively, the degrees of *uniaxial* and *biaxial* order. (*Hint*: multiply both sides of the above equation by $\left(3n_\alpha n_\beta - \delta_{\alpha\beta}\right)$ and $\left(l_\alpha l_\beta - m_\alpha m_\beta\right)$ in turn, then take the trace.)

References

Boschi, S., Zannoni, C., Chiccoli, C., Pasini, P., and Brunelli, M.C.P. 2001. *Science and Supercomputing at CINECA – Report 2001* CINECA.

Chandrasekhar, S. 1992. *Liquid Crystals*, 2nd edition. Cambridge University Press.

Chiccoli, C., Pasini, P., and Zannoni, C. 1988. A Monte-Carlo investigation of the planar Lebwohl–Lasher lattice model. *Physica A* **148**, 298–311.

de Gennes, P. G., and Prost, J. 1992. *The Physics of Liquid Crystals*, 2nd edition. Oxford Science Publications.

Deloche, B., Cabane, B., and Jérôme, D. 1971. Effect of pressure on mesomorphic transitions in para-azoxyanisole (PAA). *Mol. Cryst. Liq. Cryst.* **15**, 197–209.

Dunmur, D. A., and Sluckin, T. J. 2011. *Soap, Science and Flat-Screen TVs*. Oxford University Press.

Fabbri, U. and Zannoni, C. 1986. A Monte Carlo investigation of the Lebwohl–Lasher lattice model in the vicinity of its orientational phase transition. *Molec. Phys.* **58**, 763–788.

Goldenfeld, N. 1992. *Lectures on Phase Transitions and the Renormalization Group*. Addison-Wesley.

Humphries, R. L., James, P. G., and Luckhurst, G. R. 1972. Molecular field treatment of nematic liquid-crystals. *J. Chem. Soc. Faraday Trans.* **2**, 68, 1031–1044.

Katriel, J., Kventsel, G. F., and Sluckin, T. J. 1986. Free energies in the Landau and molecular-field approaches. *Liq. Cryst.* **1**, 337–355.

Luckhurst, G. R., and Zannoni, C. 1977. Why is Maier–Saupe theory of nematic liquid-crystals so successful? *Nature* **267**, 412–414.

Maier, W. and Saupe, A. 1958. Eine einfache molekulare Theorie des nematischen kristallinflüssigen Zustandes. *Z. Naturforsch.* **13a**, 564–566.

Maier, W. and Saupe, A. 1959. Eine einfache molekular-statistische Theorie der nematis-chen kristallinflüssigen Phase. 1. *Z. Naturforsch.* **14a**, 882–889.

Maier, W. and Saupe, A. 1960. Eine einfache molekular-statistische Theorie der nematis-chen kristallinflüssigen Phase. 2. *Z. Naturforsch.* **15a**, 287–292.

Mondal, E., and Roy, S. K. 2003. Finite size scaling in the planar Lebwohl–Lasher model. *Phys. Lett. A* **312** 397–410.

Onsager, L. 1949. The effects of shape on the interaction of colloidal particles. *Ann. N. Y. Acad. Sci.* **51**, 627–659.

Papon, P., Leblond, J., and Meijer, P. H. E. 2002. *Physique des Transitions de Phases*, 2nd edition. Dunod.

Sluckin, T. J., and Poniewierski, A. 1985. In *Fluid Interfacial Phenomena*, ed. C. A. Croxton. Wiley. pp. 215–253.

Vroege, G. J., and Lekkerkerker, H. N. W. 1992. Phase transitions in lyotropic colloidal and polymer liquid crystals. *Rep. Prog. Phys.* **55**, 1241–1309.

10 Phase transitions and critical phenomena

10.1 Introduction

In this chapter we provide a comprehensive and unified treatment of the physical concepts and quantities relevant to phase transitions and critical phenomena. Some of these concepts and quantities have already been discussed in the particular contexts of magnetism (Chapter 7), or liquid crystals (Chapter 9, We conclude the chapter with a brief reference to the application of renormalisation group ideas to critical phenomena. The resulting reconciliation of the predictions of classical theories with experiments and exact or quasi-exact solutions of model systems is a major triumph of twentieth-century physics.

10.2 Phases and phase transitions

Thermodynamics habitually deals with equilibrium systems that are uniform on the macroscopic scale – i.e., where the structure, density and composition are the same throughout the bulk of the system, which is many times greater than a single atom or molecule. Because the system is in equilibrium, intensive thermodynamic variables such as the temperature or chemical potential are also spatially uniform. Such a system is said to consist of a single *phase* – it is a one-phase system. Phases are what used to be called the states of matter. *Solids*, *liquids*, *gases* and *plasmas*[1] are four commonly occurring phases – they correspond to the four elements of Aristotelian physics: earth, water, air and fire, respectively. There are, however, many other, less easily identified phases, some of which we have already encountered in this book, such as the ferromagnetic, antiferromagnetic, nematic or superconducting phases.

On varying the temperature, pressure and chemical potential(s) of a system, properties such as its structure, density, composition, compressibility, specific heat, etc., in general change smoothly. Yet under certain conditions the changes may be abrupt – some of these properties may exhibit discontinuities, or even divergences, at a *phase transition*. Phase transitions are among the most spectacular phenomena of nature. Consider, for instance, the phase sequence of water on heating: from ice, an opaque, slippery, rock-like solid; to colourless, incompressible, liquid water, which nevertheless easily flows under

---

[1] A plasma is not universally acknowledged to be a distinct phase from a gas, as there is no obvious order parameter for the transition between the two.

gravity; to water vapour, an invisible and very low-density (greenhouse!) gas, hence of ill-defined volume. It is therefore hardly surprising that in earlier times the three phases of water should have been regarded as three chemically distinct substances, albeit inter-convertible through heating and cooling. And if this was the case with water, why would it not be with other substances as well, leading to the alchemic belief in the transmua-tion of the elements? These incorrect ideas, we now realise, arose from ignorance of the atomic nature of matter, as well as of the concept of *energy*. It is energy that allows us to distinguish between *physical* changes, such as phase transitions, which only involve a spatial rearrangement of the constituent molecules, and *chemical* changes, or chemical reactions, in which the molecules themselves change their composition. In general, the energy exchanged between the system and the environment is smaller in a physical change than in a chemical change, although this is not always the case: for example, many organic compounds will decompose before boiling – just think of caramelised sugar.

Thus far we have restricted ourselves to one-component systems, i.e., pure substances. Mixtures are, however, fertile ground for this type of phenomenon. Pour two immiscible liquids (the paradigm is oil and water) into the same vessel and attempt to mix them – for instance by shaking vigorously. As we all know, this is doomed to failure: the system will readily separate into two phases, each of which is almost one pure component (in the case of water and oil, the water-rich phase will fall to the bottom of the vessel). Phase separation is also common in solids, such as alloys: for example, at low temperatures austenitic steels will phase-separate into iron- and carbon-rich phases.

10.3 The order of a phase transition

At a phase transition, some quantities are observed to vary abruptly. Let us consider the condensation of a simple fluid. The appropriate thermodynamic potential is the Gibbs free energy, whose natural variables are the pressure and the temperature – which, as we now, do not change in the course of the transition. However, the volume and the entropy will vary discontinuously: a liquid has a much smaller volume than a gas at the same pressure and number of particles, and there is an associated latent heat. It then follows from Eqs. (2.36) and (2.37) that the Gibbs free energy has discontinuous first derivatives at the transition. Furthermore, the latent heat is related to the entropy jump through Equation (3.99). If there is no volume change at the transition, then the Gibbs free energy can be replaced by the Helmholtz free energy, but everything else stays the same: the first derivatives of the thermodynamic potential that is minimised at equilibrium are discontinuous. A phase transition displaying this property is said to be *first order*: Figure 10.1, parts (a) and (c), shows how G or F and the entropy vary in the vicinity of such a transition.

If, on the other hand, the first derivatives of the relevant thermodynamic potential are continuous, but its higher-order derivatives are not, the transition is said to be *contin-uous*: there is no latent heat associated with a continuous transition. Parts (b) and (d)

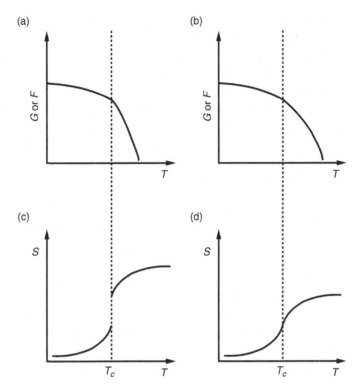

Figure 10.1 Gibbs free energy G at constant pressure, or Helmholtz free energy F at constant volume, vs temperature in the vicinity of (a) a first-order transition, or (b) a continuous transition, at $T = T_c$. (c) and (d) show the temperature dependence of the entropy, $S = -(\partial G/\partial T)_p$ or $S = -(\partial F/\partial T)_T$, obtained by differentiating the G or F vs T curve in (a) or in (b), respectively. The entropy has a step discontinuity at a first-order transition, (c), but not at a continuous transition, (d). (Figure adapted from Stanley (1971), by permission of Oxford University Press.)

of Figure 10.1 illustrate the case in which the second derivative of G or F has a step discontinuity, at the transition.

We note in passing that, in the original classification of phase transitions, due to Paul Ehrenfest, the order of a transition was that of the lowest-order derivative of the Gibbs or Helmholtz free energy that had a step discontinuity at the transition. This classification later proved inadequate, as in some cases, e.g., at the Curie point of a ferromagnet, a derivative of the free energy may have an infinite discontinuity or even diverge, but no lower-order derivatives have step discontinuities. In modern parlance, all transitions that are not first-order are said to be continuous or *second-order*.

It is easy to understand the differences between first-order and continuous phase transitions by looking at the free energy as a function of the order parameter. For a continuous transition, such as the paramagnetic–ferromagnetic transition discussed in Section 7.4, the non-dimensional Helmholtz free energy f^*_{mag} is as in Figure 7.7a. For $T < T_c, f^*_{\mathrm{mag}}$ has two symmetric minima at $m = \pm m_s$, corresponding to magnetisations of the same magnitude but opposite signs, hence physically equivalent. When $T \rightarrow T_C^-$, the two minima move towards the origin of the axes and at $T = T_c$ they merge into a single minimum at $m = 0$. Figure 7.7b shows the magnetisation m_s vs temperature: it is a continuous function, albeit with an infinite derivative at $T = T_c$. A continuous transition therefore occurs between contiguous points in the space of thermodynamic variables.

Consider now the isotropic–nematic transition of a liquid crystal studied in Section 9.2. The non-dimensional Helmholtz free energy is plotted in Figure 9.8. For $T \ll T_{NI}$ this free energy has a single minimum at $Q \neq 0$, corresponding to the nematic phase. At a temperature $T_I < T_{NI}$ – the limit of absolute stability of the isotropic phase – a new, shallower minimum appears at $Q = 0$, which corresponds to a (metastable) isotropic phase. Between this minimum and the deeper one, the free energy has a maximum. At $T = T_{NI}$ – and, unlike in the magnetic solid, only at this temperature – the two minima have the same depth and coexistence is possible between a non-zero-Q nematic phase and a zero-Q isotropic phase. If the temperature is further increased, the minimum at $Q \neq 0$ will become shallower than that at $Q = 0$, and the nematic phase will be only metastable – it resides in a local minimum of the free energy – and will eventually disorder, as it can lower its Helmholtz free energy by doing so. Finally, above $T_N > T_{NI}$ – the limit of absolute stability of the nematic phase – this local minimum disappears and only the isotropic phase can be realised: any system still in the nematic phase would disorder very rapidly if heated above T_N. Unlike the paramagnetic–ferromagnetic transition, the isotropic–nematic transition is discontinuous at $T = T_{NI}$ – thermal energy allows the system to jump from a non-zero-order-parameter phase to a zero-order-parameter phase. Note, however, that the transition would be continuous at either $T = T_I$ or $T = T_N$.

10.4 Critical points and critical exponents

As we saw earlier in this chapter, at a phase transition many properties of a system undergo abrupt changes. For instance, when water boils its density is reduced some 1000 times.[2] Yet it is often possible, by varying an appropriate thermodynamic parameter such as the temperature. to weaken the transition – i.e., to make the two phases more and more similar, until they become identical at a *critical point*. Beyond this critical point there is only one phase, and no phase transition: it is therefore possible continuously to transform, say, a liquid into a gas with no discontinuous change of its density. In other words, at a critical point the latent heat vanishes and a first-order transition becomes continuous. This was first observed experimentally, for carbon dioxide, by Andrews in 1869 (Andrews, 1869).

The commonest critical point is that associated with the condensation of a pure fluid. This is the point at which the liquid–vapour coexistence curve terminates, and at which the pressure, temperature and density take well-defined values for each substance, p_c, T_c and ρ_c respectively. In a binary mixture or an alloy, one usually finds two phases of different compositions at low temperature, but a single phase at a higher temperature. The

[2] Freezing also entails a change in density, albeit much less spectacular: the density of ice is about 10% less than that of liquid water.

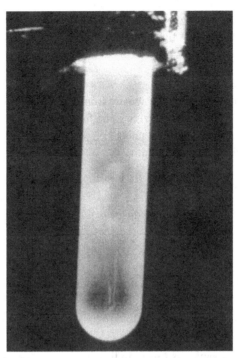

Figure 10.2 Critical opalescence in a mixture of cyclohexane and aniline just below the critical point. Reproduced from Stanley, H. E., Buldyrev, S. V., Canpolat, M., Mishima, O., Sadr-Lahijany, M. R., Scala, A., and Starr, F. W. 2000. The puzzling behaviour of water at very low temperature. *Phys. Chem. Chem. Phys.* 2, 1551–1558, with permission from the Royal Society of Chemistry.

critical temperature is then the temperature above which the two phases become identical and the interface between them disappears. Figure 10.2 illustrates the phenomenon of critical opalescence, which is due to the scattering of light by large-wavelength density (hence refractive index) fluctuations close to T_c. The critical point of a binary mixture is also characterised by well-defined values of the pressure and composition, p_c and x_c respectively.

In a ferromagnetic crystal, the critical point coincides with the Curie temperature, at which the spontaneous magnetisation vanishes in the absence of an applied magnetic field. One final example is a superconductor: below its critical temperature the resistivity drops to zero, again for zero applied magnetic field.

10.4.1 Definition of critical exponents

We saw earlier in this chapter that many thermodynamic quantities are either discontinuous or diverge at a critical point. These behaviours are described by a set of numbers called *critical exponents* or *critical indices*, which we shall now define rigorously (Stanley, 1971). Let us start by introducing a non-dimensional reduced temperature,

$$t \equiv \frac{T - T_c}{T_c} = \frac{T}{T_c} - 1, \tag{10.1}$$

which measures how far we are from the critical point. Clearly, below T_c we have $t < 0$, and above T_c, $t > 0$. If $f(t)$ is some function of t, we shall be interested in its behaviour in the vicinity of the critical point, i.e., for $|t| \ll 1$. If we further suppose that $f(t)$ is continuous and positive, and if the limit

$$\lambda = \lim_{t \to 0} \frac{\ln f(t)}{\ln |t|}, \tag{10.2}$$

exists, then λ is said to be the *critical exponent* associated with $f(t)$. This is usually written in abbreviated form as

$$f(t) \sim |t|^{\lambda}. \tag{10.3}$$

Note that this is in general **not** the same as

$$f(t) = A|t|^{\lambda}, \tag{10.4}$$

where A is a constant, since $f(t)$ usually contains additional terms, known as *subdominant terms*, that vanish as $t \to 0$. In the simplest case we have

$$f(t) = A|t|^{\lambda} \left(1 + B|t|^{y} + \cdots \right) \qquad (y > 0), \tag{10.5}$$

where A is the *critical amplitude*. It is much easier experimentally to find the critical exponents than the full functional form of $f(t)$: λ is the slope of the plot of $f(t) \sim |t|^{\lambda}$ vs t on a log-log scale in the region $|t| \ll 1$. As we shall see, the critical exponents are the same for many different systems, since they depend only on very general properties such as a sysem's spatial dimension. In addition, there exist a large number of relations between critical exponents, some of which we shall derive in the next section.

If λ is positive, then $f(t)$ will go to zero as $t \to 0$, whereas if λ is negative, $f(t)$ will diverge at $t = 0$. $\lambda = 0$ is a special case, as it may describe three very different behaviours, as we shall now see.

1. *Logarithmic divergence.* To show that $\lambda \to 0$ may correspond to a logarithmic divergence, let us consider a function of the form

$$f_{\lambda}(t) = A \frac{|t|^{\lambda} - 1}{\lambda} + B, \tag{10.6}$$

where A and B are constants. It is easily shown (see Problem 10.1) that

$$f(t) \equiv \lim_{t \to 0} f_{\lambda}(t) = A \ln |t| + B. \tag{10.7}$$

2. *Cusp singularity.* Let us now consider a function of the form

$$f(t) = A - B|t|^{\mu} \qquad (0 < \mu < 1), \tag{10.8}$$

where A and B are constants By definition (10.2), the critical exponent is again $\lambda = 0$: the function is finite at $t = 0$, $f(t = 0) = A$, but its first derivative diverges.

3. *Analytic function*, with or without a jump discontinuity of the first derivative This would again be a function of the form (10.8), but for $\mu \geq 1$. Now $f(t)$ is analytic for all t if $\mu > 1$; if $\mu = 1$ the first derivative is jump-discontinuous at $t = 0$.

We are thus led to generalise the definition of critical exponent, Eq. (10.2), in order to take into account possible 'singular behaviour' of $f(t)$. One possible solution is to start by finding the lowest-order derivative of $f(t)$ that diverges at $t = 0$. If j is the order of that derivative, an alternative definition of critical exponent is then

$$\lambda = j + \lim_{t \to 0} \frac{\ln |d^j f(t)/dt^j|}{\ln |t|}. \tag{10.9}$$

Applied to, e.g., $f(t) = A - B|t|^\mu$, this gives $\lambda = \mu$.

10.4.2 The most important critical exponents

There are many critical exponents associated with a number of thermodynamic quantities. We now introduce the most important ones.

Experimentally, in the vicinity of the critical point of a pure substance one has

$$\frac{\rho_l(T) - \rho_v(T)}{\rho_c} \sim (-t)^\beta, \tag{10.10}$$

where $\rho_l(T)$ and $\rho_v(T)$ are, respectively, the coexistence liquid and vapour densities at temperature T, and ρ_c is the critical density. Analogously, for a ferromagetic material in the vicinity of its Curie temperature and in zero applied field, one also has

$$M(T) \sim (-t)^\beta \qquad (H = 0), \tag{10.11}$$

where $M(T)$ is the spontaneous magnetisation at temperature T. Finally, consider a binary mixture that exhibits two phases, α and β, at low temperature, but only one at higher temperature. Close to the critical point we again have

$$X_\alpha(T) - X_\beta(T) \sim (-t)^\beta, \tag{10.12}$$

where $X_\alpha(T)$ and $X_\beta(T)$ are the compositions of the coexisting phases at temperature T. In all cases, $\beta > 0$. Further, note that the quantities $[\rho_l(T) - \rho_v(T)]/\rho_c$, $M(T)$ and $X_\alpha(T) - X_\beta(T)$ are all non-zero below T_c and zero above T_c, approaching zero continuously as $T \to T_c^-$ ($-t \to 0^-$). A quantity with this property allows a phase transition to be located: Landau called it the *order parameter* of the transition. β is, then, the critical exponent associated with the order parameter of a transition. Its value is approximately the same for the above three transitions, which is somewhat unexpected, as the three systems are quite different. As we shall see later, explaining this was one of the key motivations for the study of critical phenomena.

Besides the order parameter, the thermodynamic response functions also exhibit interesting behaviours in the critical region. In the case of a fluid, the isothermal compressibility κ_T, Eq. (2.61), is

$$K_T \sim \begin{cases} (-t)^{-\gamma'} & \text{if } T < T_c, \rho = \rho_l(T) \text{ or } \rho = \rho_v(T) \\ t^{-\gamma} & \text{if } T > T_c, \rho = \rho_c \end{cases} . \qquad (10.13)$$

This means that, if we approach the critical point from the low-temperature region ($T \to T_c^-$) along either branch of the liquid–vapour coexistence curve, the isothermal susceptibility will diverge as an inverse power law of exponent γ'. If, on the other hand, we approach the critical point from the high-temperature region ($T \to T_c^+$), at a constant density equal to the critical density, then the isothermal compressibility will diverge as an inverse power law of exponent γ. The equivalent response function for magnetic solids is the isothermal susceptibility χ_T, given by Eq. (2.76). We likewise have

$$\chi_T \sim \begin{cases} (-t)^{-\gamma'} & \text{if } T < T_c, H = 0 \\ t^{-\gamma} & \text{if } T > T_c, H = 0 \end{cases} . \qquad (10.14)$$

Two other exponents are associated with the specific heat of a fluid at constant volume, given by Eqs. (2.57) or (2.59):

$$C_V \sim \begin{cases} (-t)^{-\alpha'} & \text{if } T < T_c, \rho = \rho_l(T) \text{ or } \rho = \rho_v(T) \\ t^{-\alpha} & \text{if } T > T_c, \rho = \rho_c \end{cases} , \qquad (10.15)$$

and the specific heat of a magnetic solid in zero applied field, given by Eq. (2.73):

$$C_H \sim \begin{cases} (-t)^{-\alpha'} & \text{if } T < T_c, H = 0 \\ t^{-\alpha} & \text{if } T > T_c, H = 0 \end{cases} . \qquad (10.16)$$

Note that we have violated the mapping between fluid and magnetic quantities, Eq. (2.71), which is suitable for describing monodomain ferromagnets. This would have led us to consider the specific heat at constant magnetic moment, C_M, rather than the specific heat at constant magnetic field, C_H. These two quantities are in fact equivalent (Fisher, 1967): for $T > T_c$, we obviously have $H = 0 \Rightarrow M = 0$, since by definition a ferromagnet is paramagnetic above its Curie temperature; for $T < T_c$, a real ferromagnet will split up into domains of different orientations, in such a way that its total magnetic moment is only zero when $H = 0$.

Two other critical exponents, ν' and ν, describe the critical behaviour of the order parameter correlation length ξ, which we shall define more precisely later. Suffice it to say for now that ξ is a measure of the range of order parameter fluctuations, i.e., the lengthscale on which the order parameter varies. For T much less than T_c, ξ is of the order of the mean interatomic or intermolecular separation, i.e., it is microscopic. Close to T_c, however, order parameter fluctuations become long-ranged, extending across the whole system. In a fluid, the associated strong spatial variations of the refractive index lead to the phenomenon known as *critical opalescence*: see figure 10.2. In a fluid, one has

$$\xi \sim \begin{cases} (-t)^{-\nu'} & \text{if } T < T_c, \rho = \rho_l(T) \text{ or } \rho = \rho_v(T) \\ t^{-\nu} & \text{if } T > T_c, \rho = \rho_c \end{cases} , \qquad (10.17)$$

Table 10.1 Critical exponents α (specific heat), β (order parameter), γ (response function), ν (correlation length), and δ (critical isotherm), from experiment, theory (classical theories, two-dimensional Ising model, spherical model) and numerical simulation (three-dimensional Ising model) [Goldenfeld, 1992, chap. 3.7; Huang, 1987, chap. 16; Stanley, 1971, chap. 8]. Note that the values of any given critical exponent below (primed) and above (unprimed) T_c are the same, which fact is not explained by the classical theories.

Exponent	Experiment	Classical theories	Ising model $(d=2)$	Ising model $(d=3)$	Spherical model $(d=3)$
α	$0.110 - 0.116$	0	0	0.110	-1
β	$0.316 - 0.327$	1/2	1/8	0.325	1/2
γ	$1.23 - 1.25$	1	7/4	1.24	2
ν	$0.6 - 0.7$	1/2	1	0.64	1
δ	$4.6 - 4.9$	3	15	4.82	5

whereas in a magnetic solid,

$$\xi \sim \begin{cases} (-t)^{-\nu'} & \text{if } T < T_c, H = 0 \\ t^{-\nu} & \text{if } T > T_c, H = 0 \end{cases}. \tag{10.18}$$

One last exponent,[3] δ, describes the shape of the critical isotherm: unlike all previous ones, it relates the pressure (for a fluid) or the magnetic field (for a magnetic solid) to the order parameter on the critical isotherm, i.e., for $T = T_c$. For a fluid, one has

$$P - P_c \sim \left(\frac{\rho - \rho_c}{\rho_c}\right)^\delta \qquad (T = T_c), \tag{10.19}$$

and, for a magnetic solid,

$$H \sim M^\delta \qquad (T = T_c). \tag{10.20}$$

Values of the above exponents are collected in Table 10.1, for a number of physical systems and some statistical mechanical models.

From more comprehensive compilations of data, two regularities are apparent:

1. Physical systems can be grouped into *universality classes*. Within each universality class, each critical exponent has a well-defined value for all systems, however disparate these might be.
2. The values of a given exponent below ($T < T_c$) and above ($T > T_c$) the critical temperatures are very approximately the same. We shall retain the distinction between primed and unprimed exponents so that it is always clear whether we are above or below T_c.

The *classical* theories, such as Weiss's theory and the Van der Waals theory that we shall encounter shortly, predict that a given critical exponent should have the same value for all systems, in clear disagreement with experimental evidence. A valid theory of criticality should be able to explain why there are universality classes and predict the numerical

---

[3] The last we shall consider here. There are in fact many more;

values of the critical exponents within each class. As it happens, this goal can only be achieved through application of the renormalisation group, whose key ideas we shall sketch at the end of this chapter but which is otherwise beyond the scope of the present book.

10.4.3 Critical exponent inequalities

On the basis of purely thermodynamic considerations, two inequalities can be derived that some of the critical exponents must satisfy, regardless of their actual values. These provide a powerful test of any putative theory of critical phenomena. We shall present them now. Only the first one will be derived, as the derivation of the second is somewhat lengthy; the interested reader may consult Stanley (1971). First, however, we need to state the following lemma.

Lemma 10.1: Let $f(x) \sim x^{\lambda}$ and $g(x) \sim x^{\mu}$ be two positive real-valued functions. If $f(x) \leq g(x)$ when $0 < x \ll 1$, then $\lambda \geq \mu$.

Proof   If $f(x) \leq g(x)$, then $\ln f(x) \leq \ln g(x)$. On the other hand, $0 < x \ll 1$ implies that $\ln x < 0$, hence

$$\frac{\ln f(x)}{\ln x} \geq \frac{\ln g(x)}{\ln x}, \tag{10.21}$$

Taking the limit of both sides of inequality (10.21) when $x \to 0$ and using the definition of critical exponent, Eq. (10.2), it follows that $\lambda \geq \mu$.

The first inequality relating critical exponents is known as *Rushbrooke's inequality*. We shall prove it using the language of magnetic systems for simplicity.[4] Let us start by noting that, from Eq. (2.82), one has

$$C_M = C_H - \frac{T\alpha_H^2}{\chi_T} \geq 0 \Leftrightarrow C_H \geq \frac{T\alpha_H^2}{\chi_T}, \tag{10.22}$$

where the inequality is a consequence of the thermodynamic stability of the ferromagnetic phase, and α_H, given by Eq. (2.81), is the magnetic conterpart of the coefficient of thermal expansion α, Eq. (2.63). Using Eqs. (10.16), (10.11) and (10.14), we obtain

$$C_H \sim (-t)^{-\alpha'}, \quad \alpha_H = \left(\frac{\partial M}{\partial T}\right)_H \sim (-t)^{\beta-1}, \quad \chi_T \sim (-t)^{-\gamma'}, \tag{10.23}$$

and from Eq. (10.22) we have

$$(-t)^{-\alpha'} \geq (-t)^{2(\beta-1)+\gamma'}. \tag{10.24}$$

The lemma then implies

$$\textit{Rushbrooke's inequality} \qquad -\alpha' \leq 2(\beta-1)+\gamma' \Leftrightarrow \alpha'+2\beta+\gamma' \geq 2. \tag{10.25}$$

One other inequality relates the specific heat, order parameter and critical isotherm exponents:

$$\textit{Griffiths' inequality} \qquad \alpha'+\beta(1+\delta) \geq 2. \tag{10.26}$$

---

[4] For a derivation employing the language of fluids, see Fisher (1967).

To prove this inequality (see Stanley (1971) or Fisher (1967)) one uses the fact that the Helmholtz free energy is convex with respect to both temperature and magnetic moment, as well as the lemma stated earlier.

Many other critical exponent inequalities can be derived if one introduces additional hypotheses concerning the behaviour of thermodynamic functions; see Fisher (1967) and references therein. As we shall see later, many of these actually hold as equalities – which is not explained by thermodynmaics.

10.5 Classical theories of phase transitions. Universality

We next discuss three of the so-called 'classical' theories of critical phenomena. All predict critical exponents that do not depend on the dimension of the system, or on that of the order parameter, and which are also collected in Table 10.1.

10.5.1 Van der Waals theory of the liquid–vapour transition

The Van der Waals equation of state (1873) is an approximate equation of state for a real gas. It provided one of the earliest qualitative explanations of fluid phase transitions. Although its predictions are quantitatively incorrect near the critical point, the Van der Waals equation is credited with first allowing identification of the roles played by repulsive amd attractive intermolecular forces in phase transitions. Here we present a heuristic derivation of the Van der Waals equation, following Stanley (1971, chap. 5).

Let us start by recalling the equation of state of an ideal gas, written in terms of the particle number N, or, alternatively, of the mole number $n = N/N_A$:

$$PV = Nk_BT = nRT. \qquad (10.27)$$

This equation, which neglects all interparticle interactions, describes reasonably well the behaviour of gases at low density and/or high temperature. To treat a dense gas one needs to take into account: (i) intermolecular repulsions, i.e., the fact that molecules have finite sizes; and (ii) intermolecular attractions.

According to Eq. (10.27), on increasing the pressure of an ideal gas at constant temperature to arbitrarily large values, its volume will shrink to arbitrarily small values. This is obviously not true of a real gas. Let us assume, as a first approximation, that the molecules are hard spheres of total volume V_{min}. The accessible volume is then that of the container (V) minus that occupied by the gas molecules (V_{min}). Because the volume is an extensive quantity, $V_{min} \propto N$. Let b/N_A be the volume occupied by a single molecule. Then $V_{min} = bN/N_A = nb$ and we are thus led to make the substitution $V \rightarrow V - nb$ in Eq. (10.27). This yields a first modification of the ideal gas equation, sometimes called the *Clausius equation of state*:

$$P = \frac{nRT}{V - nb}. \qquad (10.28)$$

Table 10.2 Van der Waals parameters of some substances, from Stanley (1971).			
Gas	a (Pa m^6 mol^{-2})	b (m^3 mol^{-1})	$Z_c \equiv P_c V_c / RT_c$
He	3.45×10^{-3}	23.7×10^{-6}	0.308
H$_2$	2.47×10^{-2}	26.6×10^{-6}	0.304
O$_2$	1.38×10^{-1}	31.8×10^{-6}	0.292
CO$_2$	3.63×10^{-1}	42.7×10^{-6}	0.275
H$_2$O	5.52×10^{-1}	30.5×10^{-6}	0.230

Now the intermolecular attractions will decrease the pressure that the gas exerts on the walls of the vessel, by an amount ΔP. This decrease is a consequence of the fact that each molecule, because it is attracted by others, carries less linear momentum when it collides with the walls, and also that, for the same reason, there are fewer such collisions. If we assume that both contributions are proportional to the density, we may write $\Delta P \propto (N/V)^2$. Taking the proportionality constant to be a/N_A^2, we subtract $\Delta P = (a/N_A^2)(N/V)^2$ from the right-hand side of Eq. (10.28), to get the *Van der Waals equation of state*:

$$P = \frac{nRT}{V - nb} - \frac{an^2}{V^2}, \tag{10.29}$$

which is usually written in the form

$$\left(P + \frac{an^2}{V^2} \right) (V - nb) = nRT. \tag{10.30}$$

a and b are phenomenological parameters. Table 10.2 (Stanley, 1971) collects a and b for a few substances, which fit the experimental data quite well when far from T_c.

For one mole of gas ($n = 1$), Eq. (10.30) can be written in the form

$$V^3 - \left(b + \frac{RT}{P} \right) V^2 + \frac{a}{P} V - \frac{ab}{P} = 0. \tag{10.31}$$

At low enough temperatures, Eq. (10.31) has three real solutions, i.e., for each pair of values of P and T there are three possible volumes V, not just one as in the case of the ideal gas. Figure 10.3 shows a few Van der Waals isotherms. On increasing the temperature, the three solutions come together and merge into a single solution, V_c, at $T = T_c$. At temperatures higher than T_c, two of the roots are complex, and at very high temperatures, Eq. (10.31) may be written approximately as $V^3 - (RT/P)V^2 = 0$, i.e., it approaches the ideal gas equation of state. As can be seen from Fig. 10.3, the critical point is simultaneously a stationary point and an inflection point of P vs V; hence we must have

$$\left(\frac{\partial P}{\partial V} \right)_T = \left(\frac{\partial^2 P}{\partial V^2} \right)_T = 0. \tag{10.32}$$

These two equations, together with Eq. (10.31), yield the critical pressure P_c, critical volume V_c, and critical temperature T_c, as functions of a and b (see Problem 10.2):

$$P_c = \frac{a}{27b^2}; \quad V_c = 3b; \quad T_c = \frac{8a}{27Rb}. \tag{10.33}$$

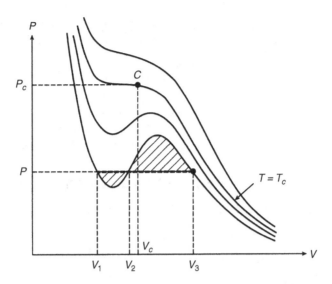

Figure 10.3 *P–V* isotherms of the Van der Waals equation of state. For $T < T_c$ the isotherms have a positive-slope portion, which is unphysical as in those regions the compressibility would be negative. This artefact is a consequence of the failure of Van der Waals theory to ensure that the equilibrium state is a global minimiser of the Gibbs free energy. It follows that the liquid phase is artificially continued into a region of the phase diagram where it is metastable, or even unstable. To rectify this deficiency one needs to impose mechanical equilibrium of the two coexisting phases, in addition to thermal equilibrium and equilibrium with respect to particle exchange (i.e., equality of chemical potentials). Mechanical equilibrium requires that the two coexisting phases should have the same pressure, i.e., that for $T < T_c$ the isotherms should be horizontal in the coexistence region (thick dashed line in figure). By Maxwell's construction, Eq. (10.47), the position of this horizontal portion should be such that the two shaded areas are equal. (Figure adapted from Huang (1987).)

If we now measure the pressure in units of P_c, the volume in units of V_c, and the temperature in units of T_c, by defining the following reduced variables:

$$\overline{P} = \frac{P}{P_c}; \quad \overline{V} = \frac{V}{V_c}; \quad \overline{T} = \frac{T}{T_c}, \tag{10.34}$$

then the Van der Vaals Eq., (10.31), becomes

$$\left(\overline{P} + \frac{3}{\overline{V}^2}\right)\left(\overline{V} - \frac{1}{3}\right) = \frac{8}{3}\overline{T}. \tag{10.35}$$

The Van der Waals equation written in terms of these variables does not contain any substance-specific parameter. It is a universal equation, valid for all substances, and known as a *law of corresponding states*. Experimentally, it has been found that even fluids that are not described by the Van der Waals equation obey a law of corresponding states: plots of the coexistence pressure, volume and temperature curves, scaled by the critical values (*Guggenheim plots*) are very nearly universal (Stanley, 1971).

Further note that, from Eqs. (10.33), one may derive the following universal relation involving the critical pressure, volume and temperature of a Van der Waals fluid:

$$Z_c \equiv \frac{P_c V_c}{RT_c} = \frac{a}{27b^2} \times 3b \times \left(\frac{8a}{27b}\right)^{-1} = \frac{3}{8} = 0.375. \tag{10.36}$$

Typical experimental values of Z_c are listed in Table 10.2. These are close to 0.375 for the simpler gases, but progressively deviate from this value as their complexity increases and the gases are less well described by the Van der Waals equation.

Critical exponents for the Van der Waals equation of state

This derivation follows Stanley (1971). Start by defining a new set of reduced variables that measure how far we are from the critical point:

$$p \equiv \overline{P} - 1 = \frac{P - P_c}{P_c}; \quad v \equiv \overline{V} - 1 = \frac{V - V_c}{V_c}; \quad t \equiv \overline{T} - 1 = \frac{T - T_c}{T_c}. \tag{10.37}$$

In terms of these new variables, the Van der Waals equation, Eq. (10.35) is now

$$\left[(1+p) + 3(1+v)^{-2}\right][3(1+v) - 1] = 8(1+t). \tag{10.38}$$

Multiplying both sides of this equation by $(1+v)^2$, expanding all powers and regrouping terms, we get

$$2p\left(1 + \frac{7}{2}v + 4v^2 + \frac{3}{2}v^3\right) = -3v^3 + 8t\left(1 + 2v + v^2\right). \tag{10.39}$$

Solving this equation with respect to p near the critical point yields

$$p = 4t - 6tv - \frac{3}{2}v^3 + \mathcal{O}\left(tv^2, v^4\right), \tag{10.40}$$

where omission of the terms denoted $\mathcal{O}\left(tv^2, v^4\right)$ will be justified *a posteriori*. We further define

$$\eta = \frac{1}{\overline{V}} - 1 = \frac{\rho - \rho_c}{\rho_c}. \tag{10.41}$$

It then follows from the third of Eqs. (10.37) that

$$v = \frac{1}{1+\eta} - 1 = -\eta + \mathcal{O}\left(\eta^2\right). \tag{10.42}$$

Substituting this into Eq. (10.40), we get

$$p = 4t + 6t\eta + \frac{3}{2}\eta^3 + \mathcal{O}\left(t\eta^2, \eta^4\right). \tag{10.43}$$

Setting $t = 0$ gives the critical isotherm exponent δ:

$$p \propto \eta^3 = \left(\frac{\rho - \rho_c}{\rho_c}\right)^3 \Rightarrow \delta = 3. \tag{10.44}$$

We next turn to the critical isochore, $\rho = \rho_c \Rightarrow v = 0$. From Eqs. (2.61), (10.37) and (10.39) we find, for the isothermal compressiblity K_T when $T \to T_C^+$,

$$(-V_c K_T)^{-1} = \frac{P_c}{V_c} \left(\frac{\partial p}{\partial v} \right)_T = \frac{P_c}{V_c}(-6t) \Rightarrow K_T \propto t^{-1} \Rightarrow \gamma = 1. \qquad (10.45)$$

To find β, the critical exponent of the order parameter, we need to know the shape of the coexistence curve when $T \to T_c^-$. This is given by $v_l(p)$ and $v_v(p)$ as determined by *Maxwell's equal-area construction* (Goldenfeld, 1992, chap. 4) (see the caption of figure 10.3), which we shall now derive. Differentiating Eq. (3.81) and comparing the result with Eq. (2.35), we find, for constant N,

$$d\mu = -\frac{S}{N}dT + \frac{V}{N}dP. \qquad (10.46)$$

Integrating this along any isotherm ($dT = 0$) between the vapour (v) and liquid (l) phases, we get

$$\mu_l - \mu_v = \int_v^l d\mu = \int_{P_v}^{P_l} \frac{V}{N}dP = 0. \qquad (10.47)$$

Equation (10.47) follows from purely thermodynamic considerations and therefore must hold for any functional form $V(P)$, not just for the Van der Waals isotherms. It implies the condition of mechanical equilibrium of the two coexisting phases, $P_l = P_v$. Hence any P–V isotherm must be flat in the coexistence region. To see what Eq. (10.47) means geometrically, consider the lowermost isotherm in Figure 10.3. This is reproduced in Figure 10.4 as V vs P. Let the endpoints of the (now vertical) thick-dashed line be A, of

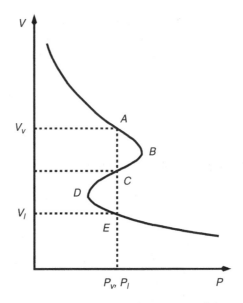

Figure 10.4 Lowest-temperature isotherm in Figure 10.3 plotted as V vs P, for the purpose of illustrating Maxwell's construction.

coordinates (P_v, V_v), and E, of coordinates (P_l, V_l). The integral can be broken into several portions:

$$\int_{P_v}^{P_l} \frac{V}{N} dP = \int_A^E \frac{V}{N} dP = \int_A^B \frac{V}{N} dP + \int_B^C \frac{V}{N} dP + \int_C^D \frac{V}{N} dP + \int_D^E \frac{V}{N} dP$$
$$= 0, \tag{10.48}$$

which may be rearranged into

$$\int_A^B \frac{V}{N} dP - \int_C^B \frac{V}{N} dP = \int_D^C \frac{V}{N} dP - \int_D^E \frac{V}{N} dP. \tag{10.49}$$

Now the integral between A and B is the area under the arc AB in Figure 10.4, and the integral between C and B is the area under the arc CB. The left-hand side of Eq. (10.49) is the difference between these integrals, which equals the area in the closed region $ABCA$. Likewise, the right-hand side of Eq. (10.49) equals the area in the closed region $CDEC$. $ABCA$ and $CDEC$ are the two shaded regions in Figure 10.3, which must therefore have the same area. This is Maxwell's equal area construction: it tells us where to draw the horizontal (in Figure 10.3) portion of the isotherm. One immediate consequence is that a first-order transition is an isobaric process ($P_l = P_v$), as noted in Section 3.4.7.

Recall that to find β we need to know the shape of the coexistence curve near T_c. From Eqs. (10.37) and (10.40) we have

$$dP = P_c \left(-6t - \frac{9}{2} v^2 \right) dv. \tag{10.50}$$

Substituting this into Eq. (10.47) and noting that t is constant along an isotherm, we find

$$\int_{v_v}^{v_l} v \left(-6t - \frac{9}{2} v^2 \right) dv = 0. \tag{10.51}$$

Now this must be true along any isotherm, i.e., for any t, in the range of validity of (10.40). Thus we conclude that, close to the critical point, we must have $v_v = -v_l$: the coexistence curve is symmetric in the critical region. To find the t-dependence of v_v or v_l, we write Eq. (10.40) for $v = v_v$ and $v_l = -v_v$:

$$p_v = 4t - 6tv_v - \frac{3}{2} v_v^3, \tag{10.52}$$

$$p_l = 4t + 6tv_v + \frac{3}{2} v_v^3. \tag{10.53}$$

Setting $p_v = p_l$ and solving with respect to v_v, we obtain

$$v_v = 2(-t)^{1/2} = \frac{1}{2} (v_v - v_l) \approx \frac{1}{2} (\eta_l - \eta_v) \Rightarrow \frac{\rho_l - \rho_v}{\rho_c} \approx 4(-t)^{1/2} \Rightarrow \beta = \frac{1}{2}, \tag{10.54}$$

where we have used Eqs. (10.41) and (10.42).

Lastly, to find the critical exponent of the specific heat at constant volume C_V, we start by deriving the Helmholtz free energy of the Van der Waals gas using Eq. (10.29) and the relation $P = -(\partial F/\partial V)_T$:

$$-F(V, T) = Nk_B T \ln (V - Nb) + \frac{aN^2}{V} + f(T), \tag{10.55}$$

where $f(T)$ is an integration constant with respect to V, fixed by the requirement that when $a = b = 0$ the above expression must reduce to the ideal gas result, Eq. (5.56):

$$f(T) = Nk_B T \ln\left(\text{const.} \times T^{3/2}\right). \tag{10.56}$$

From Eqs. (10.54) and (10.55), and the definition of specific heat at constant volume, Eq. (2.57), it then follows that

$$(C_V)_{VdW} = \frac{3}{2}Nk_B = (C_V)_{\text{ideal}} \Rightarrow \alpha = 0, \tag{10.57}$$

i.e., the specific heat has a step discontinuity at the critical point.

The Van der Waals equation of state as a mean-field theory

The critical exponents in Van der Waals theory are the same as in the Weiss theory of the paramagnetic–ferromagnetic transition, which we calculated in Chapter 7. Recall that, in the latter theory, each spin interacts with a mean field generated by its nearest neighbours. In this section we shall show, following Reif (1985, chap. 10), that the Van der Waals equation of state can be derived by assuming that each gas molecule is acted on by a mean field generated by all the other molecules in the fluid. We shall assume the molecules to be spheres, of radius $R_0/2$, so that the excluded volume of a pair of molecules is a sphere of radius R_0: see Figure 10.5. The mean-field approximation consists in assuming that the fluid is composed of non-interacting molecules subjected to an effective potential of the form

$$\Phi(r) = \begin{cases} \infty & \text{if } r < R_0 \\ \bar{u} & \text{if } r > R_0 \end{cases} \qquad (\bar{u} < 0). \tag{10.58}$$

Note that the attractive part of the potential has infinite range, i.e., it is felt by all the molecules in a macroscopic sample. The partition function for this system is just the

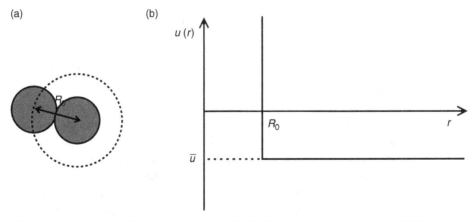

Figure 10.5 (a) Molecules approximated as hard spheres of radius $R_0/2$; (b) effective interaction potential, Eq. (10.58).

product of the partition functions of the N non-interacting molecules:

$$Z = \frac{Z_N}{N!}, \qquad Z_N = Z_1^N, \tag{10.59}$$

where

$$Z_1 = \int\int e^{-\beta[p^2/2m+\Phi(r)]}\frac{d^3r\,d^3p}{h^3} = \int e^{-\beta p^2/2m}\frac{d^3p}{h^3}\int e^{-\beta\Phi(r)}d^3r. \tag{10.60}$$

The integral over d^3p is given by Eq. (5.46), leading to

$$Z_1 = \left(\frac{2\pi mk_BT}{h^2}\right)^{3/2}\left[V_{exc}e^{-\infty} + (V - V_{exc})e^{-\beta\bar{u}}\right], \tag{10.61}$$

where $V_{exc} = 4\pi R_0^3/3$ is the excluded volume of a pair of molecular hard cores, see Eq. (10.58), and we have used the result

$$\int_{\theta=0}^{\pi}\int_{\phi=0}^{2\pi}d^3r = 4\pi r^2 dr.$$

For the pressure we thus obtain

$$\begin{aligned}
P &= -\left(\frac{\partial F}{\partial V}\right)_T = Nk_BT\left(\frac{\partial\ln Z_1}{\partial V}\right)_T \\
&= Nk_BT\frac{\partial}{\partial V}\left[\ln(V - V_{exc}) - \frac{\bar{u}}{k_BT}\right].
\end{aligned} \tag{10.62}$$

Clearly $V_{exc} \propto N$. If we let the proportionality constant be b/N_A, it follows that

$$V_{exc} = (b/N_A)N = bn. \tag{10.63}$$

Because $\bar{u}$ is the mean attraction felt by a molecule, it is not unreasonable to assume that it is proportional to the mean fluid density. Taking a/N_A^2 as the constant of proportionality, we write

$$\bar{u} = -\frac{a}{N_A^2}\frac{N}{V} = -\frac{an^2}{NV}. \tag{10.64}$$

Equation (10.62) with Eqs. (10.63) and (10.64) then yields

$$P = Nk_BT\left(\frac{1}{V - bn} - \frac{an^2}{Nk_BTV^2}\right) = \frac{nRT}{V - bn} - \frac{an^2}{V^2}, \tag{10.65}$$

which is the Van der Waals Eq., (10.29).

The above derivation again highlights the key plausible assumption at the heart of the Van der Waals theory of the liquid–vapour transition: that in a real gas the molecules interact via a strong, short-ranged repulsion (which we now know to be a consequence of Pauli's exclusion principle) and a weak, long-ranged attraction (which we now know to be of the induced-dipole type). The repulsions contribute entropy to the free energy – a molecule may occupy any position that is not excluded by the presence of other molecules; whereas attractions contribute enthalpy – the energy is lowered because the intermolecular forces perform work.

10.5.2 Weiss theory of magnetism

This was discussed in great detail in Section 7.3, so we just briefly summarise its main results. It is also a mean-field theory, and indeed from Eq. (7.47) it follows that $\gamma = 1$, from Eq. (7.52) that $\beta = 1/2$, and from Eq. (7.76) that $\alpha = 0$. These are exactly the same as for the Van der Waals gas, which is a completely different physical system. As was said earlier, these values of the critical exponents do not agree with experiment.

10.5.3 Landau theory of phase transitions. Universality

This was also discussed in great detail, in Section 7.4 for continuous transitions, and in Section 9.4 for first-order transitions. Landau theory does not make any assumptions concerning the nature of the interparticle interactions, and is therefore quite general. The one key assumption of the theory is that in the transition region the relevant free energy is an analytic function of the order parameter. The order parameter is, in turn, defined as a suitable average over the whole system of an appropriate microscopic quantity, such as molecular orientations in a liquid crystal. Although this assumption appears *a priori* reasonable, we shall see that it is in general not valid. However, because of its simplicity, Landau theory remains a powerful tool for the study of phase transitions and the structure of ordered phases, as long as we are not interested in critical properties. The values of its critical exponents are the same as in Weiss theory.

10.6 The Onsager revolution

The Ising model, which we first encountered in Chapter 8, is one of the most famous models of statistical physics. Originally it was designed to study ferromagnetism (Section 8.1), but is also applicable to the liquid–vapour transition (Section 8.4). Although the Ising model is a gross oversimplification of any physical system, it possesses the considerable advantage that it can be solved analytically in two dimensions, in zero applied field. In other words, it is possible to perform an analytically exact calculation of the partition function of the two-dimensional Ising model in zero field. As we shall see shortly, the critical exponents of the Ising model differ appreciably from the predictions of the classical theories (Van der Waals, Weiss and Landau), which fact triggered the so-called Onsager revolution in the study of critical phenomena.

10.6.1 Exact solution of the two-dimensional Ising model*

Here we shall only sketch Onsager's 1944 solution (Onsager, 1944).[5] For details see, e.g., (Huang, 1987, chap. 16). As in the one-dimensional case studied in Chapter 8, the transfer matrix method is used.

---

[5] Note that in 1936 Peierls had proved that the Ising model has a non-zero spontaneous magnetisation.

Consider a square lattice composed of n rows and n columns. The total number of spins is, of course, $N = n^2$. We again impose periodic boundary conditions: the lattice repeats itself indefinitely both along rows and along columns, i.e., it has the topology of a torus – the two-dimensional equivalent of a ring.

Let μ_i $(i = 1, 2, \ldots, n)$ be a vector whose components are the n spins in the ith row:

$$\mu_i = (\sigma_{i1}, \sigma_{i2}, \ldots, \sigma_{in}). \tag{10.66}$$

A given lattice configuration is then specified by the set of vectors $(\mu_1, \mu_2, \ldots, \mu_n)$. Because interactions are restricted to nearest neighbours, row i will interact only with rows $i - 1$ and $i + 1$. Let $E(\mu_i, \mu_{i+1})$ be the interaction energy between rows i and $i + 1$, and $E(\mu_i)$ be the interaction energy of all spins within the ith row, plus their interaction energy with an external magnetic field H:

$$E(\mu_i, \mu_{i+1}) = -J \sum_{j=1}^{n} \sigma_{ij}\sigma_{i+1j}, \tag{10.67}$$

$$E(\mu_i) = -J \sum_{j=1}^{n} \sigma_{ij}\sigma_{ij+1} - H \sum_{j=1}^{n} \sigma_{ij}. \tag{10.68}$$

Periodic boundary conditions require that, for each row,

$$\sigma_{in+1} = \sigma_{i1}. \tag{10.69}$$

The total energy of a given spin configuration $(\mu_1, \mu_2, \ldots, \mu_n)$ is then

$$E(\mu_1, \mu_2, \ldots, \mu_n) = \sum_{i=1}^{n} [E(\mu_i, \mu_{i+1}) + E(\mu_i)]. \tag{10.70}$$

This yields, for the partition function,

$$Z(H, T) = \sum_{\mu_1} \cdots \sum_{\mu_n} \exp\left\{ -\beta \sum_{i=1}^{n} [E(\mu_i, \mu_{i+1}) + E(\mu_i)] \right\}$$

$$= \sum_{\{\mu_i\}} \prod_{i=1}^{n} \exp\{-\beta [E(\mu_i, \mu_{i+1}) + E(\mu_i)]\}, \tag{10.71}$$

which is analogous to Eq. (8.4) for the one-dimensional Ising model. If we now define the *transfer matrix* $\mathbf{T}$ as

$$T_{ii+1} = \exp\{-\beta [E(\mu_i, \mu_{i+1}) + E(\mu_i)]\}, \tag{10.72}$$

it is not difficult to convince ourselves that

$$Z(H, T) = \sum_{i=1}^{n} \left(\sum_{k_1=1}^{n} \sum_{k_2=1}^{n} \cdots \sum_{k_{N-1}=1}^{n} T_{ik_1} T_{k_1 k_2} \cdots T_{k_{N-1} i} \right) = \mathrm{Tr}\,\mathbf{T}^n, \tag{10.73}$$

where Tr denotes the trace of a matrix. That Z equals the trace of the nth power of $\mathbf{T}$ given by Eq. (10.72) is a consequence of imposing periodic boundary conditions on the vectors μ_i. The problem of finding the partition function of the two-dimensional Ising model thus reduces to calculating the trace of the nth power of its transfer matrix, which can be found in any linear algebra textbook to be

$$\mathrm{Tr}\,\mathbf{T}^n = \sum_{k=1}^{p} \lambda_k^n, \tag{10.74}$$

where λ_k ($k = 1, 2, \ldots, p$), with $p = 2^n$, are the eigenvalues of $\mathbf{T}$.

As in the one-dimensional case, the partition function will be dominated by the largest eigenvalue of $\mathbf{T}$. Determining this eigenvalue is the most difficult part of the calculation. Here we just quote the final result:

$$\lim_{N\to\infty} \frac{1}{N} Z(0, T) = \ln\left[2\cosh(2\beta J)\right] + \frac{1}{2\pi} \int_0^\pi d\phi \, \ln\left[\frac{1}{2}\left(1 + \sqrt{1 - \kappa^2 \sin^2\phi}\right)\right], \tag{10.75}$$

where $\kappa = 2\sinh(2\beta J)/\cosh^2(2\beta J)$.

Once we have the partition function we can compute all thermodynamic quantities of interest, such as the Helmholtz free energy,

$$\frac{\beta F(0, T)}{N} = -\ln\left[2\cosh(2\beta J)\right] - \frac{1}{2\pi} \int_0^\pi d\phi \, \ln\left[\frac{1}{2}\left(1 + \sqrt{1 - \kappa^2 \sin^2\phi}\right)\right], \tag{10.76}$$

and the internal energy,

$$\frac{U(0, T)}{N} = -J\coth(2\beta J)\left[1 + \frac{2}{\pi}\kappa' K_1(\kappa)\right], \tag{10.77}$$

where

$$K_1(\kappa) = \int_0^{\pi/2} \frac{d\phi}{\sqrt{1 - \kappa^2 \sin^2\phi}} \tag{10.78}$$

is the complete elliptic integral of the first kind, and $\kappa' = \sqrt{1 - \kappa^2}$. Finally, for the specific heat we have

$$\frac{1}{k_B} c(0, T) = \frac{2}{\pi}\left[\beta J\coth(2\beta J)\right]^2 \left\{2K_1(\kappa) - 2E_1(\kappa) - (1 - \kappa')\left[\frac{\pi}{2} - \kappa' K_1(\kappa)\right]\right\}, \tag{10.79}$$

where

$$E_1(\kappa) = \int_0^{\pi/2} \sqrt{1 - \kappa^2 \sin^2\phi}\, d\phi \tag{10.80}$$

is the complete elliptic integral of the second kind.

The elliptic integral, $K_1(\kappa)$, has a singularity at $\kappa = 1$ (or $\kappa' = 0$); hence all thermodynamic functions will be singular at some temperature T_c such that

$$2\tanh^2\left(\frac{2J}{k_B T_c}\right) = 1 \Leftrightarrow \frac{k_B T_c}{J} \approx 2.269185, \tag{10.81}$$

which is the critical temperature of the two-dimensional Ising model in zero applied magnetic field.[6]

---

[6] To prove Eq. (10.81) note that $\kappa = 1 \Rightarrow 2\sinh(2\beta_c J) = \cosh^2(2\beta_c J) = 1 + \sinh^2(2\beta_c J) \Leftrightarrow [\sinh(2\beta_c J) - 1]^2 = 0$, leading to $\sinh(2\beta_c J) = 1$ and $\cosh(2\beta_c J) = \sqrt{2}$, where $\beta_c = 1/k_B T_c$.

Expanding the specific heat around T_c, we get

$$\frac{1}{k_B}c(0,T) \approx \frac{2}{\pi}\left(\frac{2J}{k_B T_c}\right)^2 \left[-\ln\left|1-\frac{T}{T_c}\right| + \ln\left(\frac{k_B T_c}{2J}\right) - \left(1+\frac{\pi}{4}\right)\right], \tag{10.82}$$

which diverges logarithmically as $|T-T_c| \to 0$, implying $\alpha = 0$. On the other hand, the internal energy, Eq. (10.77), is continuous at $T = T_c$, i.e., there is no latent heat.

In order to be able to identify the singularities of the thermodynamic functions with a phase transition between ordered and disordered states, we must find the order parameter – the total magnetic moment of the lattice – and show that it is non-zero for $T < T_c$ and vanishes for $T > T_c$. This calculation is of complexity comparable to that of the partition function, so we just quote the final result:

$$M(T) = \begin{cases} 0 & \text{if } T > T_c \\ \left\{1 - \left[\sinh\left(\frac{2J}{k_B T}\right)\right]^{-4}\right\}^{1/8} & \text{if } T < T_c \end{cases}. \tag{10.83}$$

This is plotted along with our Monte Carlo simulation curves in Figure 8.4. It can be shown (see Problem 10.4) that Eq. (10.83) implies $\beta = 1/8$.

In contrast, the molecular field (or Bragg–Williams) approximation predicts $\beta = 1/2$ and a finite discontinuity of the specific heat at $T = T_c$ (Huang, 1987). One important consequence is that Landau theory, which assumes the free energy to be analytic at $T = T_c$, is not valid at all. The critical exponents of the two-dimensional Ising model are also listed in Table 10.1.

10.6.2 Other exact and approximate results

Onsager's solution of the two-dimensional Ising model and its disagreement with the predictions of classical theories (mean-field and Landau) aroused renewed interest, on the one hand, in more detailed experimental studies, and on the other hand, in other models exhibiting critical behaviour and amenable to either exact or approximate solution. Here we present a brief overview of the most important results, and refer the reader to Stanley (1971) and especially Domb (1996) for more details.

An experimental landmark was Guggenheim's 1945 analysis of the liquid–vapour coexistence curves of a large number of gases. If plotted in terms of the reduced units (10.34), all data collapse onto a single master curve, i.e., the law of corresponding states holds (see Figure 10.6). However, close to the critical point we have

$$\frac{\rho_l(T) - \rho_v(T)}{\rho_c} \sim (-t)^{1/3} \Rightarrow \beta = \frac{1}{3}, \tag{10.84}$$

at variance with the classical prediction $\beta = 1/2$. This same $\beta = 1/3$ was also found in studies of low-Curie-temperature ferromagnets. The normal–superfluid transition of helium-4, on the other hand, appeared to exhibit a logarithmic divergence of the specific heat. These and other slowly accumulating results therefore suggested that the classical theories are not applicable. Note that they also do not agree with the exact results for the two-dimensional Ising mode, which was, however, not a cause for much concern, given that real systems are three-dimensional.

Most theoretical investigations were performed on variants of the Ising model in $d = 1, 2$ or 3 spatial dimensions, but where the spins are vectors in n-dimensional space – which

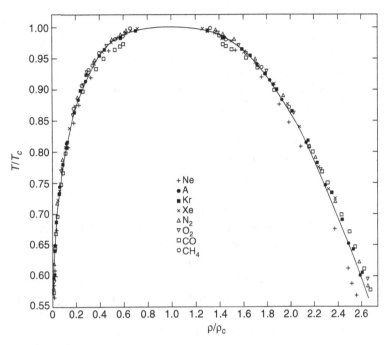

Figure 10.6 Reduced liquid and vapour coexistence densities, ρ_l/ρ_c and ρ_v/ρ_c, respectively, plotted against the reduced temperature $t = T/T_c$, for a number of simple fluids. The data are consistent with the law of corresponding states with a non-classical exponent, $\beta = 1/3$. (From Guggenheim (1945). Copyright 1945 AIP Publishing LLC.)

is, if anything, even less experimentally relevant. One interesting exception is the case $n \to \infty$, which is equivalent – in the sense that it has the same partition function – to the so-called *spherical model* of Kac and Uhlenbeck. The spherical model is a generalisation of the $n = 1$ Ising model in which the spins are not constrained to take the values ± 1, but may instead take any value, the only restriction being that the sum of the squares of all spins must equal the number of lattice sites. For all other models, only approximate results could be obtained, using perturbative methods – essentially more or less sophisticated series expansions. The general conclusion was that the critical exponents were strongly dependent both on the dimensionality of space, d, and on that of the order parameter, n, but for each (d, n) pair seemed insensitive to the lattice geometry (square, triangular, ...) or to the detailed functional form of the interactions, provided these were short-ranged. Some of these results are collected in Table 10.1; for more details, and references to the original works, see Binney *et al.* (1993), Huang (1987), Stanley (1971).

10.7 Reconciliation of the classical theories, experimental results, and exact results: the scaling hypothesis

We should now ask ourselves, why do such reliable tools of statistical physics as mean-field theory and Landau theory fail when applied to critical phenomena? As we saw earlier, these theories predict values for the critical exponents that are always ratios of small integers, and depend neither on d nor n. In other words, these theories do not discriminate between scalar and vector order parameters. Moreover, they cannot explain why the inequalities involving critical exponents, some of which we encountered in Section 10.4.3, are valid as equalities. We shall next briefly discuss the basic ideas that allowed reconciliation of the classical picture with experimental and exact model results, following Callen (1985) and Huang (1987). For a historical perspective, see Domb (1996); for more technical details, see Binney *et al.* (1993), Goldenfeld (1992) and Stanley (1971).

The key to solving this problem is an appreciation of the role of fluctuations. We saw in Section 10.4.2 that the range of order parameter fluctuations diverges at T_c. In other words, there are fluctuations on all lengthscales, from that of atoms or molecules, to that of the whole system, which has macroscopic size. In a fluctuation-dominated regime, we can no longer identify the mean value (i.e., the macroscopic value) of a thermodynamic quantity with its 'most probable value', because the distribution functions are no longer sharply peaked at their mean value.

In the critical region there is one diverging length: the correlation length of the order parameter. This suggests that it will be the only relevant lengthscale in this region, as it is larger than any other length – the lattice constant, the typical size of atoms, molecules or ions, their mean free path, etc. We are thus led to the *scaling hypothesis* – or, more precisely, the *static scaling hypothesis* – which can be formulated as follows:

The correlation length ξ being the only relevant lengthscale in the critical region, then any quantity with dimensions of (length)$^{-D}$ will be proportional to ξ^{-D} when $|t| \ll 1$.

Let us now see how the scaling hypothesis explains the relations between critical exponents, as well as the existence of universality classes. We shall follow Kadanoff's original argument, as sketched by Stanley (1971).

Consider again the d-dimensional Ising model in a magnetic field. Its Hamiltonian, given by Eq. (8.18), contains two parameters: the strength J of the interaction between spins, and the magnetic field H (denoted h in Eq. (8.18); we have changed the notation to bring it into line with the remainder of this chapter). Let us now divide up the lattice into blocks of edge length La, where a is the lattice constant and $L \gg 1$ is an arbitrary scale factor. If there are N spins in the lattice, there will be N/L^d blocks of L^d spins each. Close enough to the critical point, i.e., for $|t| \ll 1$, the correlation length will be much larger than the lattice constant and we can therefore choose L such that $\xi \gg La \gg a$. It follows that each block will contain a very large number of spins, $L^d \gg 1$. Now within each block, all spins have approximately the same value, because the block is much smaller than the spin correlation length. We can then assign a well-defined spin to a whole block. It should also be possible to write the Hamiltonian of the system in terms of the block spins, rather than the individual spins. One would expect this 'new' Hamiltonian to have the same functional form as the 'old' one, only with the old parameters J and H replaced by new parameters $\tilde{J}$ and $\tilde{H}$ describing, respectively, the interaction between two blocks of spins, and between a block of spins and the magnetic field. These new parameters would obviously depend on L.

If the two Hamiltonians, that of individual spins and that of blocks of spins, have the same functional form, it is not unreasonable to assume that the same will be true of their Gibbs free energies. We are thus led to postulate that the Gibbs free energy of the system of blocks of spins has, in terms of the new variables $\tilde{t}$ and $\tilde{H}$, the same functional form as the Gibbs free energy of the system of individual spins, in terms of the original t and H, times a scaling factor:

$$g\left(\tilde{t},\tilde{H}\right) = L^d g(t,H). \tag{10.85}$$

The left-hand side of the above equation is the Gibbs free energy of a block of spins, while the right-hand side is the Gibbs free energy per spin of the same system. Because the two systems, that of blocks of spins and that of individual spins, are actually the same, $\tilde{t}$ and $\tilde{H}$ must be proportional to t and H, so as to preserve the system's critical properties. The proportionality constants will depend on L:

$$\tilde{t} = L^y t, \tag{10.86}$$

$$\tilde{H} = L^x H. \tag{10.87}$$

Combining Eqs. (10.85)–(10.87), we obtain

$$g\left(L^y t, L^x H\right) = L^d g(t,H) \Leftrightarrow g\left(L^{y/d} t, L^{x/d} H\right) = L g(t,H), \tag{10.88}$$

i.e., the Gibbs free energy is a homogeneous function of t and H.

Let us examine the consequences of this result. Differentiating both sides of Eq. (10.88) with respect to H and using the fact that $M = -(\partial G/\partial H)_T$, we get

$$L^{x/d} \frac{\partial g\left(L^{y/d} t, L^{x/d} H\right)}{\partial\left(L^{x/d} H\right)} = L \frac{\partial g(t,H)}{\partial H}$$

$$\Leftrightarrow L^{x/d} M\left(L^{y/d} t, L^{x/d} H\right) = L M(t,H). \tag{10.89}$$

Recall now that there are two critical exponents associated with the magnetisation: β, when $-t \to 0$ and $H = 0$; and δ, when $H \to 0$ and $t = 0$. Let us start with β: setting $H = 0$ in Eq. (10.89) yields

$$M(t,0) = L^{x/d-1} M\left(L^{y/d} t, 0\right). \tag{10.90}$$

Because L is an arbitrary scale factor (with the only restriction that $L \gg 1$), we can take $L = (-t)^{-d/y}$, whence

$$M(t,0) = (-t)^{(d-x)/y} M(-1,0). \tag{10.91}$$

Comparing Eqs. (10.91) and (110.11), we conclude that

$$\beta = \frac{d-x}{y}. \tag{10.92}$$

For $H \to 0$ and $t = 0$, we may likewise derive (see Problem 10.5) that

$$\delta = \frac{x}{d-x}. \tag{10.93}$$

Equations (10.92) and (10.93) can be solved with respect to x and y, with the result

$$x = d \frac{\delta}{\delta + 1}, \tag{10.94}$$

$$y = \frac{d}{\beta} \frac{1}{\delta + 1}. \tag{10.95}$$

Further relations involving the critical exponents and parameters x and y may be derived. Differentiating both sides of Eq. (10.88) twice with respect to H and using the fact that $\chi_T = (\partial M / \partial H)_T = -(\partial^2 G / \partial H^2)_T$, we obtain

$$L^{2x/d} \frac{\partial^2 g \left(L^{y/d} t, L^{x/d} H \right)}{\partial \left(L^{x/d} H \right)^2} = L \frac{\partial^2 g(t, H)}{\partial H^2}$$

$$\Leftrightarrow \chi_T(t, H) = L^{2x/d-1} \chi_T \left(L^{y/d} t, L^{x/d} H \right). \tag{10.96}$$

The zero-field susceptibility is then

$$\chi_T(t, 0) = L^{2x/d-1} \chi_T \left(L^{y/d} t, 0 \right). \tag{10.97}$$

Taking again $L = (-t)^{-d/y}$ and comparing the result with Eq. (10.14), we conclude that

$$\gamma' = \frac{2x - d}{y}. \tag{10.98}$$

Substituting Eq. (10.95) into Eq. (10.98), we arrive at the following relation involving β, δ and γ':

$$\gamma' = \beta(\delta - 1), \tag{10.99}$$

known as *Widom's scaling relation*.

If, on the other hand, we differentiate both sides of Eq. (10.88) with respect to T and use the fact that $C_H = -T(\partial^2 G / \partial T^2)_H$, then

$$L^{2y/d} \frac{\partial^2 g \left(L^{y/d} t, L^{x/d} H \right)}{\partial \left(L^{y/d} t \right)^2} = L \frac{\partial^2 g(t, H)}{\partial t^2}$$

$$\Leftrightarrow C_H(t, H) = L^{2y/d-1} C_H \left(L^{y/d} t, L^{x/d} H \right). \tag{10.100}$$

As before, we set $H = 0$ and $L = (-t)^{-d/y}$ and compare the result with Eq. (10.16), leading to

$$\alpha' = 2 - \frac{d}{y} \Leftrightarrow \alpha' + \beta(\delta + 1) = 2, \tag{10.101}$$

where we have used Eq. (10.95). Equation (10.101) is the Griffiths' inequality, (10.26), now valid as an equality. From Eqs. (10.99) and (10.101) one likewise derives Rushbrooke's inequality, (10.25), as an equality:

$$\alpha' + 2\beta + \gamma' = 0, \tag{10.102}$$

known as the *Essam–Fisher scaling relation*.

In summary, the scaling hypothesis allows us to derive relations between critical exponents, but not to predict their actual values. It is, however, able to explain why the critical exponent associated with a given quantity takes the same values above (unprimed) and below (primed) T_c: the 're-scaled' Gibbs free energy is postulated to be the same for $T > T_c$ and $T < T_c$.

10.8 The Ginzburg criterion

It remained to clarify why some phase transitions, namely the normal–superconducting transition or the paraelectric–ferroelectric transition (the electrical analogue of the paramagnetic–ferromagnetic transition), exhibit classical critical behaviour, as described by Landau theory, but others, such as the normal-superfluid transition of helium-4, or the paramagnetic–ferromagnetic transition, do not.

Ginzburg was able to reconcile these results by postulating that, although Landau theory is in general valid, it ceases to be so in the vicinity of the critical point, where the amplitude of order parameter fluctuations becomes significant. If some systems appear to exhibit classical critical exponents, this can only mean that their non-classical region is too small to be experimentally accessible.

For definiteness, let us consider a ferromagnet. As we saw earlier, classical theories predict, for the global order parameter

$$M(T) \sim (-t)^{1/2}, \tag{10.103}$$

and for its fluctuations (Huang, 1987, chap. 17)

$$\Delta M(T) \sim (-t)^{\frac{d-2}{4}}. \tag{10.104}$$

Therefore, the fluctuations will be negligible with respect to the global order parameter when

$$\frac{\Delta M(T)}{M(T)} \sim (-t)^{\frac{d}{4}-1} \ll 1 \Rightarrow d \geq 4. \tag{10.105}$$

In other words, in the limit $t \to 0$, the fluctuations will be negligible only if $d \geq 4$. Otherwise, around T_c there will always exist a range of temperatures, however narrow, in which order parameter fluctuations dominate and the critical exponents will deviate substantially from their classical values. Condition (10.105) is known as the *Ginzburg criterion*.

Alternatively, the Ginzburg criterion tells us how wide is the window ΔT of non-classical behaviour: it is that t for which $\Delta M(T)/M(T) \sim 1$ for $d < 4$. This calculation requires knowledge of the prefactors in Eqs. (10.103) and (10.104): for the normal–superconducting transition the width of the non-classical range is $\Delta T \sim 10^{-16}$ K, hence unobservable, whereas for the normal–superfluid transition of helium-4 it is $\Delta T \sim 0.6$ K, hence observable.

10.9 A short overview of the renormalisation group

As we saw earlier, the scaling hypothesis allows us to derive general relations involving the critical exponents, which were discovered either experimentally, or through analytical or numerical investigations of model systems. It is, however, unable to predict the actual values of the critical exponents. This only became possible through the application of renormalisation group ideas to critical phenomena, in what was one of the great triumphs of statistical mechanics in the second half of the twentieth century. It won Kenneth G. Wilson the Physics Nobel Prize in 1982.

Let us briefly explain what the renormalisation group is about, following Domb (1996, chap. 1). The key idea is already contained in the scaling hypothesis: because the correlation length diverges in the critical region, the system is governed by the same Hamiltonian on all lengthscales; it remains to relate the parameters of the Hamiltonian on different scales. The renormalisation group is the way to systematically map a many-variable Hamiltonian onto a fewer-variable Hamiltonian. This procedure may then be iterated, gradually reducing the number of variables. The critical point, at which all lengthscales are equivalent (i.e., at which there is *scale invariance*), is the fixed point of this mapping: at such a point, the Hamiltonian no longer changes.

Detailed renormalisation group calculations are, in general, quite complex. In the hope of stoking our readers' curiosity, we just add that these calculations reveal that, in spatial dimensions higher than four, fluctuations are irrelevant and all critical exponents assume their classical values.

Problems

10.1 Derive Eq. (10.7). *Hint:* note that $t^\lambda = e^{\lambda \ln t}$.

10.2 Find the critical pressure, volume and temperature of a Van der Waals gas in terms of a and b.

10.3 Write down the Van der Waals equation of state (10.29) in terms of the *molar density* $\rho_m = n/V$ (number of moles per unit volume).

    a. Expand the resulting equation in powers of ρ_m and find the second and third virial coefficients, B_2 and B_3 respectively, defined as

$$p = RT\rho_m + B_2\rho_m^2 + B_3\rho_m^3 + \cdots$$

    b. Find the *Boyle temperature* T_B, defined as the temperature for which $B_2 = 0$, and compare it with T_c.

10.4 Following the procedure outlined below or otherwise, show that for the two-dimensional Ising model $\beta = 1/8$.

    a. Expand $[\sinh(2\beta J)]^{-4}$ to first order in powers of $T - T_c$.

b. Note that Eq. (10.81) implies that

$$\sinh\frac{2J}{k_B T_c} = 1 \quad \text{and} \quad \cosh\frac{2J}{k_B T_c} = \sqrt{2},$$

and use this result to simplify the expansion obtained earlier.

c. Rewrite the above expansion in terms of $t \equiv (T - T_c)/T_c$ and substitute it into Eq. (10.83) to find

$$M(T) = \left(\frac{8\sqrt{2}}{k_B T_c}\right)^{1/8} (-t)^{1/8} + \text{higher-order terms}.$$

10.5 Derive Eq. (10.93) by setting $t = 0$ in Eq. (10.89) and comparing the result with Eq. (10.20). *Hint:* let $L = H^{-d/x}$.

10.6 Check that the critical exponents listed in Table 10.1 obey Widom's (Eq. (10.99)), Griffiths' (Eq. (10.101)), and Essam–Fisher's (Eq. (10.102)) scaling relations.

References

Andrews, T. 1869. The Bakerian Lecture: On the Continuity of the Gaseous and Liquid States of Matter. *Phil. Trans. R. Soc.* **159**, 575–590.

Binney, J. J., Dowrick, N. J., Fisher, A. J., and Newman, M. E. J. 1993. *The Theory of Critical Phenomena.* Oxford University Press.

Callen, H. B. 1985, *Thermodynamics and an Introduction to Thermostatistics*, 2nd edition. Wiley.

Domb, C. 1996. *The Critical Point.* Taylor and Francis.

Farrell, R. A. 1968. In *Fluctuations in Superconductors*, ed. by W. S. Goree and F. Chilton. Stanford Research Institute.

Fisher, M. E. 1967. The theory of equilibrium critical phenomena. *Rep. Prog. Phys.* **30**, 615–730.

Goldenfeld, N. 1992. *Lectures on Phase Transitions and the Renormalization Group.* Addison-Wesley.

Guggenheim, E. A. 1945. The principle of corresponding states. *J. Chem. Phys.* **13**, 253–261.

Huang, K. 1987. *Statistical Mechanics*, 2nd edition. Wiley.

Onsager, L. 1944. Crystal statistics. I. A two-dimensional model with an order–disorder transition. *Phys. Rev.* **65**, 117–149.

Reif, F. 1985. *Statistical and Thermal Physics.* McGraw-Hill.

Stanley, H. E. 1971. *Introduction to Phase Transitions and Critical Phenomena.* Oxford University Press.

11 Irreversible processes

11.1 Introduction

This chapter is an introduction to systems that are not necessarily in thermodynamic equilibrium. In a real thermodynamic process, a system occupies a succession of non-equilibrium states, hence the process as a whole is irreversible. Here we study two irreversible processes of statistical physics, Brownian motion and diffusion, which, as we shall see, are intimately related.

11.2 Diffusion

Let us consider a non-uniform suspension of solid particles in a fluid medium. Although the total number of particles in the whole volume is fixed, the mean number of particles per unit volume will be a function of both position and time; in one spatial dimension, we denote this quantity by $n(x,t)$. In zero external field this is a non-equilibrium state: the system (fluid plus particles) will therefore be driven towards a state of higher entropy, i.e., a more uniform distribution of particles throughout the volume.[1] This type of particle motion is called *diffusion*; contrast it with the motion associated with the flow of a fluid.

We define J_x, the *diffusion flux* along OX, as the mean number of particles per unit time that cross a plane of unit area placed perpendicular to OX. In the simplest approximation, the flux is proportional to the particle concentration gradient in the OX direction – this is *Fick's law*:

$$J_x = -D\frac{\partial n(x,t)}{\partial x}, \tag{11.1}$$

where D is the *diffusion coefficient* or *diffusivity*, and the minus sign means that particles always diffuse 'down' the concentration gradient. Diffusion is said to be *steady-state* if J_x is constant, and *non-steady-state* otherwise.

D is a measure of the ease with which a concentration gradient drives a diffusion flux: if D is large, then only a weaker gradient of the concentration is required to produce a given

---

[1] Although, strictly speaking, the entropy is a function of state that can only be defined at equilibrium, we know from the second law that the entropy increases in spontaneous non-equilibrium processes, and is maximised at equilibrium. In addition, there exists a whole thermodynamics of irreversible processes which uses the entropy (Boltzmann's H theorem, Prigogine's law of minimum entropy production, etc.). These topics are outside the scope of this book.

Figure 11.1 $n(x,t)Adx$ = number of particles in the slab of thickness dx and cross-sectional area A, at time t. $AJ_x(x)$ = number of particles per unit time that enter the slab through its bottom surface, and $AJ_x(x+dx)$ = number of particles per unit time that exit the slab through its top surface, both also at time t.

value of the diffusive flux than if D is small. Dimensional analysis of Eq. (11.1) reveals that D has dimensions of

$$[D] = L^2 T^{-1}. \tag{11.2}$$

(Note that this does not depend on our choice of units of concentration.)

Because the total number of particles is conserved, the change per unit time in the number of particles contained in a slab of fluid of thickness dx must equal the number of particles per unit time that enter the slab through its bottom surface at x, minus the number of particles per unit time that exit the slab through its top surface at $x + dx$, as sketched in Figure 11.1.

We may then write

$$\frac{\partial}{\partial t}[n(x,t)Adx] = AJ_x(x) - AJ_x(x+dx)$$

$$= AJ_x(x) - A\left[J_x(x) + \frac{\partial J_x}{\partial x}dx\right] \tag{11.3}$$

$$\Rightarrow \frac{\partial n(x,t)}{\partial t} = -\frac{\partial J_x}{\partial x}.$$

If we now insert Eq. (11.1) into Equation (11.3) and assume constant D, we obtain the (one-dimensional) *diffusion equation*:[2]

$$\frac{\partial n(x,t)}{\partial t} = D\frac{\partial^2 n(x,t)}{\partial x^2}. \tag{11.4}$$

The above derivation can be generalised to more than one dimension. In three dimensions, the diffusion equation reads

$$\frac{\partial n(\boldsymbol{r},t)}{\partial t} = D\nabla^2 n(\boldsymbol{r},t), \quad \nabla^2 = \frac{\partial^2}{\partial x^2} + \frac{\partial^2}{\partial y^2} + \frac{\partial^2}{\partial z^2}. \tag{11.5}$$

As with all partial differential equations, the solution of Eq. (11.4) will depend on its initial conditions and boundary conditions. It is instructive to consider a special case for which

---

[2] There are many situations of practical interest where D is non-constant. Think, for instance, of moisture (undesirably) diffusing through a *Kit Kat* bar: clearly, the water molecules will diffuse differently through the chocolate coating and the wafer core. Hence D depends on composition, which in turn depends on position.

an analytical solution exists: at time $t = 0$, all particles are located at the origin. We then have (Sommerfeld, 1964, chap. II, paragraph 12):

$$n(x,t) = \frac{n_0}{\sqrt{4\pi Dt}} \exp\left(-\frac{x^2}{4Dt}\right), \tag{11.6}$$

where

$$n_0 = \int_{-\infty}^{+\infty} n(x,t)dx \tag{11.7}$$

is the (constant) total number of particles per unit volume: see Figure 11.2. Strictly speaking, this solution is only valid if the system is infinite. It is, however, approximately valid if $n(x,t)$ at the boundaries is much smaller than at the origin, i.e., if $n(\pm L, t) \ll n(0,t)$, where $2L$ is the linear size of the system. Of course, in any finite system this will no longer be true at sufficiently late times, as $n(x,t) \rightarrow n_0 = $ constant when $t \rightarrow \infty$.

Using Eqs. (11.6) and (11.7), we can interpret

$$f(x,t) = \frac{n(x,t)}{n_0} \tag{11.8}$$

as the *probability density* of finding a particle at a position between x and $x + dx$, at a time between t and $t + dt$. This probability density is Gaussian: see Figure 11.2 and Chapter 1; its mean and variance, are, respectively,

$$\bar{x} = 0, \tag{11.9}$$

$$\sigma^2(x) = \overline{x^2} - \bar{x}^2 = 2Dt, \tag{11.10}$$

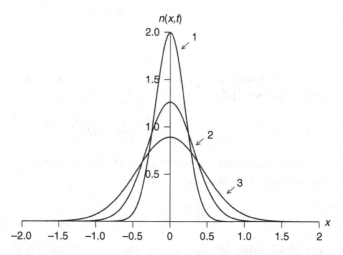

Figure 11.2 Concentration $n(x,t)$ given by Eq. (11.6) vs x, at times (in units of D) $t = 0.02$ (curve 1), $t = 0.05$ (curve 2) and $t = 0.1$ (curve 3). At time $t = 0$, all particles are at the origin, $x = 0$. The area under all three curves is the same and equals the total number of particles per unit volume, n_0. In this particular example $n_0 = 1$, hence $n(x,t)$ coincides with the probability density $f(x,t)$, Eq. (11.8). Note that as time increases the distribution widens, meaning that particles become more and more evenly dispersed.

whence the mean square displacement of a particle is

$$\overline{x^2} = 2Dt \Rightarrow \sqrt{\overline{x^2}} \approx x \propto t^{1/2}. \tag{11.11}$$

Contrast this with the law of motion of a particle starting from the origin and moving along OX at constant speed ('ballistic motion'):

$$\overline{x^2} = x^2 = v^2 t^2 \Rightarrow x \propto t. \tag{11.12}$$

The ballistic particle is thus much faster than the diffusing particle, which suggests that a diffusing particle does not move at constant velocity. Indeed a diffusing particle undergoes many collisions with the constituent particles of the medium wherein it is diffusing, e.g., solvent molecules, as a result of which both the direction and the magnitude of the particle's velocity repeatedly change, as we shall see in the next section.

Equations (11.2) and (11.11) further allow us to define a *characteristic timescale for diffusion*, t_D, as the time a particle takes to diffuse a distance equal to its size. If, as in many practical situations, the particle can be approximated as a sphere of diameter d, then

$$t_D = \frac{d^2}{2D}. \tag{11.13}$$

(The factor of 2 in the denominator is, of course, inessential and is sometimes omitted.)

Let us now consider the effect of an applied field – e.g., Earth's gravitational field, or an electric field if the particles are charged. We follow Kubo (1985). Assume the field to be spatially uniform, as is Earth's gravitational field near Earth's surface, or the electric field between the parallel plates of a planar capacitor. Such an external field will exert on each diffusing particle a spatially uniform force K, which will accelerate the particle along the field direction. This motion will be opposed by the viscous friction force $\overline{F}$ of the solvent. If the mean particle velocity $\overline{v}$ is not too large, we can take $\overline{F}$ to be proportional to $\overline{v}$. Clearly, a steady state will be reached in which the force due to the field is balanced by the viscous friction force and $\overline{v} = \overline{v}_0$, the *terminal drift velocity*: [3]

$$K + \overline{F} = 0 \Leftrightarrow K - \alpha \overline{v}_0 = 0 \Leftrightarrow \overline{v}_0 = \frac{K}{\alpha}. \tag{11.14}$$

In addition to the diffusion flux J_x given by Eq. (11.1), there is now a flux J_K due to the external field:

$$J_K = n(x,t)\overline{v}_0 = n(x,t)\frac{K}{\alpha}. \tag{11.15}$$

The total flux is then

$$J = J_x + J_K = -D\frac{\partial n(x,t)}{\partial x} + \frac{K}{\alpha}n(x,t). \tag{11.16}$$

---

[3] We are implicitly assuming *strong damping*, i.e., that the steady state is attained in a time much shorter than the typical time for diffusion t_D. If it were not so, then $\overline{v}$ would not be constant (and equal to $\overline{v}_0$) and an additional equation would be needed to determine it.

Substituting this into Eq. (11.2), we find the equation for diffusion in an external field to be

$$\frac{\partial n(x,t)}{\partial t} = D\frac{\partial^2 n(x,t)}{\partial x^2} - \frac{K}{\alpha}\frac{\partial n(x,t)}{\partial x}. \tag{11.17}$$

This equation can easily be solved by making the following change of variables:

$$\xi = x - \frac{K}{\alpha}t, \tau = t \Rightarrow \begin{cases} \dfrac{\partial n(x,t)}{\partial x} = \dfrac{\partial n(\xi,\tau)}{\partial \xi}, \quad \dfrac{\partial^2 n(x,t)}{\partial x^2} = \dfrac{\partial^2 n(\xi,\tau)}{\partial \xi^2} \\[2mm] \dfrac{\partial n(x,t)}{\partial t} = \dfrac{\partial n(\xi,\tau)}{\partial \tau} - \dfrac{K}{\alpha}\dfrac{\partial n(\xi,\tau)}{\partial \xi} \end{cases}. \tag{11.18}$$

In terms of the new variables, Eq. (11.17) is written

$$\frac{\partial n(\xi,\tau)}{\partial \tau} = D\frac{\partial^2 n(\xi,\tau)}{\partial \xi^2}, \tag{11.19}$$

which is exactly the same as the diffusion equation, Eq. (11.4), and therefore has the same solutions – only in terms of the new variables (ξ,τ). In the special case considered earlier, where all particles are located at the origin at $t = \tau = 0$, we thus find

$$n(\xi,\tau) = \frac{n_0}{\sqrt{4\pi D\tau}}\exp\left(-\frac{\xi^2}{4D\tau}\right) \Leftrightarrow n(x,t) = \frac{n_0}{\sqrt{4\pi Dt}}\exp\left[-\frac{\left(x - \frac{K}{\alpha}t\right)^2}{4Dt}\right], \tag{11.20}$$

which is again a Gaussian distribution, cf. Figure 11.2, centred at $x_0 = \bar{v}_0 t = (K/\alpha)t$, i.e., there is a collective drift of all the diffusing particles due to the external field.

It is instructive to consider the long-time limit, $t \to \infty$, of Eq. (11.19). Now we can no longer use the above solution, Eq. (11.20), as it may no longer be true that $n(x = \pm L) \ll n(0,t)$ However, at very long times the systems will likely have reached equilibrium, so the total particle flux J given by Eq. (11.16) must vanish, with the result

$$n(x, t \to \infty) \approx n_0 \exp\left(\frac{K}{D\alpha}x\right). \tag{11.21}$$

In zero field at long times, the particles would be distributed uniformly throughout the system, giving $n(x,t) = n_0 = $ constant. We conclude that the effect of an external field is to set up a non-uniform distribution of the diffusing particles.

Let us now take the external field to be Earth's gravitational field near the Earth's surface. If the OX axis is along the vertical direction, then $K = -mg$ and

$$n(x) \approx n_0 \exp\left(-\frac{mg}{D\alpha}x\right), \tag{11.22}$$

where m is the mass of the diffusing particle. Comparing this with the barometric law, Eq. (5.91), we find

$$\frac{mg}{D\alpha} = \frac{mg}{k_B T} \Rightarrow D = \frac{k_B T}{\alpha}. \tag{11.23}$$

i.e., the diffusion coefficient is proportional to the thermodynamic temperature and inversely proportional to the viscous friction coefficient. We shall show in the next section that this result is valid for any particle performing Brownian motion, Eq. (11.51), in any external field or even in zero external field.

11.3 Brownian motion

The random motion of particles suspended in a fluid (liquid or gas) is known as *Brownian motion*, after the Scottish botanist Robert Brown, who in 1828 published an account of his observations of the irregular motions of various types of small particles suspended in water. At first Brown thought this was a sign that the grains were alive – he had used fresh pollen grains – so he repeated the experiment with other grains that were unquestionably dead, such as pollen from the British Museum, guaranteed to be more than a hundred years old, and even a finely powdered chunk of the sphinx, always with the same result.

Observations carried out over long times revealed that Brownian particles appear to perform a random walk; see Figure 11.3. Now from Chapter 1 we know that the mean square displacement of a random walker scales with the number of steps N as

$$\overline{r^2} = N\sigma^2(s) \Rightarrow \overline{r^2} \propto N, \tag{11.24}$$

where $\sigma^2(s)$ is the variance of the displacement s at every step. Let τ be the mean time that one step takes; then the number of steps N in the time t is $N = t/\tau$, whence

$$\overline{r^2} \propto t. \tag{11.25}$$

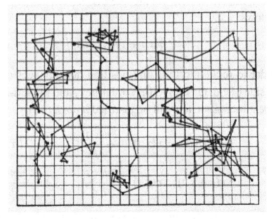

Figure 11.3 Facsimile of Perrin's sketch of his microscopic observation of the motion of three colloidal particles of radius 0.53 μm suspended in a liquid. The dots are the particle positions at 30-second intervals. The straight lines connecting the dots are not the particle trajectories, as between each two consecutive recorded positions the particles underwent a similarly erratic motion. The grid size is 3.2 μm. (From Perrin (1948), ©Presses Universitaires de France. Reprinted with permission. Data from Perrin, J. B. 1909. Mouvement brownien et réalité moléculaire. *Ann. de Chimie et de Physique (VIII)* **18**, 5–114.)

Comparing this result with Eq. (11.11), we are led to interpret Brownian motion as a *diffusion process*: each suspended particle undergoes many collisions with the solvent molecules. This is the basis of Einstein's 1905 treatment of Brownian motion, which established a connection between the random displacement of a single particle and the diffusion of a large number of particles, cf. the reasoning that leads to Eqs. (11.8) and (11.11).

11.3.1 Statistical interpretation of diffusion*

In 1905 Einstein gave a new derivation of the diffusion equation that allows diffusion to be interpreted statistically. This was motivated by his reflections on Brownian motion (Pais, 1982, chap. 5). If the concentration of diffusing particles is low, we may assume that each of them moves independently of all others. Let τ be a time that is much shorter than the typical observation time t, but much larger than the timescale of any microscopic processes, so that the motion of a particle during time τ does not depend on its earlier history. Then particle motion on the macroscopic scale may be described as a *Markov process*: this is a stochastic process, i.e., a sequence of random events such that the probability of any given event depends only on the event immediately preceding it. For simplicity, we shall consider the one-dimensional case.

Let $\phi(s)$ be the probability that a particle will suffer a displacement between s and $s+ds$ in the time τ. $\phi(s)$ is normalised to unity, and if all directions are equivalent (e.g., if there are no external fields favouring some directions over others) then $\phi(s)$ is also symmetric:

$$\int_{-\infty}^{+\infty} \phi(s)ds = 1, \quad \phi(s) = \phi(-s). \tag{11.26}$$

Because the particles all move independently, we can relate the number of particles per unit length at time t, $n(x,t)$, with the number of particles per unit length at time $t+\tau$, $n(x,t+\tau)$, as follows:

$$n(x,t+\tau)dx = dx \int_{-\infty}^{+\infty} n(x+s,t)\phi(s)ds. \tag{11.27}$$

In order to relate the time and position dependences of $n(x,t)$, we shall expand the left-hand side of Eq. (11.27) to first order in τ, and the left-hand side to second order in s (this apparent inconsistency will be justified *a posteriori*.

$$n(x,t+\tau) = n(x,t) + \frac{\partial n(x,t)}{\partial t}\bigg|_{\tau=0}\tau + \mathcal{O}\left(\tau^2\right), \tag{11.28}$$

$$n(x+s,t) = n(x,t) + \frac{\partial n(x,t)}{\partial x}\bigg|_{s=0}s + \frac{1}{2}\frac{\partial^2 n(x,t)}{\partial x^2}\bigg|_{s=0}s^2 + \mathcal{O}\left(s^3\right). \tag{11.29}$$

Substituting Eqs. (11.28) and (11.29) into Eq. (11.27), and using Eq. (11.26) and the fact that $\int_{-\infty}^{+\infty} s\phi(s)ds = 0$, we obtain

$$\frac{\partial n(x,t)}{\partial t} = \frac{\overline{s^2}}{2\tau}\frac{\partial^2 n(x,t)}{\partial x^2}, \tag{11.30}$$

where

$$\overline{s^2} = \int_{-\infty}^{+\infty} s^2\phi(s)ds. \tag{11.31}$$

Equation (11.30) is just the diffusion Eq. (11.4) with diffusion coefficient

$$D = \frac{\overline{s^2}}{2\tau}. \tag{11.32}$$

Einstein's treatment of diffusion thus provides a statistical interpretation of the phenomenological coefficient D: it is the mean square displacement per unit time of a diffusing particle. If we assume D is constant, we shall further conclude that the mean square displacement in a time τ is proportional to τ. It is therefore consistent to carry the series expansions leading to Eq. (11.30) to second order in the spatial coordinate, but only to first order in the time coordinate.

Moreover, note that, from Eq. (11.32) and $t/\tau = N$, the mean square displacement given by Eq. (11.11) of a particle obeying the diffusion Eq. (11.4) becomes

$$\overline{x^2} = N\overline{s^2}, \tag{11.33}$$

which has the same form as Eq. (11.24). We thus conclude that, *when all directions are equivalent (i.e., when $\bar{s} = 0$), the diffusion of a particle may be understood as an unbiased ramdom walk.*

11.3.2 Equation of motion

From a macroscopic point of view, the set of all solvent molecules may be regarded as a heat reservoir at temperature T. Likewise, the collisions between these molecules and the diffusing particle may be interpreted as an effective force $F_{eff}(t)$ acting on the particle. If in addition there is an external force $F_{ext}(t)$ acting on the particle which is due to some external field (gravitational, electric, etc.), then the equation of motion for a particle of mass m and velocity v is (in one spatial dimension):

$$m\frac{dv}{dt} = F_{ext}(t) + F_{eff}(t). \tag{11.34}$$

The effective force clearly depends on the positional coordinates and velocities of the many solvent molecules; hence $F_{eff}(t)$ is a rapidly fluctuating function of time and it makes sense to formulate the problem in *statistical* terms. Let us consider an ensemble of many identical systems to the one under study, i.e., each consisting of a diffusing particle and its surrounding medium. For each of these, *the effective force varies randomly in time.* It then follows from Eq. (11.34) that also the particle velocity $v(t)$ will fluctuate randomly with time. On the other hand, the ensemble-averaged velocity, $\overline{v}(t)$, will vary with time much more slowly than $v(t)$ itself. Therefore it makes sense to split the velocity into a sum of two contributions:

$$v(t) = \overline{v}(t) + v'(t). \tag{11.35}$$

Here, $\overline{v}(t)$ is the slowly-varying contribution, which governs the long-time behaviour; and $v'(t)$ is the fast-varying component, with zero mean value:

$$\overline{v'}(t) = 0. \tag{11.36}$$

Now $F_{eff}(t)$ depends on particle motion, so it must also comprise a slow and a fast component:

$$F_{eff}(t) = \overline{F}(t) + F'(t), \tag{11.37}$$

with

$$\overline{F'}(t) = 0. \tag{11.38}$$

The slowly varying component of $F_{eff}(t)$ must be a function of the ensemble-averaged velocity, such that $\overline{F}(\overline{v}) = 0$ at equilibrium, i.e., when $\overline{v} = 0$ (we shall henceforth omit the explicit time dependence of most quantities for simplicity). If $\overline{v}$ is not too large, $\overline{F}(\overline{v})$ can be expanded in a power series of $\overline{v}$ truncated at first order:

$$\overline{F} = -\alpha \overline{v}, \tag{11.39}$$

where α is a positive constant and the minus sign means that the effective force opposes particle motion. We may then identify the effective force of Eq. (11.39) with a friction force, with α the friction coefficient. Equation (11.34) for the slowly varying component of the particle velocity is written:

$$m \frac{d\overline{v}}{dt} = F_{ext} + \overline{F} = F_{ext} - \alpha \overline{v}. \tag{11.40}$$

For the full particle velocity, we have

$$m \frac{dv}{dt} = m \frac{d\overline{v}}{dt} + m \frac{dv'}{dt} = F_{ext} - \alpha \overline{v} + F'. \tag{11.41}$$

To find a closed equation of motion, we make the approximation $\alpha \overline{v} \simeq \alpha v$. This amounts to neglecting $\alpha v'$, which is a fast-varying function of time and much smaller than F', provided the particle mass m is not too small. We thus arrive at the *Langevin equation*.

$$m \frac{dv}{dt} = F_{ext} - \alpha v + F'. \tag{11.42}$$

This equation contains the friction term, $-\alpha v$, which subsumes the interactions between the diffusing particle and the heat reservoir. Because friction is a *dissipative* process, the motion described by the Langevin equation is *irreversible*.

We have gone from reversible to irreversible motion using qualitative arguments. A more detailed analysis shows (Reif, 1985, chap. 15.7) that the friction coefficient and the random force F_{eff} are related at a deeper level.

11.3.3 Electric circuit analogy

In this subsection we explore an analogy between the Langevin equation and the equation for a resistor-inductor (RL) alternating-current (AC) circuit. Let $\epsilon(t)$ be the electromotive force applied to the circuit. The electrons that make up the electric current interact with the ions in the crystal lattices of the resistor. The effect of these interactions on the current $I(t)$ can be modelled by an effective fluctuating electromotive force $V(t)$. In turn, $V(t)$ can be decomposed into a slowly varying component, $\overline{V}(t) = -RI(t)$ (where the equality follows from Ohm's law), and a fast-varying component $V'(t)$ with zero mean value.[4] Kirchhoff's voltage law (the loop rule) then yields

$$\epsilon + V - L\frac{dI}{dt} = 0 \Leftrightarrow L\frac{dI}{dt} = \epsilon(t) - RI + V'. \tag{11.43}$$

Comparison with the Langevin Eq. (11.42) shows that the electrical resistance R plays the role of friction. This is consistent with our intuitive idea of resistance as something that opposes the motion of electrons in conductors. Interestingly, the inductance L plays the role of particle mass: the larger L, the greater the 'inertia' of the circuit, i.e., the less sensitive it is to fluctuations.

11.3.4 Mean square displacement

In the absence of any external forces, the Langevin equation is written

$$m\frac{dv}{dt} = -\alpha v + F'. \tag{11.44}$$

In thermal equilibrium, we must have $\bar{x} = 0$, i.e, the mean displacement vanishes. What about the mean *square* displacement? To find it, following Reif (1985, chap. 15.6) start by multiplying both sides of Eq. (11.44) by x:

$$mx\frac{d\dot{x}}{dt} = m\left[\frac{d(x\dot{x})}{dt} - \dot{x}^2\right] = -\alpha x\dot{x} + xF', \tag{11.45}$$

where we have defined $v = dx/dt \equiv \dot{x}$. We next take the ensemble average of both sides:

$$m\left\langle\frac{d(x\dot{x})}{dt}\right\rangle = m\langle\dot{x}^2\rangle - \alpha\langle x\dot{x}\rangle + \langle xF'\rangle, \tag{11.46}$$

where $\langle\ldots\rangle$ denotes the mean value of a quantity. Now on the basis of the ergodic hypothesis we assume that the mean of the derivative equals the derivative of the mean:

$$\left\langle\frac{d(x\dot{x})}{dt}\right\rangle = \frac{d}{dt}\langle x\dot{x}\rangle. \tag{11.47}$$

Furthermore, from the equipartition theorem, we have

$$\langle m\dot{x}^2\rangle = k_B T. \tag{11.48}$$

Finally, it is not unreasonable to assume that x and F' are uncorrelated quantities, leading to

$$\langle xF'\rangle = \langle x\rangle\langle F'\rangle = 0. \tag{11.49}$$

Combining Eqs. (11.46)–(11.49), we find

$$m\frac{d}{dt}\langle x\dot{x}\rangle = k_B T - \alpha\langle x\dot{x}\rangle, \tag{11.50}$$

with solution

$$\langle x\dot{x}\rangle = Ce^{-t/\tau} + \frac{k_B T}{\alpha}, \tag{11.51}$$

---

[4] Here we have departed from the conventions of electrical circuits and used capital letters to denote time-dependent voltages and currents. This is because small v already denotes the velocity.

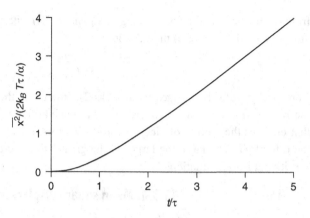

Figure 11.4 Mean square displacement (in reduced units) vs time, Eq. (11.55). It is non-linear at early times, Eq. [11.56] and linear at late times, Eq. (11.57).

where

$$\tau = \frac{m}{\alpha} \tag{11.52}$$

is a characteristic timescale of the system, and C is a constant of integration. If we assume that all particles in the ensemble are at position $x = 0$ at $t = 0$, then $C = -k_B T/\alpha$, and Eq. (11.51) becomes

$$\langle x\dot{x} \rangle = \frac{k_B T}{\alpha} \left(1 - e^{-t/\tau} \right). \tag{11.53}$$

Noting that

$$\langle x\dot{x} \rangle = \frac{1}{2} \frac{d}{dt} \langle x^2 \rangle, \tag{11.54}$$

we can integrate Eq. (11.53) to find the mean square displacement (Figure 11.4):

$$\langle x^2 \rangle = \overline{x^2} = \frac{2k_B T}{\alpha} \left[t - \tau \left(1 - e^{-t/\tau} \right) \right]. \tag{11.55}$$

Let us consider two limiting behaviours:

(a) *Early times*

$$t \ll \tau \Rightarrow e^{-t/\tau} \simeq 1 - \frac{t}{\tau} + \frac{1}{2} \left(\frac{t}{\tau} \right)^2 \Rightarrow \overline{x^2} = \frac{k_B T}{m} t^2. \tag{11.56}$$

Now from Eq. (5.86), $\overline{v^2} = k_B T/m$ is the mean square velocity for motion in one spatial dimension. It thus follows that, over short times, the particle moves as a *free particle*, with constant velocity $v = \sqrt{\overline{v^2}} = \sqrt{k_B T/m}$. Note that, in this case, friction is irrelevant.

(b) *Late times*

$$t \gg \tau \Rightarrow e^{-t/\tau} \to 0 \Rightarrow \overline{x^2} = \frac{2k_B T}{\alpha} t. \tag{11.57}$$

This means that, over long times, the particle performs an *unbiased random walk*, i.e., it behaves as a *diffusing particle*, with diffusion coefficient

$$D = \frac{k_B T}{\alpha}. \tag{11.58}$$

Note that, in this case, the particle mass is irrelevant. If we further assume that the particle is a sphere of radius a and (realistically) that its velocity is not too large, we may use *Stokes's law*:

$$\alpha = 6\pi \eta a, \tag{11.59}$$

where η is the solvent viscosity. Inserting this into Eq. (11.57) yields the *Einstein formula* for Brownian motion:

$$\overline{x^2} = \frac{k_B T}{3\pi \eta a} t, \tag{11.60}$$

which is yet another example of a fluctuation-dissipation relation: fluctuations in the position $\overline{(x^2)}$ are hereby related to dissipation described, at a phenomenological level, by the viscosity. If we take the solvent to be water at 17°C, we have $\eta = 11.3 \times 10^{-4}$ Pa s, and for a particle of radius $a = 1$ μm, Eq. (11.60) gives $\left(\overline{x^2}\right)^{1/2} \simeq 5$ μm for $t = 1$ minute.

From the Einstein formula, Eq. (11.60), Boltmann's constant k_B may be found by measuring the mean square displacement of particles of known radius in a medium of known viscosity. Using this and the known value of the universal gas constant R, Jean Perrin in 1910 was able to determine Avogadro's number with remarkable accuracy: $N_A = R/k_B = 6.02 \times 10^{23}$ mol^{-1}.[5,6]

11.4 The Fokker–Planck equation*

In this section we derive a generalised diffusion equation which describes the time evolution of the probability distribution of the velocity of a Brownian particle. This result will then be used to find the particle's velocity correlation functions. We restrict ourselves to the one-dimensional case and follow Haken (1983), to which we also refer readers for the generalisation to higher spatial dimensions.

Let us consider a particle suspended in a fluid medium and constrained to move along a straight line, which we take as the OX axis. This particle suffers random collisions with fluid molecules coming from the left or from the right. As discussed in the preceding section, the total effective force acting on the particle is composed of a friction force, which we assume to be of the form $-\alpha v$, plus a random, fluctuating, force $F'(t)$. We shall

---

[5] When Perrin succeeded in determining the number of molecules in a mole, he named it after the Italian Scientist Amedeo Avogadro (1776–1856). For details see Kubinga (2013).

[6] Note that the ideal gas equation of state (5.58)) can be written in the form $PV = nRT$. where $n = N/N_A$ is the number of moles of gas.

now specify what this random force is. Consider the effect of a single collision, or 'kick', denoted ϕ_j, at time $t = t_j$. Because the kick has a very short duration, we can approximate it as a Dirac δ function (see Appendix 11.5) of strength φ:

$$\phi_j = \varphi\delta(t - t_j), \tag{11.61}$$

According to Newton's second law, the change in particle velocity due to this one kick is

$$m\frac{dv}{dt} = \varphi\delta(t - t_j). \tag{11.62}$$

Integrating both sides of the above equation over a short time interval (of width $2\epsilon \ll t$) centred at $t = t_j$, we find

$$\int_{t_j-\epsilon}^{t_j+\epsilon} m\frac{dv}{dt}dt = \int_{t_j-\epsilon}^{t_j+\epsilon} \varphi\delta(t - t_j)dt \Rightarrow m\left[v(t_j + \epsilon) - v(t_j - \epsilon)\right] \equiv m\Delta v = \varphi, \tag{11.63}$$

i.e., 'at' $t = t_j$ the velocity jumps by an amount φ/m. The total random force is the sum of all kicks, each given by Eq. (11.61), over all collision times t_j, noting that a molecule may impact the particle either coming from its left (i.e., moving in the positive sense of OX) or coming from its right (i.e., moving in the negative sense of OX):

$$F'(t) = \varphi\sum_j (\pm 1)_j\delta(t - t_j), \tag{11.64}$$

where, of course, the sequence of plus and minus signs is random.

We now ask ourselves what the probability is that the particle will follow a given trajectory in the (v,t) plane.[7] If the particle has velocity $u_1(t)$, then the probability that at time t its velocity is $v \neq u_1(t)$ is obviously nil. On the other hand, the probability that its velocity is $v = u_1(t)$ is unity. Consequently, the probability density for the particle velocity is a Dirac δ function:

$$P_1(v,t) = \delta(v - u_1(t)). \tag{11.65}$$

If on another trajectory the particle velocity is $u_2(t)$, we likewise have

$$P_2(v,t) = \delta(v - u_2(t)), \tag{11.66}$$

and so on. We now introduce a new probability density $f(v,t)$ by averaging over all trajectories:

$$f(v,t) = \langle P(v,t)\rangle = \langle\delta(v - u(t))\rangle. \tag{11.67}$$

$f(v,t)dv$ is the probability that the particle velocity takes a value between v and $v + dv$ at time t. This is as yet an unknown quantify, so we shall next derive an evolution equation for it. Over a time Δt, the change in $f(v,t)$ is

$$\Delta f(v,t) = f(v,t + \Delta t) - f(v,t) = \langle\delta(v - u(t + \Delta t))\rangle - \langle\delta(v - u(t))\rangle. \tag{11.68}$$

---

[7] Recall that the particle moves along a straight line.

We set $u(t+\Delta t) = u(t) + \Delta u(t)$ and expand the first δ function in Eq. (11.68) in a power series of $\Delta u(t)$. To second order, we have

$$\Delta f(v,t) = \left\langle \left[\frac{d}{du(t)} \delta(v - u(t)) \right] \Delta u(t) \right\rangle$$
$$+ \frac{1}{2} \left\langle \left[\frac{d^2}{du(t)^2} \delta(v - u(t)) \right] \Delta u(t)^2 \right\rangle. \tag{11.69}$$

Further progress requires knowledge of $\Delta u(t)$ and of the derivatives of the δ functions. $\Delta u(t)$ can be found from the Langevin Eq., (11.42), which for simplicity we rewrite as

$$\frac{du(t)}{dt} = -\gamma u(t) + \Psi(t), \tag{11.70}$$

where $\gamma = \alpha/m$ and $\Psi(t) = F'(t)/m$. We now assume that, in a time Δt, the particle suffers many kicks from the fluid particles (so that it is meaningful to take the mean) but its velocity does not change very much (which, by Eq. (11.63), is reasonable if the particle mass is not too small). Integration of Eq. (11.70) between t and $t+\Delta t$ then yields:

$$\int_t^{t+\Delta t} \frac{du(t')}{dt'} dt' = u(t+\Delta t) - u(t) \equiv \Delta u(t)$$
$$= -\int_t^{t+\Delta t} \gamma u(t') dt' + \int_t^{t+\Delta t} \Psi(t') dt'$$
$$\simeq -\gamma u(t) \Delta t + \Delta \Psi(t). \tag{11.71}$$

Inserting $\Delta u(t)$ given by Eq. (11.71) into the first term on the right-hand side of Eq. (11.69), we get

$$\left\langle \left[\frac{d}{du(t)} \delta(v - u(t)) \right] \Delta u(t) \right\rangle = -\gamma \Delta t \left\langle \frac{d}{du(t)} [\delta(v - u(t))] u(t) \right\rangle$$
$$+ \left\langle \frac{d}{du(t)} \delta(v - u(t)) \Delta \Psi(t) \right\rangle. \tag{11.72}$$

Now using the property of the δ function, Eq. (11.106),

$$\left\langle \frac{d}{du(t)} [\delta(v - u(t))] u(t) \right\rangle = \int f(u(t),t) \frac{d}{du(t)} [\delta(v - u(t))] u(t) du(t)$$
$$= -\frac{d}{dv} [vf(v,t)]. \tag{11.73}$$

On the other hand, $u(t)$ is determined (via the Langevin Eq. (11.70)) by the sequence of kicks at times $t_j < t$, whereas $\Delta \Psi(t)$ is determined (via Eq. (11.71)) by the sequence of kicks at times t_j such that $t < t_j < t+\Delta t$. It is then reasonable to assume that in a δ function $u(t)$ and $\Delta \Psi(t)$ are uncorrelated, leading to the approximation

$$\left\langle \frac{d}{du(t)} \delta(v - u(t)) \Delta \Psi(t) \right\rangle \approx \left\langle \frac{d}{du(t)} \delta(v - u(t)) \right\rangle \langle \Delta \Psi(t) \rangle. \tag{11.74}$$

Moreover, by Eqs. (11.64) and (11.70) $\Psi(t)$ consists of a sequence of kicks in the positive and negative directions, with equal probabilities. Then if the sequence is sufficiently long, its mean will be close to zero; hence

$$\langle \Psi(t) \rangle = 0 \Rightarrow \langle \Delta \Psi(t) \rangle = 0, \tag{11.75}$$

and the second term on the right-hand side of Eq. (11.72) vanishes.

We now turn to the second term on the right-hand side of Eq. (11.69). Noting that $\Delta u(t)$ results from kicks at times $t_j > t$, we likewise approximate

$$\left\langle \left[\frac{d^2}{du(t)^2} \delta(v - u(t)) \right] \Delta u(t)^2 \right\rangle \approx \left\langle \frac{d^2}{du(t)^2} \delta(v - u(t)) \right\rangle \left\langle \Delta u(t)^2 \right\rangle. \tag{11.76}$$

Inserting Eq. (11.71) for $\Delta u(t)^2$ into the preceding equation yields a sum of three terms, proportional to Δt^2, $\Delta t \Delta \Psi(t)$ and $\Delta \Psi(t)^2$. The first of these is a higher-order infinitesimal, and therefore negligible; whereas the second and the third require calculation of $\langle \Delta \Psi(t) \rangle$ and $\langle \Psi(t)^2 \rangle$, respectively. We argued earlier that $\langle \Delta \Psi(t) \rangle = 0$. On the other hand,

$$\left\langle \Delta \Psi(t)^2 \right\rangle = \langle \Delta \Psi(t) \Delta \Psi(t) \rangle = \int_t^{t+\Delta t} \int_t^{t+\Delta t} \langle \Psi(t_1) \Psi(t_2) \rangle dt_1 dt_2. \tag{11.77}$$

We define $\langle \Psi(t_1) \Psi(t_2) \rangle$ as the *correlation function* of the random force: physically, this measures how the random forces at times t_1 and t_2 are related. In the simplest case we have

$$\langle \Psi(t_1) \Psi(t_2) \rangle = Q \delta(t_1 - t_2), \tag{11.78}$$

where Q is a constant. This means that the random force is not correlated in time, i.e., that its value at some time t does not depend on its value at any other time t'. In this case Eq. (11.77) straightforwardly gives

$$\langle \Delta \Psi(t)^2 \rangle = Q \Delta t, \tag{11.79}$$

which, upon substitution into Eq. (11.76), yields

$$\left\langle \frac{d^2}{du(t)^2} \delta(v - u(t)) \right\rangle \langle \Delta u(t)^2 \rangle = \int f(u(t)) \frac{d^2}{du(t)^2} \delta(v - u(t)) du(t) \times Q \Delta t$$

$$= Q \Delta t \frac{d^2 f(v,t)}{dv^2}, \tag{11.80}$$

where we have again used the property of the δ function, Eq. (11.106).

If we now take Eq. (11.69) with Eqs. (11.72)–(11.75) and (11.80), divide both sides by Δt and take the limit $\Delta t \to 0$, we obtain the *Fokker–Planck* equation:

$$\frac{df(v,t)}{dt} = \frac{d}{dv} [\gamma v f(v,t)] + \frac{1}{2} Q \frac{d^2 f(v,t)}{dv^2}. \tag{11.81}$$

This equation describes the time evolution of the probability density function of the velocity, in this particular case a Brownian particle. $K \equiv -\gamma v$ is known as the *drift coefficient* and Q as the *diffusion coefficient*. It can be shown that K and Q are the two first moments of $f(v,t)$ (Reif, 1985, chap. 15.11).

We shall now find Q for a Brownian particle, by integrating the Langevin Eq., (11.70). Using, e.g., the variation of constants method, we find

$$v(t) = \int_0^t e^{-\gamma(t-\tau)}\Psi(\tau)d\tau + v(0)e^{-\gamma t}. \tag{11.82}$$

Now Q characterises the steady state, which is only achieved for times $t \gg 1/\gamma$. So we are justified in neglecting the fast-decaying second term in Eq. (11.82), and compute

$$\left\langle v(t)^2 \right\rangle = \left\langle \int_0^t e^{-\gamma(t-\tau)}\Psi(\tau)d\tau \int_0^t e^{-\gamma(t-\tau')}\Psi(\tau')d\tau' \right\rangle. \tag{11.83}$$

Using Eq. (11.78), this becomes

$$\langle v(t)^2 \rangle = Q \int_0^t \int_0^t e^{-2\gamma t + \gamma(\tau+\tau')}\delta\left(\tau - \tau'\right)d\tau d\tau' = Qe^{-2\gamma t} \int_0^t e^{2\gamma\tau}d\tau$$

$$= \frac{Q}{2\gamma}\left(1 - e^{-2\gamma\tau}\right). \tag{11.84}$$

For $t \gg 1/\gamma$ (late times), we can neglect the exponentially decaying term and write the kinetic energy of the Brownian particle as

$$\frac{m}{2}\langle v^2 \rangle = \frac{mQ}{4\gamma}. \tag{11.85}$$

Because the particle is in thermal equilibrium with the solvent at temperature T, we can use the equipartition theorem, with the result

$$\frac{mQ}{4\gamma} = \frac{1}{2}k_B T \Rightarrow Q = \frac{2\gamma k_B T}{m}. \tag{11.86}$$

Recall that Q is the strength of time correlations, cf. Eq. (11.78). If follows that Eq. (11.86) is another example of a fluctuation-dissipation relation, which connects the amplitude of fluctuations ($\sim Q$) to the strength of dissipation ($\sim \gamma$) and to absolute temperature

It is also of interest to calculate the velocity correlation function at two different times, t and t', which measures how fast the system loses memory of its velocity; or, in other words, how long it takes for the velocity at time t' to differ appreciably from what it was at time t. Rewriting Eq. (11.82) for $t \neq t'$, we find

$$\langle v(t)v(t') \rangle = Q \int_0^t \int_0^{t'} e^{-\gamma(t-\tau)-\gamma(t'-\tau')}\delta\left(\tau - \tau'\right)d\tau d\tau'$$

$$= \frac{Q}{2\gamma}\left[e^{-\gamma(t-t')} - e^{-\gamma(t+t')}\right]. \tag{11.87}$$

For a steady-state process, again the relevant limit is $t \gg 1/\gamma$ and $t' \gg 1/\gamma$, but $|t - t'|$ need not be large, leading to the final result:

$$\langle v(t)v(t')\rangle = \frac{Q}{2\gamma}e^{-\gamma(t-t')}. \tag{11.88}$$

This result may also be derived by solving the time-dependent Fokker–Planck equation, which we will not do here; the interested reader is referred to Haken (1983, chap. 6) or Reif (1985, chap. 15.11). We thus conclude that the time after which the particle loses memory of its velocity is

$$\tau = \frac{1}{\gamma} = \frac{m}{\alpha}. \tag{11.89}$$

This is precisely the characteristic time introduced in the preceding section, Eq. (11.52), and whose physical meaning is now apparent. If $\alpha \to 0$ the velocity correlation function will diverge: in the absence of friction, the fluctuations will be very large (on the scale of the system size), which we call *critical fluctuations*.

Steady-state solutions of the Fokker–Planck equation

The theory of Brownian motion plays a key role in other fields of physics besides mechanics, as well as in chemistry and biology. Consequently, in Eq. (11.81) v may be any variable q appropriate to the description of a system subjected to both deterministic and random forces, which may be internal or external to the system (Haken, 1983).

In the most general case, the diffusion coefficient may also depend on q, and the Fokker–Planck equation is written

$$\frac{df(q,t)}{dt} = -\frac{d}{dq}[K(q)f(q,t)] + \frac{1}{2}\frac{d^2}{dq^2}[Q(q)f(q,t)]. \tag{11.90}$$

We shall restrict ourselves to constant Q only. In this case, Eq. (11.90) may be rewritten as

$$\frac{df}{dt} + \frac{d}{dq}\left(Kf - \frac{1}{2}Q\frac{df}{dq}\right) = 0, \quad f = f(q,t), \quad K = K(q). \tag{11.91}$$

If we define a *probability flux* as

$$j = Kf - \frac{1}{2}Q\frac{df}{dq}, \tag{11.92}$$

then Eq. (11.91) becomes

$$\frac{df}{dt} + \frac{dj}{dq} = 0, \tag{11.93}$$

which is a one-dimensional continuity equation. In higher dimensions it can be shown that the time derivative of the probability density f equals minus the divergence of the probability flux j (Haken, 1983, chap. 6). If f is time-independent, we find after integration of Eq. (11.93) that

$$j = \text{constant}. \tag{11.94}$$

We now impose the so-called 'natural boundary conditions', i.e., that f should vanish when $q \to \pm\infty$. Equation (11.92) then implies that $j = 0$, and

$$\frac{1}{2}Q\frac{df}{dq} = Kf, \tag{11.95}$$

which can be readily integrated to yield

$$f(q) = \mathcal{N}e^{-2V(q)/Q}, \tag{11.96}$$

where $\mathcal{N}$ is a normalisation constant, and

$$V(q) = -\int_{q_0}^{q} K(q')dq' \tag{11.97}$$

has the physical meaning of a generalised potential.

Let us first consider the case where the drift coefficient is linear in q. We have:

$$K(q) = -\gamma q \Rightarrow V(q) = \frac{\gamma}{2}q^2, \tag{11.98}$$

where we have dropped an arbitrary constant in the definition of $V(q)$; this may be absorbed into $\mathcal{N}$. For a Brownian particle, q is the velocity, Q is given by Eq. (11.86), and the probability density is therefore

$$f(v) = \mathcal{N}e^{-mv^2/2k_BT}, \tag{11.99}$$

which is Gaussian: it is actually the one-dimensional Maxwell distribution, Eq. (5.79).

Figure 11.5 shows the generalised potential $V(q)$, equation (11.98), and the corresponding probability density, Eq. (11.99). We may imagine that the random force pushes the particle up the potential well, and that between any two consecutive kicks it slides back down along the slope. The most probable value of q is therefore $q = 0$, but because of all the random kicks the particle may also be found with $q \neq 0$. If we make γ smaller, the potential well opens up and the probability density spreads out: a smaller friction coefficient allows q to take values farther away from zero.

Consider now a drift coefficient that is non-linear in q:

$$K(q) = -\gamma q - \beta q^3 \Rightarrow V(q) = \frac{\gamma}{2}q^2 + \frac{\beta}{4}q^4, \tag{11.100}$$

where $\beta > 0$, but γ may be positive or negative. This potential and the corresponding probability density are plotted in Figure 11.6. If $\gamma > 0$, everything is qualitatively the same as for the linear friction coefficient, Eq. (11.98). If $\gamma < 0$, however, the situation is qualitatively different: if there were no fluctuations, the particle would fall into either of the two potential minima, at positive or negative q of the same absolute value, so the symmetry would be broken, as at a *continuous phase transition* (cf. Figure 7.7). In the presence of fluctuations, the probability density is symmetric, meaning that the particle may be found with equal probability at either minimum.

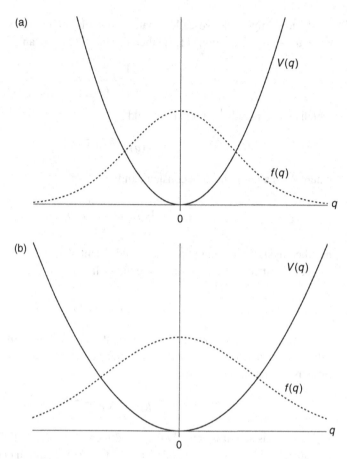

Figure 11.5 Generalised potential (solid line) and probability density (dashed line) for the steady-state Fokker–Planck equation with a friction coefficient linear in q, Eq. (11.98). Arbitrary units have been used along all axes. In (b) the friction coefficient is half of what it is in (a).

Finally, let us consider an even more general form of the friction coefficient:

$$K(q) = -\gamma q - \lambda q^2 - \beta q^3 \Rightarrow V(q) = \frac{\gamma}{2}q^2 + \frac{\lambda}{3}q^3 + \frac{\beta}{4}q^4, \qquad (11.101)$$

where $\beta > 0$, $\lambda < 0$, and γ may be positive or negative. This potential and the corresponding probability density can be seen in Figure 11.7. When γ goes from positive to negative, the potential acquires two asymmetric minima, at positive and negative q of in general different absolute values. Start by assuming that the particle rests at the bottom of the shallower minimum. If there are no fluctuations, it will just stay there. If however we change γ in such a way that this minimum disappears, then the particle can 'tumble' into the deeper minimum – a 'mechanical' phase transition. If there are fluctuations, then the asymmetric probability density means that the particle may jump the barrier between the two minima and occupy the deeper minimum even if γ is not varied, in which case the transition between the two states is 'thermodynamic', analogous to a *first-order phase transition* (cf. Figure 7.8). As we saw in Chapter 10, a phase transition is first-order or

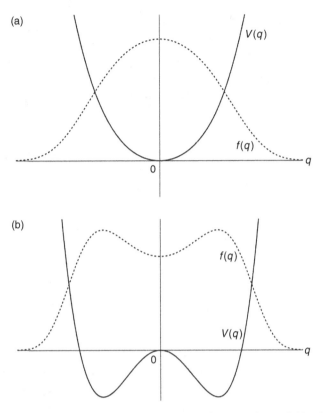

Figure 11.6 Generalised potential (solid line) and probability density (dashed line) for the steady-state Fokker–Planck equation with the friction coefficient of Eq. (11.100). Arbitrary units have been used along all axes. (a) $\gamma > 0$; (b) $\gamma < 0$.

continuous depending on whether the free energy written as a power series of the order parameter contains a cubic term or not. Here the potential $V(q)$ plays the role of the free energy; compare Eqs. (11.100) and (11.101).

11.5 The Dirac δ function

The Dirac δ function is defined as follows:

$$\delta(x - x_0) = 0 \quad \text{if} \quad x \neq x_0, \tag{11.102}$$

$$\int_{x_0 - \epsilon}^{x_0 + \epsilon} \delta(x - x_0) dx = 1 \quad (\epsilon > 0). \tag{11.103}$$

The δ function may be seen as an infinitely sharp, infinitely tall Gaussian function:

$$\delta(x - x_0) = \lim_{\alpha \to 0} \left\{ \frac{1}{\alpha \sqrt{\pi}} \exp\left[-\frac{(x - x_0)^2}{\alpha^2} \right] \right\}. \tag{11.104}$$

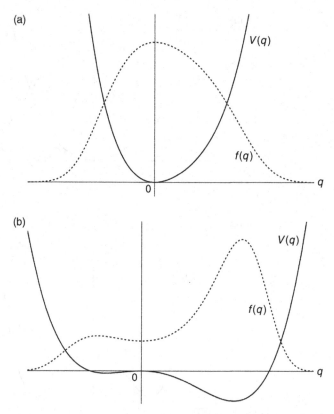

Figure 11.7   Generalised potential (solid line) and probability density (dashed line) for the steady-state Fokker–Planck equation with the friction coefficient of Eq. (11.101). Arbitrary units have been used along all axes. (a) $\gamma > 0$; (b) $\gamma < 0$.

The δ function has the following properties:

$$\int_{x_0-\epsilon}^{x_0+\epsilon} f(x)\delta(x-x_0)dx = f(x_0), \tag{11.105}$$

$$\int_{x_0-\epsilon}^{x_0+\epsilon} f(x)\frac{d^n}{dx^n}\delta(x-x_0)dx = (-1)^n\frac{d^n f(x)}{dx^n}\bigg|_{x=x_0}, \tag{11.106}$$

where $f(x)$ is any n-times differentiable function.

Problems

11.1 Diffusion.

     a. Compute the diffusion coefficient D of a spherical particle of diameter $d = 1\ \mu$m placed in water at $17°C$ ($\eta = 11.3 \times 10^{-4}$ Pa s).

b. Assume that at $t = 0$ the particle is at the origin of coordinates, and that there are no privileged directions for diffusion. Compute the root mean square displacement $r_{rms} = \left(\overline{r^2}\right)^{1/2}$ of the particle after 10 hours.

11.2 For long observation times, compute the probability that a Brownian particle is farther than one root mean square displacement from the origin. You may need to use the result $\text{erfc}(\sqrt{3/2}) \simeq 0.083$, where $\text{erfc}(x) = 1 - \text{erf}(x)$ and $\text{erf}(x)$ is the error function, defined as

$$\text{erf}[x] = \frac{2}{\sqrt{\pi}} \int_0^x e^{-u^2} du.$$

11.3 Consider a gas of molecules of mass m enclosed in a volume V at temperature T.

a. Assume that the density n varies along the OX axis, i.e., $n = n(x)$. Let λ_x and $\overline{v_x}$ be, respectively, the mean free path and the mean velocity along OX. Further assume that, on average, any molecules crossing a surface of unit area placed perpendicular to OX start from a distance λ_x from that surface on either side, where the density is either $n(x - \lambda_x)$ or $n(x + \lambda_x)$. Show that the flux of molecules across the surface is

$$J_x = -\overline{v_x}\lambda_x \frac{dn}{dx}.$$

b. Now assume that motion in all three dimensions is equally probable. Show that the mean value of $\overline{v_x}\lambda_x$ equals

$$\langle \overline{v_x}\lambda_x \rangle = \frac{1}{3}\overline{v}\lambda,$$

and use this to show that the diffusion coefficient is

$$D = \frac{1}{3}\overline{v}\lambda.$$

11.4 Electrical conductivity of a metal.

a. Consider a Brownian particle of mass m and charge q in an electric field of strength E. For simplicity, asssume motion along one dimension only. Using the appropriate Langevin equation show that, in the steady state,

$$\overline{v} = \frac{qE\tau}{m}.$$

b. The conduction electrons in a metal can be viewed as Brownian particles. Assuming that the metal obeys Ohm's law, use the preceding result to find the following expression for the electric conductivity:

$$\sigma = \frac{ne^2\tau}{m}.$$

c. Compute the electric conductivity of copper at $T = 300$ K and $T = 4$ K. For very pure samples, $\tau(300\,\text{K}) \simeq 2 \times 10^{-14}$ s, $\tau(4\,\text{K}) \simeq 2 \times 10^{-9}$ s, and $n = 8.45 \times 10^{28}$ m^{-3} (Kittel, 1995, chap. 6).

d. The mean free path of a conduction electron is

$$\lambda = v_F \tau,$$

where v_F is the Fermi velocity of the electron. Compute λ for copper at $T = 300$ K and at $T = 4$ K, taking $v_F = 1.57 \times 10^6$ ms^{-1} (Kittel, 1995, chap. 6).

11.5 Thermal conductivity of a metal.

a. Heat propagation may be treated as a diffusion process. Let $\bar{\varepsilon}$ be the mean energy per particle. The energy density is therefore $\rho = n\bar{\varepsilon}$. Show, by multiplying both sides of Fick's law in three dimensions, $\mathbf{J} = -D\nabla n(\mathbf{r},t)$, by $\bar{\varepsilon}$, that the energy density flux is

$$\mathbf{J} = -D\frac{\partial \rho}{\partial T}\nabla T.$$

On the other hand, Fourier's law of heat propagation states that

$$\mathbf{J} = -\kappa \nabla T,$$

where κ is the heat conductivity. Noting that

$$c_v = \frac{\partial \rho}{\partial T},$$

where c_V is the specific heat per unit volume, show that the thermal conductivity κ and the diffusion coefficient D are related by

$$\kappa = Dc_V.$$

b. The specific heat per unit volume of a free-electron gas is, from Eq. (6.73),

$$c_V = \frac{\pi^2}{2}nk_B^2\frac{T}{\varepsilon_F}, \quad T \ll T_F.$$

Using $D = \bar{v}\lambda/3$, $\lambda = v_F\tau$, $v_F = \sqrt{2\varepsilon_F/m}$, and $\kappa = Dc_V$, show that

$$\kappa = \frac{\pi^2 nk_B^2 \tau T}{3m}.$$

Hint: assume that $\bar{v} = v_F$.

c. Assuming that the relaxation times are approximately the same for electric and thermal conduction, derive the *Wiedemann–Franz* law:

$$\frac{\kappa}{\sigma} = \frac{\pi^2 k_B^2}{3e^2}T, \quad \frac{\pi^2 k_B^2}{3e^2} = 2.443 \times 10^{-8}\,\text{J}^2\text{K}^{-1}\text{C}^{-2}.$$

According to this law, at not too low temperatures the ratio of thermal to electrical conductivity of a metal is proportional to its temperature, with a universal proportionality constant. At very low temperatures ($T \ll \Theta_{\text{Debye}}$), this constant decreases, possibly because the electrical and thermal relaxation times are no longer identical. See Kittel (1995, chap. 6) for details.

References

Haken, H. 1983. *Synergetics*, 3rd edition. Springer Verlag.

Kittel, C. 1995. *Introduction to Solid State Physics*, 7th edition. Wiley.

Kubo, R., Toda, M., and Hashitsume, N. 1985. *Statistical Physics II – Nonequilibrium Statistical Mechanics*. Springer-Verlag.

Kubinga, H. 2013. A tribute to Jean Perrin. *Europhys. News* **44/5**, 16–18.

Pais, A. 1982. *Subtle is the Lord*. Oxford University Press.

Perrin, J. B. 1948. *Les Atomes*. Presses Universitaires de France.

Reif, F. 1985. *Fundamentals of Statistical and Thermal Physics*. McGraw-Hill.

Sommerfeld, A. 1964. *Partial Differential Equations in Physics* (Lectures on Theoretical Physics, vol. VI). Academic Press.

Appendix A Values of some fundamental physical constants

Quantity	Symbol, equation	Value
Speed of light in vacuum	c	$299792458 \text{ m s}^{-1}$
Planck's constant	h	$6.62606957(29) \times 10^{-34} \text{ J s}$
Planck's constant, reduced	$\hbar = h/2\pi$	$1.054571726(47) \times 10^{-34} \text{ J s}$
Electron charge magnitude	e	$1.602176565(35) \times 10^{-19} \text{ C}$
Electron mass	m_e	$9.10938291(40) \times 10^{-31} \text{ kg}$
Proton mass	m_p	$1.672621777(74) \times 10^{-27} \text{ kg}$
		$= 1836.15267245(75) \, m_e$
Atomic mass constant	$m_u = (\text{mass } {}^{12}\text{C atom})/12$	$1.660538921(73) \times 10^{-27} \text{ kg}$
Permittivity of free space	$\epsilon_0 = 1/\mu_0 c^2$	$8.854187817\ldots \times 10^{-12} \text{ F m}^{-1}$
Permeability of free space	μ_0	$4\pi \times 10^{-7} \text{ N A}^{-2}$
Bohr magneton	$\mu_B = e\hbar/2m_e$	$9.27400968(20) \times 10^{-24} \text{ J T}^{-1}$
Nuclear magneton	$\mu_N = e\hbar/2m_p$	$5.05078353(11) \times 10^{-27} \text{ J T}^{-1}$
Avogadro's number	N_A	$6.02214129(27) \times 10^{23} \text{ mol}^{-1}$
Boltzmann's constant	k_B	$1.3806488(13) \times 10^{-23} \text{ J K}^{-1}$
Ideal gas constant	$R = N_A k_B$	$8.3144621(75) \text{ J mol}^{-1} \text{ K}^{-1}$
Molar volume, ideal gas at STP	$V_m = N_A k_B$	$22.413968(20) \times 10^{-3} \text{ m}^3 \text{ mol}^{-1}$
	$\times (273.15 \text{ K})/(101\,325 \text{ Pa})$	
Wien displacement law constant	$b = \lambda_{\max} T$	$2.8977721(26) \times 10^{-3} \text{ m K}$
Stefan–Boltzmann constant	$\sigma = \pi^2 k_B^4/60\hbar^3 c^2$	$5.670373(21) \times 10^{-8} \text{ W m}^{-2} \text{ K}^{-4}$

Table A.1

The figures in parentheses after the values give the 1-standard-deviation uncertainties in the last digits. This set of constants is recommended for international use by CODATA (the Committee on Data for Science and Technology). The full 2010 CODATA set of constants can be found at http://physics.nist.gov/constants.

Appendix B Some useful unit conversions

$1 \text{ cal} = 4.1840 \text{ J}$

$1 \text{ in} = 0.0254 \text{ m}$

$1 \text{ Å} = 0.1 \text{ nm}$

$1 \text{ G} = 10^{-4} \text{ T}$

$1 \text{ dyne} = 10^{-5} \text{ N}$

$1 \text{ erg} = 10^{-7} \text{ J}$

$1 \text{ eV} = 1.602176565(35) \times 10^{-19} \text{ J}$

$2.99792458 \times 10^{9} \text{ esu} = 1 \text{ C}$

$k_B T \text{ at } 300 \text{ K} = [38.681731(35)]^{-1} \text{ eV}$

$0°\text{C} = 273.15 \text{ K}$

$1 \text{ atmosphere} = 101\,325 \text{ Pa}$

Index

Printed in the United States
By Bookmasters